Based on
NFPA 70E®
2009 Edition

ELECTRICAL SAFETY-RELATED WORK PRACTICES

SECOND EDITION

NATIONAL JOINT APPRENTICESHIP AND TRAINING COMMITTEE

This material is copyright by the NJATC. All rights are reserved. No part of this material shall be reproduced, stored in a retrieval system, or transmitted by any means; whether electronic, mechanical, photocopying, recording or otherwise; without the expressed written permission of the NJATC.

© Copyright 2009 by the National Joint Apprenticeship and Training Committee for the Electrical Industry

JONES AND BARTLETT PUBLISHERS
Sudbury, Massachusetts
BOSTON   TORONTO   LONDON   SINGAPORE

World Headquarters

Jones and Bartlett Publishers
40 Tall Pine Drive
Sudbury, MA 01776
978-443-5000
info@jbpub.com
www.jbpub.com

Jones and Bartlett's books and products are available through most bookstores and online booksellers. To contact Jones and Bartlett Publishers directly, call 800-832-0034, fax 978-443-8000, or visit our website www.jbpub.com.

Substantial discounts on bulk quantities of Jones and Bartlett's publications are available to corporations, professional associations, and other qualified organizations. For details and specific discount information, contact the special sales department at Jones and Bartlett via the above contact information or send an email to specialsales@jbpub.com.

Copyrights

Copyright © 2009 by the National Joint Apprenticeship and Training Committee for the Electrical Industry.

All rights reserved. No part of this book may be reproduced in any form or by any means, stored in a retrieval system, transmitted by any means, electronic, mechanical, photocopying, recording, or otherwise; without permission in writing from the National Joint Apprenticeship and Training Committee (NJATC).

Copyrights, Trademarks, and Registered Marks:

Portions of this book are reproduced with permission from NFPA 70E – 2009 Standard for Electrical Safety in the Workplace, Copyright © 2008, National Fire Protection Association, Inc. This reprinted material is not the complete and official position of the NFPA on the referenced subject, which is represented only by the standard in its entirety.

Electrical Safety in the Workplace® and NFPA-70E® are registered trademarks of the National Fire Protection Association, Inc., Quincy, MA. *National Electrical Code*® and *NEC*® are registered trademarks of the National Fire Protection Association, Inc. *The National Electrical Code*®, or *NEC*® (NFPA 70), is copyright by the National Fire Protection Association, Inc.

Bussmann® and *Safety* BASICS™ are trademarks of Cooper Bussmann, Inc. *Safety* BASICS™ is copyright by Cooper Bussmann, Inc.

Arc-flash hazard equations in Chapter 12 reprinted with permission from IEEE 1584 Guide for Performing Arc-Flash Hazard Calculations Copyright 2002, by IEEE. The IEEE disclaims any responsibility or liability resulting from the placement and use in the described manner.

Notice Concerning Liability

Publication of this work is for the purpose of circulating information and opinion among those concerned for electrical safety and related subjects. While every effort has been made to achieve a work of high quality, neither the NJATC nor the authors and contributors to this work guarantee the accuracy or completeness of or assume any liability in connection with the information and opinions contained in this work. The NJATC and the authors and contributors shall in no event be liable for any personal injury, property, or other damages of any nature whatsoever, whether special, indirect, consequential, or compensatory, directly or indirectly resulting from the publication, use of or reliance upon this work.

This work is published with the understanding that the NJATC and the authors and contributors to this work are supplying information and opinion but are not attempting to render engineering or other professional services. If such services are required, the assistance of an appropriate professional should be sought.

Some images in this book feature models. These models do not necessarily endorse, represent, or participate in the activities represented in the images.

Production Credits

Publisher: Kimberly Brophy
Acquisitions Editor—Electrical: Martin Schumacher
Managing Editor: Carol B. Guerrero
V.P., Manufacturing and Inventory Control: Therese Connell
Production Manager: Jenny L. Corriveau
Production Assistant: Tina Chen
Associate Marketing Manager: Meagan Norlund

Cover Design: Kristin E. Parker
Interior Design: Anne Spencer
Photo Research Manager and Photographer: Kimberly Potvin
Assistant Photo Researcher: Jessica Elias
Cover Image: Courtesy of Salisbury Electrical Safety L.L.C.
Printing and Binding: RR Donnelly
Cover Printing: RR Donnelly

Additional illustration and photographic credits appear on page 365, which constitutes a continuation of the copyright page.

Library of Congress Cataloging-in-Publication Data
Hickman, Palmer.
 Electrical safety-related work practices / Palmer Hickman ; National Joint
Apprenticeship Training Committee. — 2nd ed.
   p. cm.
 ISBN-13: 978-0-7637-5428-0 (pbk.)
 ISBN-10: 0-7637-5428-5 (pbk.)
 1. Electric engineering—United States—Safety measures. I. National
Joint Apprenticeship and Training Committee for the Electrical Industry. II.
Title.
 TK152.H53 2008
 621.3028'9—dc22
                          2008036398
6048
Printed in the United States of America
15 14 13 12 10 9 8 7 6 5 4

Acknowledgments

Principal Writer

Palmer Hickman
Director of Safety, Codes and Standards
NJATC

Contributing Writers

Mary Capelli Schellpfeffer, MD, MPA
CapSchell, Inc.

Tim Crnko
Manager, Training & Tech Services
Cooper Bussmann®

Jane G. Jones
Technical Writer

Stephen M. Lipster
Training Director
Columbus Joint Apprenticeship
    and Training Committee for the
    Electrical Industry

Dan Neeser
Senior Technical Sales Engineer
Cooper Bussmann®

Dustin Priemer
District Engineer—Houston, Texas
Cooper Bussmann®

Vincent J. Saporita
Vice President Technical Marketing
    and Services
Cooper Bussmann®

James R. White
Training Director
Shermco Industries

Text Contributors

The NJATC wishes to thank the following individuals for permission to reprint materials in this edition.

Vince A. Baclawski
Technical Director
Power Distribution Products
National Electrical Manufacturers
Association (NEMA)

John Gaston
Sales and Marketing Manager
Westex Inc.

Dennis K. Neitzel, CPE
Director
AVO Training Institute

Ellen Parson
Managing Editor
EC&M Magazine

Kristen K. Schmidt
Technical Services Coordinator
InterNational Electrical Testing
Association (NETA)

Contents

CHAPTER 1 Electrical Safety Culture 1

Introduction 1

Culture 2

   Workplace Pressure 2

   Historical Perspective 2

Hazards 3

Requirements 8

Choices 9

   Appropriate Priorities 9

   Training and Personal Responsibility . . . 9

   Costs of Energized Work Versus

      Shutting Down 9

   Time Pressure 9

   Employee Qualification 10

Implementation Choices 10

Summary 11

Lessons Learned 13

Current Knowledge 16

CHAPTER 2 Electrical Hazard Awareness . . 19

Introduction 19

Workplace Hazards 19

Electrical Shock 20

   Skin Resistance 20

   Extent of Injury 21

   Prevention 22

Arcing Faults: Arc Flash and Arc Blast 23

   Arcing Fault Basics 23

   Arc Flash 23

   Arc Blast 24

   How Arcing Faults Can Affect

      Humans 24

   Staged Arc Flash Tests 25

   The Role of Overcurrent Protective

      Devices in Electrical Safety 27

Summary 29

Lessons Learned 30

Current Knowledge 34

CHAPTER 3 OSHA Requirements 37

Introduction 37

OSHA: History and Purpose 37

   OSHA Requirements 38

   29 CFR Part 1926 38

   29 CFR Part 1910 42

Summary 49

Lessons Learned 50

Current Knowledge 54

CHAPTER 4 Design Considerations 57

Introduction 57

Design Considerations 57

Current Limitation of Overcurrent

  Protective Devices 58

   Non-Current-Limiting Overcurrent

      Protective Devices 58

   Current-Limiting Overcurrent

      Protective Devices 58

Devices and Measures That Enhance

  Electrical Safety 65

   Consistency Over Time of

      Overcurrent Protective Devices 65

   Current-Limiting Fuses 66

   Circuit Breakers 66

Summary 78

Lessons Learned 79

Current Knowledge 84

**CHAPTER 5 History, Evolution, Layout,
Style, and Scope of
NFPA 70E®** 89

Introduction 89

History of NFPA 70E 89

   The First Version 90

Evolution of NFPA 70E 90

   The Fifth Edition 90

   The Sixth Edition 91

The Seventh Edition 91

Layout of *NFPA 70E* 91

Purpose 91

Scope 91

Arrangement 92

Organization 92

Numbering Style 92

***NFPA 70E* Today** 95

Summary 95

Lessons Learned 96

Current Knowledge 99

**CHAPTER 6 Overview of *NFPA 70E*®
Concepts** **101**

Introduction 101

Organization 102

Scope 103

Purpose and Responsibility 104

Host and Contractor Responsibilities . . . 105

Host Employer Responsibilities 105

Contractor Responsibilities 105

Training Requirements 105

Electrical Safety Program 105

**Working While Exposed to Electrical
Hazards** 106

General Requirements 106

Requirements for Working Inside the
Limited Approach Boundary 106

Use of Equipment 107

Summary 109

Lessons Learned 110

Current Knowledge 113

CHAPTER 7 Electrical Safety Program . . . **117**

Introduction 117

**Establishing an Effective Electrical
Safety Program** 118

Electrical Safety Program
Objectives 118

Program Cost Considerations 119

***NFPA 70E* and the Electrical Safety
Program** 119

General Program Requirements 120

Awareness and Self-Discipline 120

Program Principles, Controls, and
Procedures 120

Hazard/Risk Evaluation 120

Job Briefing 124

Auditing 125

Groundbreaking Development 126

Summary 126

Lessons Learned 127

Current Knowledge 130

CHAPTER 8 Training **133**

Introduction 133

Is Having a Program in Place Enough? . . 134

Training Requirements 135

Safety Training 136

Type of Training 137

Emergency Procedure Training 137

Qualified and Unqualified Employee
Training 138

Documentation of Training 141

Summary 141

Lessons Learned 143

Current Knowledge 146

**CHAPTER 9 The Control of Hazardous
Energy** **149**

Overview 149

Introduction 150

**The Control of Hazardous Energy
(Lockout/Tagout)** 151

Scope, Application, and Purpose 151

Definitions 152

General 153

Application of Control 159

Release from Lockout or Tagout 160

Additional Requirements 161

Applicable Part 1910 Subpart S
Requirements 162

Typical Minimal Lockout Procedures 163

**Achieving an Electrically Safe Work
Condition 163**

The Process of Achieving an
Electrically Safe Work Condition . . .163

Lessons Learned 166

Current Knowledge 169

**CHAPTER 10 Justification, Assessment,
and Documentation 175**

Introduction 175

Justification 175

Energized Electrical Work Permit 176

Approach Boundaries 178

Shock Hazard Analysis 178

Shock Protection Boundaries 178

Approach by Qualified Persons 179

Approach by Unqualified Persons . . .180

Arc Flash Hazard Analysis 180

Arc Flash Protection Boundary 181

Personal Protective Equipment 181

Equipment Labeling 182

Summary 182

Lessons Learned 183

Current Knowledge 186

**CHAPTER 11 Calculation of Short-Circuit
Currents 189**

Introduction 189

**Short-Circuit Calculation
Requirements** 190

**Effect of Short-Circuit Current on Arc
Flash Hazards** 190

Short-Circuit Calculation Basics 191

Sources of Short-Circuit Current 191

Short-Circuit Current Factors 192

Effect of Transformers on
Short-Circuit Current 193

Effect of Conductors on
Short-Circuit Current 193

Procedures and Methods 194

Point-to-Point Method 194

Basic Short-Circuit Calculation
Procedure .195

Calculation of Short-Circuit
Currents on Transformer Secondary
with Primary Short-Circuit Current
Known 197

**Short-Circuit Calculation Software
Programs 202**

Summary 203

Lessons Learned 204

Current Knowledge 207

**CHAPTER 12 Arc Flash Hazard Analysis
Methods 211**

Overview 211

Introduction 212

Arc Flash Hazard Analysis 213

Fundamental Information
Required 213

Methods Available 217

Determining the AFPB 218

Determining the Incident Energy
or PPE 222

Arc Flash Hazard Analysis
Summary 230

**Arcing Currents in the Long Time
Characteristic of Overcurrent
Protective Devices** 231

Additional Considerations 231

Summary 232

Lessons Learned 233

Current Knowledge. 236

CHAPTER 13 Testing, Overhead Lines, Other Precautions, and Other Protective Equipment 239

Introduction. 239

Performing Testing Work. 239

Uninsulated Overhead Lines 240

Work Within the Limited Approach
Boundary 241

Other Precautions for Personnel
Activities 242

Alertness 243

Blind Reaching 243

Illumination 243

Conductive Articles, Materials,
Tools, and Equipment 243

Confined or Enclosed Spaces 243

Housekeeping. 244

Flammable Materials 244

Anticipating Failure 244

Routine Opening and Closing of
Circuits. 245

Reclosing Circuits After Protective
Device Operation. 245

Other Protective Equipment 246

General Requirements 246

Equipment Care 246

Other Protective Equipment 246

Alerting Techniques 248

Standards for Other Protective
Equipment 248

Summary 249

Lessons Learned 250

Current Knowledge. 252

CHAPTER 14 Personal Protective Equipment 257

Introduction. 257

Personal Protective Equipment. 258

General Requirements 258

Equipment Care 258

PPE Selection 258

Arc-Flash Protective Equipment
(FR Clothing). 258

General PPE Requirements 258

PPE Maintenance and Use
Requirements 262

Foot Protection. 262

Standards for PPE. 262

Factors in Selection of Protective
Clothing 263

Arc Flash Protective Equipment. 264

Clothing Material Characteristics . . . 266

Prohibited Apparel 266

Care and Maintenance of FR
Clothing and FR Arc Flash Suits . . . 266

HRC Classifications to Select PPE 268

HRC Classifications and Use of Rubber
Insulating Gloves and Insulated and
Insulating Hand Tools 268

Protective Clothing and PPE
Selection. 271

Protective Clothing Characteristics . . 271

Summary 272

Lessons Learned 273

Current Knowledge. 276

CHAPTER 15 Existing Electrical Equipment: Maintenance and Safety . . . 279

Introduction. 279

Work Practices and System Upgrades . . 279

Resetting Circuit Breakers or
Replacing Fuses 279

Circuit Breaker Evaluation 280

Fuses 280

Moving People Outside the Arc
Flash Protection Boundary 281

Marking Equipment with the Arc
Flash Hazard. 281

Updating Arc Flash Hazard
Information 283

Evaluating Overcurrent Protective
Devices in Existing Facilities for
Proper Interrupting Rating 284

Reduce Arc Flash Hazard for Existing
Fusible Systems. 286

Adjusting the Instantaneous Trip
Setting on Existing Circuit Breakers 286

Sizing Under-Utilized Circuits with
Lower Ampere Rated Fuses or
Circuit Breakers 286

Relaying Schemes 287

**Maintenance: How It Relates to
Electrical Safety** 287

How a Lack of Maintenance Can
Increase Hazards 288

Maintenance Can Affect Safety. 290

General Maintenance
Requirements 291

Maintenance Programs and
Frequency of Maintenance. 292

Maintenance and the Hazard/Risk
Analysis. 293

Qualifications for Performing
Maintenance on Electrical
Equipment 294

Legal Repercussions 295

Safe Versus Unsafe Example 296

Summary 296

Lessons Learned 298

Current Knowledge. 301

CHAPTER 16 Review and Implementation . 305

Introduction 305

Article 110 Review 306

Host and Contractor
Responsibilities. 306

Training Requirements. 306

Electrical Safety Program 306

Working While Exposed to
Electrical Hazards. 306

Use of Equipment. 307

**Lockout/Tagout and an Electrically Safe
Work Condition Review** 307

Article 130 Review 307

Justification. 307

Energized Electrical Work Permit . . . 308

Approach Boundaries. 309

Arc Flash Hazard Analysis 309

Performing Testing Work 310

Uninsulated Overhead Lines 310

Other Precautions for Personnel
Activities. 310

General Requirements for Personal
and Other Protective Equipment. . . 311

Personal Protective Equipment 311

Other Protective Equipment 311

Alerting Techniques 312

Practice Examples Review. 312

Example 1, Test 1 313

Example 2, Test 4 313

Example 3, Test 3 316

Summary 320

Lessons Learned 322

Current Knowledge. 324

APPENDIX A OSHA Citation Examples . . . 327

**Example 1: Violation of 29 CFR
1926.21(b)(2).** 327

**Example 2: Violation of 29 CFR
1926.416(a)(1).** 330

**Example 3: Violation of 29 CFR
1926.95(a)(1).** 331

Example 4: Violation of 29 CFR
1910.333(a)(2). 333

Example 5: Violation of 29 CFR
1910.335(a)(1)(i) 335

APPENDIX B Partnership Agreement 337

Excerpts from the Occupational Safety
and Health Act of 1970: *29 USC 651*. . . .337
  I. Introduction and Identification
    of Partners. 337
  II. Purpose and Scope 338
  III. Goals, Strategies and Measures 339
  IV. Performance Measures 340
  V. Annual Evaluation. 341
  VI. NECA/IBEW Partnership
    Management and Operation. 341
  VII. OSHA's Commitment/Role. 342
  VIII. Participating Employer
    Verification 342
  IX. Employer/Employee Rights. 344
  X. Termination of this Charter 344

**APPENDIX C ANSI/NETA Frequency of
Maintenance 345**

APPENDIX B: Frequency of
Maintenance Tests. 345

**APPENDIX D Maintenance Tests for Molded-
Case and Insulated-Case
Circuit Breakers 348**

7. INSPECTION AND TEST PROCEDURES
7.6.1.1 Circuit Breakers, Air, Insulated-
Case/Molded-Case 348

Tables Referenced by 7.6.1.1. 350

Glossary 353

Index 359

Credits 365

Resource Preview

This new edition brings several new and exciting features designed to enhance learning and synthesis of the material.

Beginning-of-Chapter Features

Chapter Outline A chapter outline lists the chapter's main topics, providing an overview of what will be learned.

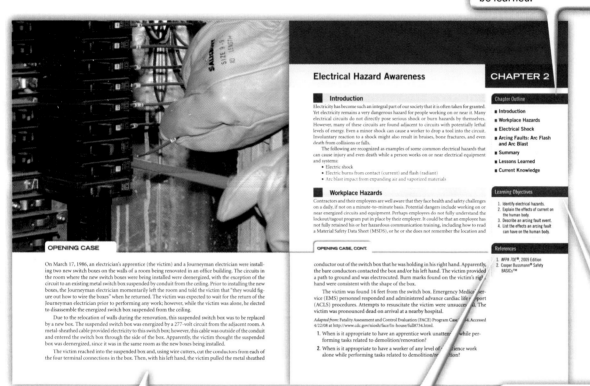

Opening Case Each chapter opens with an official National Institute for Occupational Safety and Health (NIOSH) Fatality Assessment and Control Evaluation (FACE) case study, accompanied by two critical thinking questions to focus the reader's attention.

References Key references that the reader may choose to consult are listed here.

Learning Objectives Learning objectives outline the main goals of the chapter—what the reader should understand upon completion.

Chapter Features

Figure 8.7 Certain tasks may require the worker to interact with enclosed energized equipment, as shown in this photo, increasing the hazard.

Figure 8.8 Electrical safety training may occur in the classroom or on the job, or both.

Figures A dynamic 4-color art program showcases photos and illustrations from leading electrical safety product manufacturers, reflecting current products and equipment.

energized work could include many different kinds of energy. She further reports that the results of trauma following electrical incidents include many different types of injuries.

In addition to being trained to understand the hazards of energized electrical work, training must include the practices and procedures required to guard against those hazards and that injury or death could be the result of being exposed to electrical hazards. Those practices and procedures could include, but are not limited to, lockout/tagout, establishing an electrically safe work condition, and performing an electrical hazards analysis (including a shock and arc flash hazard analysis including labeling, determination of boundaries, and PPE).

OSHA Tip

1910.332(c) Type of training.
The training required by this section shall be of the classroom or on-the-job type. The degree of training provided shall be determined by the risk to the employee.

Type of Training

Like OSHA, NFPA 70E Section 110.6(B) that the required training could be accom variety of settings and that not all training for all tasks. Classroom training or on-the only might be appropriate in some case However, for many tasks, both classroom job training are necessary for the training

It clearly takes a different amount and complexity of training to work safely on a doorbell circuit than it does to work on a 4160-volt branch circuit. The more complex the task becomes, the more thorough and detailed the training must be.

Emergency Procedure Training

Section 110.6(C) addresses emergency procedure requirements. The first of the Section 110.6(C) emergency procedure requirements addresses training on how to release victims from contact. A shock hazard is defined as the possible release of energy caused by contact or approach to energized electrical conductors or circuit parts. Therefore, workers who are exposed to the possible release of energy, either by contact with or approach to energized electrical equipment, must be trained in how to safely remove workers from that exposure. Consider the relationship between this requirement to Section 110.7(F), where it is recommended to identify when a second person could be required and the training and equipment that person should have.

(F) Hazard/Risk Evaluation Procedure. An electrical safety program shall identify a hazard/risk evaluation procedure to be used before work is started within the limited approach boundary of energized electrical conductors and

Key Excerpts Important excerpts from key sources are clearly identified by indented, smaller type.

Tables Tables display numerical information in an easy-to-read format.

Table 2.1   Human Resistance Values for Skin-Contact Conditions

Condition	Resistance (ohms)	
	Dry	Wet
Finger touch	40,000 to 1000	4000 to 15,000
Hand holding wire	15,000 to 50,000	3000 to 6000
Finger-thumb grasp	10,000 to 30,000	2000 to 5000
Hand holding pliers	5000 to 10,000	1000 to 3000
Palm touch	3000 to 8000	1000 to 2000
Hand around 1½-inch pipe	1000 to 3000	500 to 1500
Two hands around 1½-inch pipe	500 to 1500	250 to 750
Hand immersed	N/A	200 to 500
Foot immersed	N/A	100 to 300
Human body, internal, excluding skin	200 to 1000	

N/A, not applicable.
Data source: Kouwenhoven, W. B., and Milnor, W. R., "Field Treatment of Electric Shock Cases—I," AIEE Trans. Power Apparatus and Systems, Volume 76, pp. 82–84, April 1957; discussion pp. 84–87

electrolytes and are highly conductive with limited resistance to alternating electric current. As the skin is broken down by electrical current, resistance drops, and current levels increase.

Consider an example of a person with hand-to-hand resistance of only 1000 ohms. The voltage determines the amount of current passing through the body.

While 1000 ohms might appear to be low, even els can be approached by someone with sweat[...]oninsulating gloves on both hands and a full[...]p of an energized conductor and a grounded[...]n conduit. Moreover, cuts, abrasions, or blisters on[...] negate skin resistance, leaving only internal[...]tance to oppose current flow. A circuit in the[...]50 volts could be dangerous in this instance.

[...]g Ohm's law, we can calculate the current (I[...]es) in a circuit based on the circuit voltage (V[...]) resistance (R). **Ohm's law** states that current[...]) equals voltage (volts) divided by resistance[...]n equation form:

$$I \text{ (amperes)} = \frac{V \text{ (volts)}}{R \text{ (ohms)}}$$

[...]mple 1: $I = \frac{480}{1000} = 0.480$ amps (480 mA)

[...]mple 2: $I = \frac{120}{1000} = 0.120$ amps (120 mA)

[...]rical currents can cause muscles to lock up, [...]in the inability of a person to release his or her

grip from the current source. This is known as the **let-go threshold** current. At 60 Hz, most people have a "let-go" limit of 10 to 40 milliamperes (mA) (Table 2.2).

Potential injury (current flow) also increases with time. A victim who cannot "let go" of a current source is much more likely to be electrocuted than someone whose reaction removes him or her from the circuit more quickly. A victim who is exposed for only a fraction of a second is less likely to sustain an injury.

Extent of Injury

The most damaging path for electrical current is through the chest cavity and head (Figure 2.7). In short, any prolonged exposure to 60 Hz current of 10 mA or more might be fatal. Fatal ventricular fibrillation of

Table 2.2   Effects of Electrical Shock (60 Hz AC)

Response	60 Hz, AC Current (mA)
Tingling sensation	0.5 to 3
Muscle contraction and pain	3 to 10
"Let-go" threshold	10 to 40
Respiratory paralysis	30 to 75
Heart fibrillation; might clamp tight	100 to 200
Tissue and organs burn	More than 1500

Data source: Kouwenhoven, W. B., and Milnor, W. R., "Field Treatment of Electric Shock Cases—I," AIEE Trans. Power Apparatus and Systems, Volume 76, pp. 82–84, April 1957; discussion pp. 84–87.

Vocabulary Vocabulary terms are bolded and underlined in the chapter prose. Terms are defined at the end of the chapter and in the book's glossary.

196   Electrical Safety-Related Work Practices

Step 2: Determine the transformer multiplier:

$$\text{Multiplier} = \frac{100}{\%Z}$$

Note 2: The marked transformer impedance (% Z) value may vary ±10% from the actual values determined by the American National Standards Institute/Institute of Electrical and Electronics Engineers (ANSI/IEEE) test. See UL Standard 1561. For worst-case conditions, multiply transformer %Z by 0.9.

Step 3: Determine the short-circuit current (ISCA) at the transformer secondary:
a) Table 11.7
b) Formula:
3ø transformer (see Notes 3 through 5):
$$I_{SCA\,(L-L-L)} = I_{FLA} \times \text{Multiplier}$$
1ø transformer (see Notes 3 and 4):
$$I_{SCA\,(L-L)} = I_{FLA} \times \text{Multiplier}$$

Note 3: Utility voltage may vary ±10% for power, and ±5.8% for 120-volt lighting services. For worst-case conditions, multiply values calculated in step 3 by 1.1 and/or 1.058, respectively.

Note 4: Motor short-circuit contribution, if significant, may be determined at all fault locations throughout the system. A practical estimate of motor short-circuit contribution is to multiply the total load current by 4 or 6. For worst-case, calculate total motor load current, multiply by 4 and add to step 3.

Note 5: For 3-phase systems, line-to-line-to-line (L-L-L), line-to-line (L-L), line-to-ground (L-G), and line-to-neutral (L-N) bolted faults and arcing faults can be found in Table 11.8 based on the calculated 3-phase line-to-line-to-line (L-L-L) bolted fault.

Use the following procedure to determine the short-circuit current (I_{SCA}) at the end of a run of conductor or busway:

Step 4: Calculate the "f" factor:
a) 3ø line-to-line-to-line (L-L-L) fault (see Note 6):
$$f = \frac{1.732 \times L \times I_{SCA(L-L-L)}}{C \times E_{L-L}}$$
b) 1ø line-to-line (L-L) fault (see Notes 5 and 6):
$$f = \frac{2 \times L \times I_{SCA(L-L)}}{C \times E_{L-L}}$$

c) 1ø line-to-neutral (L-N) fault (see Notes 5, 6, and 7):
$$f = \frac{2 \times L \times I_{SCA(L-N)}}{C \times E_{L-N}}$$

Note 6: See below for an explanation of variables:
L = length (feet) of conduit to the fault.
C = constant from Table 11.3, Table 11.4, or Table 11.5. For parallel runs, multiply C values by the number of conductors per phase.
I_{SCA} = available short-circuit current in amperes (A) at beginning of circuit.
E_{L-N} = line-to-neutral voltage

Note 7: The L-N short-circuit current is higher than the L-L short-circuit current at the secondary terminals of a single-phase center-tapped transformer. The short-circuit current available (I_{L-N}) at the transformer terminals is typically adjusted as follows:
$$I_{L-N} = 1.5 \times I_{L-L} \text{ at Transformer Terminals}$$
The 1.5 multiplier is an approximation and will theoretically vary from 1.33 to 1.67. At some distance from the terminals, depending upon wire size, the L-N short-circuit current is lower than the L-L short-circuit current.

Step 5: Determine the "M" (multiplier):
a) Table 11.6
b) Formula: $M = \dfrac{1}{1+f}$

Step 6: Calculate the available short-circuit current (I_{SCA}) at end of circuit:
3ø line-to-line-to-line (L-L-L) fault:
$$I_{SCA\,(L-L-L)} = I_{SCA\,(L-L-L)} \times M$$
AT END OF CIRCUIT   AT BEGINNING OF CIRCUIT

Note 8: For 3-phase systems, line-to-line-to-line (L-L-L), line-to-line (L-L), line-to-ground (L-G), and line-to-neutral (L-N) bolted faults and arcing faults can be found in Table 11.8 based on the calculated 3-phase line-to-line-to-line (L-L-L) bolted fault.
1ø line-to-line (L-L) fault:
$$I_{SCA\,(L-L)} = I_{SCA\,(L-L)} \times M$$
AT END OF CIRCUIT   AT BEGINNING OF CIRCUIT
1ø line-to-neutral (L-N) fault:
$$I_{SCA\,(L-N)} = I_{SCA\,(L-N)} \times M$$
AT END OF CIRCUIT   AT BEGINNING OF CIRCUIT

Calculations Calculations are displayed in an easy-to-read design and explained step-by-step, facilitating comprehension of equations and their application. Formulas, steps, and examples are shown.

Resource Preview, continued

Boxed Features

70E Highlights The 70E Highlights boxes emphasize important points and excerpts from *NFPA 70E* that directly relate to the material being discussed at this point in the chapter.

Background Boxes These include additional background information that may be beyond the chapter's scope, but helpful to the reader.

OSHA Tips OSHA tips boxes discuss OSHA requirements that relate to material being discussed at this point in the chapter.

CAUTION Boxes Caution boxes note specific safety tips that are important for workers to remember on the job.

End-of-Chapter Features

Lessons Learned Each chapter concludes with Lessons Learned, a FACE case study accompanied by multiple-choice questions about the case—allowing the student to synthesize the information learned and apply it to a real-life case.

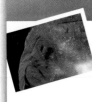

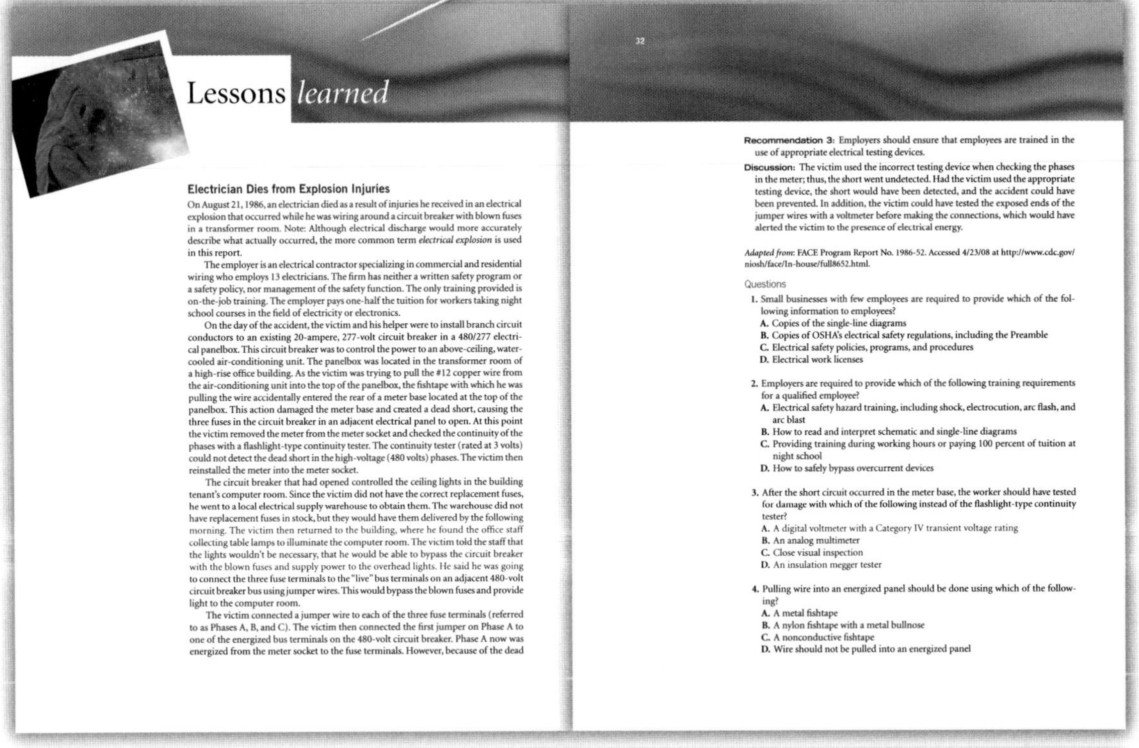

Lessons *learned*

Electrician Dies from Explosion Injuries

On August 21, 1986, an electrician died as a result of injuries he received in an electrical explosion that occurred while he was running around a circuit breaker with blown fuses in a transformer room. Note: Although electrical discharge would more accurately describe what actually occurred, the more common term *electrical explosion* is used in this report.

The employer is an electrical contractor specializing in commercial and residential wiring who employs 13 electricians. The firm has neither a written safety program or a safety policy, nor management of the safety function. The only training provided is on-the-job training. The employer pays one-half the tuition for workers taking night school courses in the field of electricity or electronics.

On the day of the accident, the victim and his helper were to install branch circuit conductors to an existing 20-ampere, 277-volt circuit breaker in a 480/277 electrical panelbox. This circuit breaker was to control the power to an above-ceiling, water-cooled air-conditioning unit. The panelbox was located in the transformer room of a high-rise office building. As the victim was trying to pull the #12 copper wire from the air-conditioning unit into the top of the panelbox, the fishtape with which he was pulling the wire accidentally entered the rear of a meter base located at the top of the panelbox. This action damaged the meter base and created a dead short, causing the three fuses in the circuit breaker in an adjacent electrical panel to open. At this point the victim removed the meter from the meter socket and checked the continuity of the phases with a flashlight-type continuity tester. The continuity tester (rated at 3 volts) could not detect the dead short in the high-voltage (480 volts) phases. The victim then reinstalled the meter into the meter socket.

The circuit breaker that had opened controlled the ceiling lights in the building tenant's computer room. Since the victim did not have the correct replacement fuses, he went to a local electrical supply warehouse to obtain them. The warehouse did not have replacement fuses in stock, but they would have them delivered by the following morning. The victim then returned to the building, where he found the office staff collecting table lamps to illuminate the computer room. The victim told the staff that the lights wouldn't be necessary, that he would be able to bypass the circuit breaker with the blown fuses and supply power to the overhead lights. He said he was going to connect the three fuse terminals to the "live" bus terminals on an adjacent 480-volt circuit breaker bus using jumper wires. This would bypass the blown fuses and provide light to the computer room.

The victim connected a jumper wire to each of the three fuse terminals (referred to as Phases A, B, and C). The victim then connected the first jumper on Phase A to one of the energized bus terminals on the 480-volt circuit breaker. Phase A now was energized from the meter socket to the fuse terminals. However, because of the dead

---

32

Recommendation 3: Employers should ensure that employees are trained in the use of appropriate electrical testing devices.

Discussion: The victim used the incorrect testing device when checking the phases in the meter; thus, the short went undetected. Had the victim used the appropriate testing device, the short would have been detected, and the accident could have been prevented. In addition, the victim could have tested the exposed ends of the jumper wires with a voltmeter before making the connections, which would have alerted the victim to the presence of electrical energy.

Adapted from: FACE Program Report No. 1986-52. Accessed 4/23/08 at http://www.cdc.gov/niosh/face/In-house/full8652.html.

Questions

1. Small businesses with few employees are required to provide which of the following information to employees?
   - A. Copies of the single-line diagrams
   - B. Copies of OSHA's electrical safety regulations, including the Preamble
   - C. Electrical safety policies, programs, and procedures
   - D. Electrical work licenses

2. Employers are required to provide which of the following training requirements for a qualified employee?
   - A. Electrical safety hazard training, including shock, electrocution, arc flash, and arc blast
   - B. How to read and interpret schematic and single-line diagrams
   - C. Providing training during working hours or paying 100 percent of tuition at night school
   - D. How to safely bypass overcurrent devices

3. After the short circuit occurred in the meter base, the worker should have tested for damage with which of the following instead of the flashlight-type continuity tester?
   - A. A digital voltmeter with a Category IV transient voltage rating
   - B. An analog multimeter
   - C. Close visual inspection
   - D. An insulation megger tester

4. Pulling wire into an energized panel should be done using which of the following?
   - A. A metal fishtape
   - B. A nylon fishtape with a metal bullnose
   - C. A nonconductive fishtape
   - D. Wire should not be pulled into an energized panel

35

CURRENT Knowledge

Ready for Review

- Electrical circuits can be serious shock or burn hazards, either in and of themselves or because they are adjacent to a circuit with potentially lethal levels of energy.
- Common electrical hazards include electric shock, electric burns, and arc blast. These can cause injury and death.
- Working on or near energized circuits and equipment is dangerous. Workers must understand and follow their employers' lockout/tagout program.
- Very little current is required to cause injury or death. Even the current drawn by a 7½ watt, 120-volt lamp can be enough to cause electrocution.
- The effects of electric current on the human body depend on circuit characteristics, contact resistance, internal resistance of the body, the current's pathway through the body, duration of contact, and environmental conditions affecting the body's resistance.
- Skin resistance is the electrical resistance introduced by human skin in a circuit in which the body is a conducting pathway.
- Skin resistance can change with moisture, temperature changes, humidity, fright, and anxiety. Cuts, abrasions, or blisters can negate the skin's resistance.
- Ohm's law states that current equals voltage divided by resistance. This principle can be used to calculate the current in a circuit.
- Electrical currents can cause muscles to lock up, [...]
- [...] cal treatment.

- The damage from an electrical shock injury is largely internal and unseen. Damage in the limbs might be the greatest due to the higher current flux per unit of cross-sectional area. Prompt medical attention can minimize the possible loss of blood circulation, the potential for amputation, and death.
- With an arcing fault, the fault current flows through the air between conductors and a grounded part. The product of the fault current and the arc voltage in a concentrated area is tremendous energy.
- An arcing fault results in an intense release of heat (as high as 35,000°F), copper vapor, molten copper, shrapnel, intense light, rapid expansion (hot air), and intense pressure and sound waves. Obviously, these exposures can be fatal to anyone near the arcing fault.
- Arcing fault current magnitude and time duration are among the most critical variables in determining how much energy is released. Other factors include available bolted short-circuit current, the speed of the OCPD, arc gap spacing, the size of the enclosure, the power factor of the fault, the system's voltage, whether the arc can sustain itself, the type of grounding scheme, and the distance of the worker's body from the arc.
- Arc flash is the hazard associated with the energy release during an arcing fault. Arc flash is measured in cal/cm²; protective clothing has an arc rating measured in these units as well.
- Arc blast is the hazard associated with the pressure released during an arcing fault.
- Arcing faults can be extremely hazardous. Contact with the circuit is not necessary for a serious or deadly burn to occur; such burns can occur at distances of more than 10 feet from energized conductors.
- The three types of burns that [...] electrical incidents are [...] flow, burns due to [...] contact burns.
- Even if the burn is [...] sight, hearing, lung [...] spiratory system, th[...] nervous system can [...]

- An arcing fault event can occur in a fraction of a second; a human cannot detect and react to it quickly enough.
- Serious accidents can occur on systems of 600 volts or less. Do not assume that a "low voltage" means low risk. The same precautions are needed when working on these systems as when working on higher voltage systems.
- If an arcing fault occurs, the survivability of the worker depends mostly on the characteristics of the OCPDs, the arc fault current, and the precautions the worker has taken prior to the event.
- The two characteristics of an OCPD that relate to the amount of energy released during an arcing fault include the time it takes the device to open and the amount of fault current the device lets through. The lower the energy released, the better.
- Staged tests conducted by the IEEE Petroleum and Chemical Industry Committee investigated fault hazards. The primary conclusions of these tests were as follows:
  - Arcing faults can release tremendous amounts of energy in many forms in a very short period of time.
  - The OCPD's characteristics can have significant impacts on the outcome.
  - Wearing PPE is extremely important and beneficial.

Vocabulary

arc blast‡: Dangerous condition caused by the release of energy in an electric arc, usually associated with electrical distribution equipment.

arc flash‡: Dangerous condition caused by the release of energy in an electric arc, usually associated with electrical distribution equipment.

arc rating*: [...] attributed to materials that de-[...] an elec-[...] expressed in [...] ined value [...] (ATPV) or [...] hould a ma-[...] onse below

the ATPV value) derived from the determined value of ATPV or E₍BT₎.
   FPN: *Breakopen* is a material response evidenced by the formation of one or more holes in the innermost layer of flame-resistant material that would allow flame to pass through the material.

arcing fault‡: Fault characterized by an electrical arc through the air.

impedance‡: Ratio of the voltage drop across a circuit element to the current flowing through the same circuit element, measured in Ohms (Ω).

incident energy*: The amount of energy impressed on a surface, a certain distance from the source, generated during an electrical arc event. One of the units used to measure incident energy is calories per centimeter square (cal/cm²).

let-go threshold: The electrical current level at which the brain's electrical signals to muscles can no longer overcome the signals introduced by an external electrical system. Because these external signals lock muscles in the contracted position, the body may not be able to let go when the brain is telling it to do so.

lockout§: The placement of a lockout device on an energy isolating device, in accordance with an established procedure, ensuring that the energy isolating device and the equipment being controlled cannot be operated until the lockout device is removed.

Ohm's law‡: Mathematical relationship between voltage, current, and resistance in an electric circuit. Ohm's law states that the current flowing in a circuit is proportional to electromotive force (voltage) and inversely proportional to resistance: I = E/R.

overcurrent protective device (OCPD): General term that includes both fuses and circuit breakers.

tagout§: The placement of a tagout device on an energy isolating device, in accordance with an established procedure, to indicate that the energy isolating device and the equipment being controlled may not be operated until the tagout device is removed.

*This is the NFPA 70E definition.
‡This definition is from the National Fire Protection Association's (NFPA's) *Pocket Dictionary of Electrical Terms*.
§This is the OSHA definition.

Ready for Review A bulleted summary of the chapter's key points; a useful tool for study and review.

Vocabulary Vocabulary terms that were bolded and underlined within the chapter are defined here, and in the book's glossary. Definitions are from key sources such as *NFPA 70E* and OSHA.

Foreword

Electrical Safety-Related Work Practices: NJATC's Guide based on *NFPA 70E*®

For years, many in the industry only considered electrical shock when contemplating worker protection and safe work practices. Today, however, the industry recognizes numerous additional hazards associated with working on or near energized electrical equipment or circuit parts. These include, for example, the hazards associated with an arcing fault including arc flash and arc blast. Consideration must be given to the devastating forces generated when molten copper expands to 67,000 times its original volume as it vaporizes, arc temperatures that can reach 35,000° F, pressures that can reach thousands of pounds per square foot, and shrapnel expelled from ruptured equipment at speeds that may exceed 700 miles per hour. Workers are exposed to these and other hazards even during a seemingly routine task such as voltage testing.

Working on circuits and equipment deenergized and in accordance with established lockout and tagout procedures has always been the primary safety-related work practice and a cornerstone of electrical safety. Only after it has been demonstrated that deenergizing is infeasible or would create a greater hazard, may equipment and circuit parts be worked on energized, and then only after other safety-related work practices, such as insulated tools and appropriate personal protective equipment have been implemented. Examples of additional concerns that should be considered include worker, contractor, and customer attitude regarding energized work, comprehensiveness of an electrical safety plan, appropriate training, the role of overcurrent protective devices in electrical safety, equipment maintenance, and design and work practice considerations. These are a few of the issues that play an important role in worker safety, and are among the topics examined in this publication.

Electrical Safety-Related Work Practices has been developed in an effort to give those in the industry a better understanding of a number of the hazards associated with working on or near energized electrical equipment and circuits and the manner and conditions under which such work may be performed. These work practices and protective techniques have been developed over many years and are drawn from industry practice, national consensus standards, and federal electrical safety requirements. In many cases these requirements are written in performance language. This program will also explore *NFPA 70E, Standard for Electrical Safety in the Workplace* as a means to comply with the electrical safety-related work practice requirements of the Occupational Safety and Health Administration.

Photo Contributors

The NJATC wishes to thank the following companies and individuals for submitting photos for inclusion in this edition.

Boltswitch
Jim Erickson
President

Cooper Bussmann®
Tim Crnko
Manager, Training & Tech Services

Dan Neeser
Senior Technical Sales Engineer

DuPont Engineering
Daniel Doan

H. Landis Floyd, II
Corporate Electrical Safety Competency
    Leader
Principal Consultant—Electrical Safety &
    Technology

Eaton Corporation
Alan Colorito
Senior Marketing Communications
    Manager

Kevin J. Lippert
Manager Codes & Standards

Fluke Corporation
Larry Wilson
Education Program Manager

Ideal Corporation
Rob Conrad
Sales Training & Development Manager

**Institute of Electrical and
Electronics Engineers (IEEE)**
Jacqueline Hansson
IEEE Intellectual Property Rights
    Coordinator

**International Brotherhood of
Electrical Workers (IBEW) Local 98**
James T. Dollard, Jr.
Safety Coordinator

Lower Colorado River Authority
Corby D. Weiss
IC&E Crew Leader
WPS/Lost Pines Power Park

**National Electrical Contractors
Association (NECA)**
Michael Johnston
Executive Director Standards and Safety
NECA

Salisbury Electrical Safety, L.L.C.
Vladimir Ostrovsky
V.P. of International Sales and Marketing

Bill Rieth
Director of Industrial Sales and Training

Shermco Industries
James R. White
Training Director

Thomas & Betts Corporation
David Kendall
Manager—Industry Affair

Greg Steinman
Technical Liaison Engineering Project
    Manager

Tyndale Company
Guy Driesbach
Senior Sales Executive

Barbara Fitzgeorge
Marketing Director

Robert Whittenberger
President

The NJATC also wishes to thank the following NECA photo participants:

Wilson Electric Co.
Ryan Hand
Michael Maffiolli

Morse Electric, Inc.
Brad Munda
Kyle Borneman

Larry McCrae, Inc.
Jeff Costello

J.P. Rainey Company, Inc.
Dave Ganther
Bill Inforzato

Carr and Duff, Inc.
George Novelli
Tom McCusker

Northern Illinois Electrical JATC
Todd Kindred

On August 1, 1995, a 36-year-old electrician's helper was electrocuted after cutting an electrical wire. The incident occurred in a retail store fitting room of a department store located in a suburban shopping mall where the victim and a coworker were replacing the overhead fluorescent light tubes and ballast transformers. The victim's employer was an electrical contractor who employed 40 workers. The employer had a written safety program that included lockout/tagout procedures and employed a safety coordinator who conducted weekly safety meetings. The victim had worked for the company for 12 days.

To save on electrical costs, the store management contracted with their electrical utility company to upgrade the fluorescent lights in the store. This was done under the utility's energy conservation program, in which the utility conducted an energy audit of the customer's building and arranged with a local electrical contractor to have the lights replaced with more energy efficient units. In this case, the victim's company had been subcontracted by the utility to replace all the fluorescent light tubes and ballast transformers in two department stores. The store's management explained that they were familiar with the electrical contractor, since the company had done work at the store before.

A work crew of four men (a foreman-mechanic and three helpers) arrived for work between 8:00 and 8:30 a.m. This was the sixth day the electrical contractor was working at the site, with one more day needed to finish the job. It was also the victim's first day at work at this site, although he had previously worked on replacing the lights at the other department store being serviced under the contract. Under the direction of the foreman, the victim and a second helper started work on replacing the ballasts in the women's fitting room. The foreman reportedly shut off the power to the lights by turning off and locking out the wall switch and then checked the lights with a circuit tester. This deenergized all but one center light fixture, which was on a separate "night light" circuit and remained on. The foreman went to check the breaker box but was unable to find the switch to shut off the remaining light. He asked the victim if he had worked on live wires, and the victim said yes, he had worked on live wires the week before with another crew. The foreman told the victim to go ahead and do the job.

Electrical Safety Culture

CHAPTER 1

Introduction

Too often a culture exists in the workplace where workers are allowed and expected to work on energized electrical circuits. Many electricians believe that performing work on energized circuits is part of their job or expected of them. Many contractors have customers with unrealistic expectations when it comes to energized work. Often a culture is in place that has a tendency to work routinely on or near energized electrical circuits. This tendency to accept the risk of an electrical injury is unacceptable and must change.

This tendency might be due to ignorance of laws that have been in place for decades, not knowing the severity of the hazards, or perhaps not realizing how quickly a task situation might change and cause an energy release. It is less likely that workers, contractors,

OPENING CASE, CONT.

The victim used a 6-ft fiberglass ladder to reach the lights and began removing the tubes and ballast transformers. At about 9:20 a.m., the victim was working on the ladder in the middle fitting room when he cut the energized black wire. The power entered though his hands and exited to the grounded metal doorframe that he was leaning against. The victim's coworker, who was working in the stall beside him, heard the victim say "help me" and saw sparks flying from the wire. The coworker cut the black wire, breaking the contact and releasing the victim, who collapsed against the metal frame.

At this time, the foreman entered the room and helped move the victim to the floor. The store manager called 911 and paged another store employee to the fitting room. The store employees found the victim on the floor with the lights out. A store employee turned on the lights at the switch, which reportedly had been unlocked by the foreman. The police arrived within 5 minutes and found the victim unresponsive. They started cardiopulmonary resuscitation (CPR) until the rescue squad and paramedics arrived. The victim was transported to the local hospital, where he was pronounced dead at 10:38 a.m. The county medical examiner attributed the cause of death to cardiac arrhythmia due to electrocution. The medical examiner's report noted electrical burns on both hands and a possible exit burn at the lower midback.

Adapted from: New Jersey FACE Investigation #95NJ080. Accessed 6/24/08 at http://www.cdc.gov/niosh/face/stateface/nj/95nj080.html.

1. What role did the on-site experience of the victim play in his electrocution?
2. Under what circumstances would performing work on an energized circuit be advisable? Why or why not?

Chapter Outline

- ■ **Introduction**
- ■ **Culture**
- ■ **Hazards**
- ■ **Requirements**
- ■ **Choices**
- ■ **Implementation Choices**
- ■ **Summary**
- ■ **Lessons Learned**
- ■ **Current Knowledge**

Learning Objectives

1. Recognize that a safety culture exists in every organization regarding energized electrical work.
2. Realize the need to change the culture that is in place regarding energized electrical work in order to better comply with regulations and further reduce worker exposure to electrical hazards.
3. Understand that a number of decisions are made before and during the time a worker is exposed to electrical hazards and that some decisions reduce or eliminate electrical hazards.

References

1. National Institute for Occupational Safety and Health (NIOSH) Fatality Assessment and Control Evaluation (FACE) program
2. Occupational Safety and Health Administration (OSHA) 29 CFR Part 1910
3. OSHA 29 CFR Part 1926

and facilities owners would allow energized work once everyone involved in the decision-making process fully understands the laws, requirements, hazards, true costs, and consequences associated with energized work.

Throughout this book, numerous Fatality Assessment and Control Evaluation (FACE) reports are presented to demonstrate the very real hazards of electrical work and to emphasize the need to make fully informed decisions. See page 12 for an explanation of the NIOSH FACE program.

Culture

"I don't care what the law says, I'm going to work it energized." "I'm an electrician; working stuff hot is part of my job." "That's what the customer expects. If my people won't do it, then they'll get another contractor." "It's the president's office, I can't deenergize the circuit to change that ballast." "You can't shut that assembly line down, it will cost too much." "There are people out of work looking for a job, so if I won't work it hot someone else will." "I've been doing it this way for 30 years and nothing has ever happened to me." "I know I should be wearing personal protective equipment (PPE), but it slows me down." "There's no time to shut it down." "That protective equipment is too expensive." "What's the worst that can happen?" "It won't happen to me."

Do any of those statements sound familiar? These are the misguided thoughts that make up a culture where the bottom line, a desire for immediate results, and a false sense of security become more important than safe practice. Are statements like those the exception or do they represent the mainstream culture that is in place in the industry? How did this culture evolve? How did an individual's willingness to take such risks develop? How did the demands and expectations of those in the industry create such pressure? Some of these questions are unanswerable and are presented only for contemplation. Others will be answered in this chapter.

Perhaps the culture is due to a lack of awareness of the hazards that exist. Maybe an "it won't happen to me" mentality has caused the culture. On the other hand, while people are aware of the hazards, they do not realize how quickly a situation can change when things go wrong.

Workplace Pressure

Far too many electricians believe that working on energized circuits is part of their job or expected of them (**Figure 1.1**). A tendency to work on or near electrical circuits while energized and accept the risk of an elec-

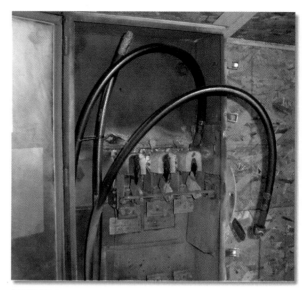

Figure 1.1   An example of the result of energized work.

trical injury is an unacceptable culture. This mind-set change must be recognized by all involved in the decision-making process.

Contractors might feel pressured by customers to work on energized equipment when a shutdown is necessary. Workers also might feel pressured by management to perform energized work when it is not justified. In reality, however, workers who accept this risk expose themselves to injury or death. They also expose the contractors and their clients to undue risks of increased insurance premiums and loss of production. Are facility owners getting the best value if they employ the services of a contractor that is not truly qualified? Chances are, customers do not really understand the total costs and risks associated with energized work.

A well-informed client understands the hazards of energized work and the financial implications associated with an electrical incident. Equipment or circuits that cannot be shut down for a few minutes ultimately might be shut down for days or weeks or more. A well-informed client is less likely to permit energized work, much less expect it.

Historical Perspective

The Background box on *Historical Methods for Testing Voltage*, which follows, demonstrates one way that acceptance of risk and the culture in place today could have evolved. The current culture's outdated work practices might be the result of the way electricians were taught in the past. Readers should contemplate the guidance that was once given to electrical workers. The recommended practice at that time actually was to check for voltage using the fingers.

BACKGROUND

Historical Methods for Testing Voltage

As late as the mid-1900s, electricians performed testing for voltage procedures, employing a variety of less desirable techniques when viewed from today's prospective. On lower voltages typically found on bell, signal, and low voltage control work, the presence of voltage (or pressure as it commonly was called) could be tested using the *tasting method*:

- A method of stripping the ends of the conductors from both sides of the circuit and placing the ends of these conductors a short distance apart on the tongue could be used to determine the presence of voltage.
- The "testee" or more appropriately, "tester" would experience a burning sensation followed by a slight salt taste. Depending on the amount of voltage present, holding one of the conductors in the bare hand and touching the other to the tongue could also be used. In this case, the body was acting as a voltage divider, lessening the burning sensation on the tongue.

Other variations of voltage testing included standing on wet ground when one end of the voltage system was grounded, while touching the tongue with the other terminal of the voltage source. This also was an "approved method." Individuals using this method often stated that once you perform these test methods, the end result was not often forgotten.

On higher voltages typically found in building power applications, the *finger method* was employed as an acceptable method of determining the presence of voltage in circuits of 250 volts or less:

- Electricians would test the wires for voltage by touching the conductors to the ends of the fingers on one hand. Often, due to skin thickness, skin dryness, and calluses, the electrician would have to first lick the fingers to wet them to be able to sense the voltage being measured.

This method was billed as easy and convenient for determining whether live wires were present. The individual electrician's threshold for pain determined whether or not this was an acceptable method for everyday use. Some electricians supposedly had the ability, depending on the intensity of the sensation, to determine the actual voltage being tested.

Source: American Electrician's Handbook: A Reference Book for Practical Electrical Workers, 5th Edition, by Terrell Croft (revised by Clifford C. Carr). New York, NY: McGraw-Hill, 1942.

Is it any wonder that today's culture does not fully understand and implement safe work practices? Safe work practices today do not allow procedures such as the "tasting method" and the "finger method" to test for the presence or absence of voltage. Much like the advances made in the medical community, significant changes in safe work practices in the electrical industry have occurred. No longer do those undergoing surgery have to bite the bullet or take a shot of whiskey. The Background box on page 4 on *Cardiopulmonary Resuscitation: Accepted Practice in the Early 1900s* contains excerpts from a procedure that once was considered an appropriate method of CPR.

CPR, like other practices and procedures, has evolved over time. Clearly, electrical safety-related work practices, like all practices and procedures, require updating over time. Which generation of practices and procedures are used in the workplace today? Are electrical workers using procedures from the early 1900s or procedures from the 21st century?

Hazards

Perhaps a full understanding of the hazards is what is missing. One would not think of petting a shark. A bomb squad worker would not disarm a bomb in a T-shirt and jeans. Performing energized work without a full understanding of the associated hazards is like entering an excavation without a trench box. Will the walls collapse? Perhaps not today, maybe not this time. How many bullets have to be out of the chamber before it is safe to play Russian roulette? The safest answer, of course, is all of them. Electrical workers must do the following as a minimum:

- Eliminate the hazard.
- Deenergize and follow all of the necessary steps of the established lockout/tagout program unless the employer demonstrates a true need for energized work.
- Perform a written hazard/risk analysis.
- Engineer out the hazards.

BACKGROUND

Cardiopulmonary Resuscitation: Accepted Practice in the Early 1900s

A review of several of the techniques used as early day methods of resuscitation show that medical technology has come a long way since the early 1900s.

The primary method of treating an individual who had experienced heart failure and/or respiratory arrest was simple. This primary method required two rescuers to perform the resuscitation procedure:

- After placing the victim on his back one rescuer would grab and wiggle the victim's tongue, while the other rescuer would work the victim's arms back and forth to help induce breathing.

While possibly resuscitating some stricken individuals, a secondary approach was to be used should the first method fail:

- In cases where manual inflation of the lungs was attempted with no success, an attempt to cause the victim to gasp for air was performed. To initiate this gasping, the rescuers would insert two fingers into the victim's rectum, pressing them suddenly and forcibly towards the back of the individual.

Needless to say, it is not hard to understand why today's CPR methods provide more favorable results for both the victim and the rescuer.

Source: The Fire Underwriters of the United States, *Standard Wiring: Electric Light and Power.* H.G. Cushing Jr, New York, NY 1911.

- Provide adequate protection against hazards when the need for energized work is demonstrated.

Consider a comparison between the hazards of driving an automobile with those associated with working on or near energized electrical equipment. Everyone is aware of the hazards of driving. The motoring public continues to drive in spite of knowledge of the potential hazards. A crash could occur due to any number of causes, such as a wet or icy roadway, a tire blowout, a distraction from a passenger, a phone call, or driving too

OSHA Tip

Occupational Safety and Health Administration (OSHA) 29 CFR 1910.333(a)(1)
Deenergized parts. Live parts to which an employee may be exposed shall be deenergized before the employee works on or near them, unless the employer can demonstrate that deenergizing introduces additional or increased hazards or is infeasible due to equipment design or operational limitations. Live parts that operate at less than 50 volts to ground need not be deenergized if there will be no increased exposure to electrical burns or to explosion due to electric arcs.

fast. Being prepared for the unexpected is important. Protective systems such as seatbelts and airbags are in place to reduce the likelihood of injury or death when things go wrong, but only if they are used. However, the protection has limitations. Motorists still might suffer severe injury or perish even if protective systems are in place and used.

It is much the same with the hazards associated with working on or near energized electrical equipment. People continue to be exposed to electrical hazards in spite of the knowledge of those hazards. These hazards include fire, falls and falling objects, electrical shock, and the hazards associated with **arcing faults**, including **arc flash** and **arc blast** (Figure 1.2). An arcing fault could be initiated by a dropped tool or by operating equipment that has not been maintained properly, for example. Will it happen? It is a well-known fact that *it happens*. The statistics speak for themselves. Protective systems, such as work practices and protective equipment, can reduce or eliminate exposure to the hazards,

Figure 1.2   An example of the results of an arcing fault.
© 1997 Institute of Electrical and Electronics Engineers. Still photo extracted from the Cooper Bussmann *Safety BASICs Handbook* with permission from Cooper Bussmann.

🔥 7OE Highlights

An electrically safe work condition is defined in *NFPA 70E®* as "A state in which an electrical conductor or circuit part has been disconnected from energized parts, locked/tagged in accordance with established standards, tested to ensure the absence of voltage, and grounded if determined necessary."

*NFPA-70E® is a registered trademark of the National Fire Protection Association, Quincy, MA.

if they are used. But again, workers may still suffer injury or death if circuits and equipment are not worked on in an **electrically safe work condition**.

The hazard of electrical shock has been recognized since the dawn of electricity. The industry has evolved and made great strides to protect against electrical shock through the use of **ground-fault circuit interrupters (GFCI)** and rubber protective goods such as insulating gloves and blankets. These products are effective when used and maintained properly. Even with these advances, injury and death still occur from electrical shock (**Figure 1.3**). The Bureau of Labor Statistics (BLS) data for electric shocks in nonfatal cases involving days away from work for the period 1992–2001 indicates that there were an average of 2726 cases annually in private industry.

But what about the lesser-known hazards of arc flash and arc blast? Unfortunately, electrical burns happen all too frequently. The BLS data for nonfatal cases involving days away from work for the period 1992–2001 indicate an average of 1710 electrical burns per year (peaking at 2200 in 1995) in private industry. That averages out to nearly one human being suffering the consequences of electrical burns every hour, based on a 40-hour work week.

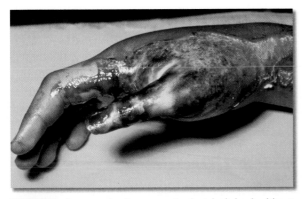

Figure 1.3 An example of exposure to electrical shock without PPE.

What are the effects of exposure to electrical hazards? An examination of multihazard electrical incidents and the incident effects on survivors is provided in a paper delivered by Mary Capelli-Schellpfeffer, MD, MPA, of CapSchell, Inc. at the 2004 IEEE IAS Electrical Safety Workshop. Excerpts from Dr. Capelli-Schellpfeffer's paper are provided in the Background box entitled *Trauma Following Electrical Events*.

BACKGROUND

Trauma Following Electrical Events

The clinical spectrum of electrical incident effects on survivors ranges from the absence of any external physical signs to severe multiple trauma. Reported neuropsychiatric difficulties can vary from vague complaints seemingly unrelated to the injury event by their distance in time or apparent severity to effects consistent with anoxic brain injury accompanying an electrical trauma. In addition to physical limitations, complaints commonly described in electrical incident survivors include hearing loss, headache, memory changes, disorientation, slowing of mental processes, agitation, confusion, irritability, affective disorders, and posttraumatic stress disorder (PTSD; severe anxiety resulting from a traumatic experience).

The evaluation and treatment of electrical incident survivors can be variable, as there is little information available to provide rigorous decision-making around the mental health care of these patients. Opinions differ about the nature and cause of patient symptoms, and the relationship between symptoms and factors like trauma severity, litigation, or premorbid personality. Not all survivors develop cognitive and emotional difficulties, and no consistent relationship has been established between characteristics, such as age, injury-related characteristics (e.g., voltage, current source, work error), and neuropsychological test performance.

Questions remain as to how electrical exposure affects central nervous system function. The pattern of neuropsychological effects suggests diffuse cerebral injury. More-

(Continued)

over, the effects of electrical incidents may produce emotional disturbance through damage to the limbic system or hypothalamic-pituitary axis. It is noteworthy that from the therapeutic perspective, it has been appreciated that the medical application of electric current in proximity to the brain during electroconvulsive therapy (ECT) affects mental status, psychiatric condition, and neuromuscular function. While the biologic mechanisms for the individual responses seen following ECT remain to be articulated, persistent alterations in patients' neuropsychiatric condition following ECT are well documented and in effect, often represent desired clinical outcomes.

Regarding electrical incident clinical effects in survivors, a study [Pliskin, Capelli-Schellpfeffer, et. al., 1998. Neuropsychological sequelae of electrical shock. *Journal of Trauma* 44 (4):709–715] analyzed the experience of the largest reported series of electrical injury survivors with neuropsychological complaints. All patients had peripheral electrical contacts (i.e., shock) with no evidence on history or examination of direct mechanical electrical contact with the head. A total of 45 males and 8 females were included in the final analysis. These individuals had a mean age of 38.5 years (range, 22 to 70 years) and a mean educational level of 13.1 years (range, 8 to 18 years).

The mean time between injury and completion of the measures used in this study was 11.2 months (range, 0.2 to 66.7 months). Twenty of the 53 patients were employed as electricians or line operators at the time of injury. There were also 7 mechanics or railroad workers, 5 office workers, 3 factory workers, 3 service technicians, 2 food service workers, 2 police officers, as well as 11 individuals with other occupations. Forty-four patients were injured on the job, and 9 were injured during nonvocational activities. At the time of follow-up contacts, 30 (56.6%) patients were working again, 18 (44.0%) patients were unemployed or retired, 1 patient was deceased, and 4 patients could not be contacted. Twenty-one of the 53 patients were injured by voltage sources less than 1000 volts (39.6%), and 27 patients sustained voltage exposures greater than 1000 volts (50.9%). Forty-four patients were hospitalized for observation or to receive initial treatment for their injuries (83.0%), while 9 patients were released after initial evaluation. Twenty patients underwent surgery for their injuries (37.7%), 32 patients received either nonsurgical treatment or no treatment, and the treatment history for 1 patient was unknown. Sixteen of the 53 patients (30.2%) sustained a loss of consciousness as a result of their electrical accident, and 4 patients (7.5%) experienced cardiac arrest. Twenty-nine patients complained of ringing in their ears (tinnitus) (18%), and 5 patients reported a loss of hearing (3%). In the 49 patients for whom complete data were available, there was no significant relationship between the reported neuropsychological symptoms, Beck Depression Inventory, self-rated memory complaints, and injury experience parameters, including voltage exposure, loss of consciousness, trauma severity, or litigation. Blast effects from the electrical incidents may help to explain why these survivors without external signs of electrical contact presented with nervous system or hearing impairment.

The potential for injury and death is directly related to the energy output from an electrical fault. However it is the actual energy exposure (i.e., the energy transferred to the individual) that provokes a biologic effect (**Figure 1.4**). Critical in predicting the extent of injury after an electrical incident are the quantity and form of energy transfer and the affected individual's biologic characteristics.

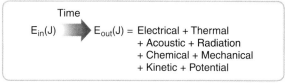

Figure 1.4 Energy transformation. Eight forms of energy are released, all of which can cause damage.

The dose or amount of energy transferred to an individual involved in an electrical incident can be conceptualized by considering the efficiency of the coupling between the energized source and the individual during the energy release. This "efficiency" is a function in part of the current, exposure duration, distance from the source, barriers used, surface area of the body exposed, and the material properties of biologic tissues, including the following:
- Tissue conductance
- Tissue impedance

- Tissue resistance
- Absorbance of human "biomaterials" (i.e., the water, lipids, fats, proteins, and minerals that constitute the body)

It is not simple to predict the possible energy transfer that may have occurred after an electrical incident. This information is routinely lacking during the immediate period of an incident investigation and may only be approximated later. Employees who may be at risk of an injury (while doing a job near energized equipment) also lack this information in concept and detail.

Arcing Faults Present Multiple Hazards

It is important to note that in an electrical incident, injury or fatality can have different causes and are not limited to a "heat" injury. In general, use of the word "burn" is often misunderstood to mean "heat injury" alone. Electrical current flow absent heating can result in a "burn." Radiation injury can also create a "burn," as for example with ultraviolet radiation damage to the corneas. There are also traumatic effects from falls, crush, and shrapnel.

In the acute or rehabilitative care setting, the relationship between survivor symptoms and the multiple effects from electrical, thermal, acoustic, and radiation forces may not be apparent. However, the magnitude of the physical forces generated during an electrical incident explains in part the resemblance of some electrical incident survivors to those with mild traumatic brain injury. In particular, the acoustic component of the electrical incidents can be a mechanism for brain injury. As electrical flash/blast reportedly occurs in almost 75 percent of survived electrical incidents, the contribution of the explosion forces is presumed similar in its injury mechanism to the acceleration-deceleration event experienced by head trauma patients in motor vehicle accidents. In the triage setting, these effects may be obscured by the lack of physical signs readily recognizable by Emergency Medical Technicians (EMTs) or triage clinicians.

Knowledge of the nature of electrical incident effects is evolving, because the electrical system itself is evolving. As power density increases across more compact spaces, more fault energy is available in electrical installations. With installations designed to be compact, smaller spatial cushions exist, with less barrier protection emerging in in-

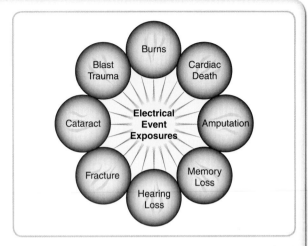

Figure 1.5 Trauma following electrical events consists of various potential injuries.

novative designs. The result may be less time and less physical distance through which an unintentional release of energy can be dissipated during an electrical incident. With higher power density and smaller, more compact spaces, risk of collateral damage from electrical incident effects increases, including injuries and fatalities (**Figure 1.5**).

High power density in a compact space has bomb-like potential, with characteristics of closed blast effects. For example, 1 megawatt (MW) is roughly equivalent to one stick of dynamite. One stick of dynamite is roughly ⅓ lb of TNT equivalent. So, a 100 MW scenario can be represented as loaded with 100 sticks of dynamite. Or, a 300 MW scenario can be represented as loaded with 100 lbs of TNT. A scenario in a closed space creates the possibility of closed blast effects, where the walls of an enclosure confine the blast, and energy can be dispersed only in one direction—toward the worker. This is unlike a scenario in an open space, such as a lineman working at the top of a utility pole in a vacant field, which allows energy to be dispersed in all directions (a free blast scenario).

With knowledge of the electrical power in a workplace scenario, hazards and possible electrical incident effects can be discussed in terms that are intuitively more obvious.

Adapted from: *Facts About Trauma About Electrical Events* by Mary Capelli-Schellpfeffer, MD, MPA, CapSchell, Inc. Reported and presented at the 2004 IEEE IAS Electrical Safety Workshop held in Oakland, California. Reprinted with permission. Courtesy of Mary Capelli-Schellpfeffer, MD, MPA.

70E Highlights

NFPA 70E includes fine print notes (FPN). Per *NFPA 70E* Section 90.5(C), these are informational only and are not enforceable.

Now that the effects of exposure to electrical hazards have been explored, it is important to recognize that electrical protective equipment provides limited protection to electrical hazards, much like seatbelts and airbags provide limited protection from the hazards that could be encountered in an automobile accident. One would not expect a hardhat to protect a person's head against a safe falling. Neither should it be expected that arc-rated **flame-resistant (FR)** clothing and other protective equipment would always allow a worker to escape an incident unscathed. All protective equipment has limitations. Although FR apparel might protect against a thermal event, many other hazards might be associated with an incident. Explosive effects, including shrapnel, could rip through protective clothing, a pressure wave could rupture eardrums, or the differential pressure that results from the wave might collapse lungs and damage other internal organs.

70E Highlights

NFPA 70E Section 130.7(A) Fine Print Note (FPN) No. 1:
The PPE requirements of 130.7 are intended to protect a person from arc flash and shock hazards. While some situations could result in burns to the skin, even with the protection selected, burn injury should be reduced and survivable. Due to the explosive effect of some arc events, physical trauma injuries could occur. The PPE requirements of 130.7 do not address protection against physical trauma other than exposure to the thermal effects of an arc flash.

▪ Requirements

Does the majority of the energized work performed today fall within what OSHA recognizes as justification to work on an energized circuit? While it would be difficult to quantify definitively, much energized work being performed probably would not be considered justified by OSHA. OSHA requires employers to furnish each employee a place of employment free from recognized hazards that are causing or are likely to cause death or serious physical harm. Live parts to which an employee might be exposed must be deenergized before the employee works on or near them, unless the employer can demonstrate that deenergizing introduces additional or increased hazards or is infeasible due to equipment design or operational limitations.

Why are some laws followed routinely while others, such as refraining from working on exposed energized electrical equipment, are all too often ignored? Perhaps it is ignorance of the laws in effect. **Figure 1.6** and **Figure 1.7** are a before and after look at electrical equipment and help illustrate why equipment must be locked out and tagged out in accordance with established policy, unless the need to work energized is demonstrated.

Figure 1.6 An example of equipment before energized work was performed.

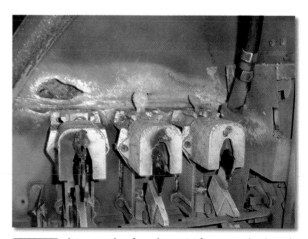

Figure 1.7 An example of equipment after energized work was performed.

What is the company policy on working on or near energized equipment? "We only work on energized equipment when we absolutely have to." That's closer to what is allowed by law. But how is "have to" defined? When it's inconvenient to deenergize? When it will cost too much to shut it down? When we know better, but it will only take a minute? Is "we never work on energized circuits" the policy in place? That's a good start but probably not entirely possible. Energized work includes voltage testing. Voltage testing is among the tasks that are infeasible to perform deenergized. As the OSHA and *NFPA 70E* requirements are fully explored in subsequent chapters, it will become apparent that the vast majority of work performed on energized equipment does not qualify as work allowed by OSHA and *NFPA 70E*.

Choices

Many choices about how to perform the task are made long before a worker is placed before a hazard. Other decisions are made along the way. Many are made just before, and even during, the performance of a task.

Many factors should be considered in creating a safe work environment. The following questions should be included in the hazard/risk analysis:

- Has an electrical safety plan been developed and implemented?
- Has appropriate training been provided?
- Are safe work practices in place and understood, including lockout/tagout and placing equipment in an electrically safe work condition?
- Has the required protective equipment been provided?
- Has an attempt been made to reduce the potential worker exposure through work practice or design considerations, such as arc-resistant switchgear or remote switching?
- Was the overcurrent protection selected solely to protect the equipment or was worker protection considered as well?
- Was a current-limiting overcurrent protective device (OCPD) selected?
- Was the OCPD applied within its rating?
- If the OCPD was replaced, was the appropriate degree of current limitation applied?
- Has the OCPD been maintained properly?
- Has the impedance on the transformer on the supply side of the service changed?

These are among a few of the concerns that must be addressed when the hazard/risk analysis is conducted.

Appropriate Priorities

A corporation's management might say they could not afford to develop a safety plan, to provide necessary training, or to provide the appropriate PPE and insulated tools. Perhaps the safety manager thinks that providing a nice gift at the company picnic is more important. In such a case, corporate priorities should be analyzed. While gift-giving is a nice gesture, it should not take precedence over spending money on crucial items such as appropriate PPE.

Training and Personal Responsibility

Some might argue that they were never trained or were never provided with the required protective equipment. Imagine a case where a worker is trained in safe work practices and then fails to implement that training. Is it lack of training? Is it lack of a safety plan? Is it lack of understanding of the hazard? The worker might know better, but the required protective equipment is left in the truck. Why was this decision made?

Costs of Energized Work Versus Shutting Down

What factors are considered when the decision is made that work must be performed on energized equipment? What is the true cost if something goes wrong? As mentioned earlier, a shutdown that "can't" be scheduled could become an unscheduled shutdown. Which type of shutdown costs more? Have the costs associated with human life, equipment, loss of production, insurance premium increases, potential exclusion from bid lists, corporate image, and worker morale been considered in the decision-making process? Can the equipment be shut down at night when very few if any people will be inconvenienced? Can equipment be shut down over the weekend? Can't a shutdown be scheduled at some point?

Consider whether supervisors are rewarded for safety shortcuts. Are the costs associated with injuries and citations charged against the job, or are they looked at as a "cost of doing business" and lumped in under overhead? If they are not charged against the job, a job site with injury after injury may appear to be more profitable than it really is. Does this type of manager make the decisions in the best interest of a company or its clients?

Time Pressure

Imagine an electrician operating out of a service truck responding to a report of a transfer switch that "blew up" at 3 a.m. at a nursing home. What is that worker thinking while driving to that site in the wee hours of the morning? That worker will be expected to get the power restored "as soon as possible." Does everyone understand "as soon as possible" in the same way? One person might interpret that to mean as soon as the electrician steps foot on the property. Another might interpret that to mean as soon as it is safely possible to do so. Does the worker know how to evaluate the magnitude of the hazards present? How and when are these decisions made?

Employee Qualification

Is the electrician qualified to perform the tasks to which they are assigned? Responding to an emergency call at 3 a.m. is not necessarily the best time to "get qualified." Is the personal and other protective equipment in the truck? Is the protection adequate? Is the electrician qualified to make those decisions at the job site, or will someone from the engineering department need to be dragged from a restful sleep as well? Has the worker been trained to know how to use the equipment properly and understand its limitations? How and when are these decisions made?

Implementation Choices

At some point a decision is made on how to implement an electrical safety program, including implementation of safety-related work practices and personal and other protective equipment.

An examination of the estimated cost of implementing an arc flash PPE program is provided in the article "Escaping Arc Danger" by H. Landis Floyd, published in the May/June 2008 issue of *IEEE Industry Applications Magazine*. The article explored the cost of arc flash clothing implementation through three examples: (1) doing nothing, (2) selecting PPE using the tables in *NFPA 70E*, and (3), hazard analysis and PPE selection based on the results of a detailed arc flash hazard analysis. Note that a comparison of the costs of implementing

the three scenarios is summarized in **Table 1.1**, which is Table 1 from the article.

Doan and Floyd [12] estimated the total cost of implementing an arc flash protective clothing by considering three options: do nothing, minimum compliance, and application of protective measures based on the state-of-the-art hazard analysis methods. This comparison is summarized in Table 1. This is based on the assumption that all recommendations are followed. In practice, human error and other factors can increase injury frequency and overall costs with any of these options.

The quality of implementing an arc hazard mitigation program can vary from doing nothing, to minimum compliance, to a state-of-the-art program with arc hazard analysis as the basis for a full range of control measures. A "do nothing" approach is very much out of step with regulatory requirements, and evolution in electrical safety knowledge, as previously noted, is likely the most costly long-term choice. Advancements in the arc flash hazards mitigation in the mining industry have been hampered somewhat by the delay in recognizing the relevance of *NFPA 70E* to mining operations.

One of the options provided in *NFPA 70E* 2004 for selecting arc flash PPE is based on tables, which provide lists of common tasks, with appropriate arc flash protective equipment noted for each task. These tables can be useful, but they can also be misapplied. The explanatory footnotes accompanying the tables may be overlooked. These notes explain that the electrical system must have certain specifications for the tables to be applicable. Failure to assure that the electrical system meets these requirements can result in either underrated or overrated PPE. Both conditions can have serious consequences. Underrated PPE can result in serious injuries from arc hazard exposure. In this case, the thermal protection rating of the multilayer system of garments was less than the exposure, resulting in severe burns, even though the inner garment layer is relatively intact. On the other extreme, overrating of PPE can lead to unnecessary heat stress.

An approach that is based on a detailed arc hazard assessment enables the identification of exposures where

Table 1.1 Cost Comparison (in U.S. $1000) of Average 5-Year Costs for Sample Company for Three Options for an Arc Flash Mitigation Program*

Method	PPE Costs	Analysis Costs	Injury Costs	Total Costs
No arc flash PPE	$0	$0	$20,800	$20,800
Two hazard level PPE choice	$1570	$100	$6150	$7820
Detailed analysis	$835	$2000	$150	$2985

*Original Table 1 from article entitled "Escaping Arc Danger."
Reprinted with permission. © 2008 Institute of Electrical and Electronics Engineers.

engineering design or administrative controls can reduce the severity or frequency of exposure, reduce the frequency of potential arc flash events, and better assure that the PPE is appropriately rated for exposures.

© 2008 Institute of Electrical and Electronics Engineers.

Summary

The statements at the beginning of this chapter, which often illustrate the electrical safety culture, might seem justified and realistic. It might be true that the worker is pressured to do a job quickly or that PPE is annoying to wear. Such a situation increases the potential for serious injury and perhaps, worse, death. When a work situation is this dangerous, issues such as time pressure and inconvenient PPE are irrelevant. The need to ensure the worker's safety overrides any other concerns that, while real and difficult, are not as important.

The statements could even be considered selfish, in that they prioritize the customer's needs over the need of the worker to remain alive and uninjured. Most workers have family and friends who will suffer the consequences of these decisions when things go wrong. Whether an incident results in an injury or a fatality, family and friends suffer emotionally and financially. Think of how many people suffer when a worker is injured or killed: A breadwinner, a companion, a best friend, a father, a mother, a son, or a daughter might be lost. If the worker does survive, who will care for him or her if he or she requires months or years of rehabilitation?

A person might get only one chance to make a decision that those left behind will regret for years, if not a lifetime. What would a loved-one recommend when asked if it is worth the risk of ignoring the rules "just this one time?" There are many reasons why risk is taken. At times, it could be calculated risk. It might be uninformed risk. Whatever the reason, it is not likely worth the risk.

There is little question that, all too often, a culture is in place where workers are allowed and expected to work on energized equipment. Too often a culture is in place that has a tendency to work routinely on or near energized electrical circuits. What is most important—that the worker performs the work without becoming injured or killed—happens only as workers, contractors, and customers become educated about the hazards and how to properly handle them. Workers, contractors, and their customers must be made aware of the requirements that are in place, the hazards that exist, the decisions that can and should be made, and the true costs associated with an incident when things go wrong. It is a multifaceted challenge that requires a multifaceted education process and a multifaceted change in culture.

BACKGROUND

Introduction to the NIOSH FACE Program

The National Institute for Occupational Safety and Health (NIOSH) FACE program is a research program designed to identify and study all fatal occupational injuries, including electrical. The goal of the FACE program is to prevent occupational fatalities across the nation by identifying and investigating work situations at high risk for injury and then formulating and disseminating prevention strategies to those who can intervene in the workplace.

FACE Program's Two Components

NIOSH in-house FACE began in 1982. Participating states voluntarily notify NIOSH of traumatic occupational fatalities resulting from targeted causes of death that have included confined spaces, electrocutions, machine-related fatalities, falls from elevation, and logging incidents. In-house FACE currently is targeting investigations of deaths associated with machinery, deaths of youths under 18 years of age, and street/highway construction work zone fatalities.

The FACE program began operating as a state program in 1989. Currently, nine state health or labor departments have cooperative agreements with NIOSH for conducting surveillance, targeted investigations, and prevention activities at the state level using the FACE model.

FACE is a research program; investigators do not enforce compliance with state or federal occupational safety and health standards and do not determine fault or blame.

Primary Activities of the FACE Program

The primary activities of the FACE program include the following:
- Conducting surveillance to identify occupational fatalities.
- Performing investigations of specific types of events to identify injury risks.
- Developing recommendations designed to control or eliminate identified risks.
- Making injury prevention information available to workers, employers, and safety and health professionals.

On-Site Investigations

On-site investigations are essential for observing sites where fatalities have occurred and for gathering facts and data from company officials, witnesses, and coworkers. Investigators collect facts and data on what was happening just before, at the time of, and right after the fatal injury. These facts become the basis for writing investigative reports.

During the on-site investigations, facts and data are collected on items such as the following:
- Type of industry involved
- Number of employees in the company
- Company safety program
- The victim's age, sex, and occupation
- The working environment
- The tasks the victim was performing
- The tools or equipment the victim was using
- The energy release that results in fatal injury
- The role of management in controlling how these factors interact

Each day, on average, 16 workers die as a result of a traumatic injury on the job. Investigations conducted through the FACE program allow the identification of factors that contribute to fatal occupational injuries. This information is used to develop comprehensive recommendations for preventing similar deaths.

FACE Information and Reports

Surveillance and investigative reports are maintained by NIOSH in a database. NIOSH researchers use this information to identify new hazards and case clusters. FACE information may suggest the need for new research or prevention efforts or for new or revised regulations to protect workers. NIOSH publications are developed to highlight high-risk work situations and to provide safety recommendations. These reports are disseminated to targeted audiences and are available on the Internet through the NIOSH homepage or through the NIOSH publications office.

The names of employers, victims, and/or witnesses are not used in written investigative reports or included in the FACE database.

Adapted from: Fatality Assessment and Control Evaluation (FACE) program web site, http://www.cdc.gov/niosh/face/brochure.html. Accessed 7/7/08.

Lessons *learned*

30-Year-Old Electrician Electrocuted in Maryland

On July 29, 1987, an electrician was electrocuted when he contacted an energized conductor while installing new wiring when he and a coworker were continuing work on the relocation of conduit and wiring for a carbon dioxide (CO_2) system in which CO_2 is used as a coolant for refrigeration. The employer is an electrical contractor who specialized in commercial and industrial electrical services and employs 40 full-time workers. The owner of the company manages the safety function. Individual foremen are responsible for on-site safety. A safety program that includes electrical safety had been developed and was administered on a routine basis.

Two electrical supply systems powered the CO_2 unit: (1) an 8 American Wire Gauge (AWG), 3-phase, 480 VAC, 50-ampere system supplied power to the compressor motor and (2) a 10 AWG, 3-phase, 480 VAC, 15-ampere system supplied power to the controls and gauges. Disconnect switches for the two systems were located in adjacent panel boxes.

On the afternoon of the incident, the victim and a building maintenance man proceeded to the panel box area to disconnect the power to the CO_2 unit. The two panel box covers were marked in pencil, and the pencil marking for the compressor motor was barely legible. The power supply for the controls and gauges was opened, but the power supply for the compressor motor apparently was overlooked, and the disconnecting means remained closed.

The victim disconnected the energized conductor from the compressor motor and taped the energized ends with electrical tape. He then pushed a metal fish tape through the conduit until its end protruded. The victim then attached the conductor to the fish tape. He pulled the fish tape back through the conduit until the end of the energized wire protruded from it. At that time, the victim removed the electrical tape from the energized wire, while part of the fish tape remained in the conduit. The victim was holding the fish tape, which was grounded against the conduit, when he contacted the energized wire. This action completed the path to ground and the victim was electrocuted.

After contacting the energized wire, the victim collapsed, breaking contact. A coworker, working in the vicinity, saw what had happened and ran to a telephone to call for assistance. The local emergency medical service responded in approximately 5 minutes and began CPR. The victim was transported to a nearby hospital, where he was pronounced dead. The medical examiner listed the cause of death as electrocution.

Recommendations/Discussion

Recommendation 1: Disconnecting means and circuits should be adequately identified.

Discussion: The control panel box cover for the compressor motor circuit disconnect had been marked in pencil. Only with close scrutiny were the markings legible. Current OSHA standard 1910.303(f)(2) states that: "Each service, feeder, and branch circuit, at its disconnecting means or overcurrent device, shall be legibly marked to indicate its purpose." An inspection and adequate marking of all the disconnecting devices should be initiated immediately.

Recommendation 2: Employers should reinforce their standard operating procedures concerning circuit testing.

Discussion: Although the victim had a voltmeter available, he failed to test the circuit, and this omission led to his death. Standard operating procedures should be reviewed, revised as needed, and consistently enforced by the employer.

Recommendation 3: Employees and/or employers should be trained in CPR.

Discussion: CPR should begin within 4 minutes (in accordance with American Heart Association guidelines) in order to achieve the best results. To meet this criteria for successful resuscitation, workers should be trained in CPR to support the victim's circulation and respiration until trained medical personnel arrive. No one at the accident site was trained in CPR, and therefore, resuscitation was delayed until the Emergency Medical Services (EMS) arrived.

Adapted from: FACE 87-63, 30-Year-Old Electrician Electrocuted in Maryland. http://www.cdc.gov/niosh/face/In-house/full8763.html. Accessed 6/24/08.

Questions

1. The electrical safety culture that existed within the contractor's organization was heavily influenced by which of the following?
   A. The wide-ranging training program provided by the employer
   B. The safety program that was administered by the line organization, including the owner and foreman
   C. The owner's detailed knowledge and understanding of electrical hazards
   D. The general acceptance of working on energized circuits

2. The worker was exposed to electrocution because of which of the following?
   A. The owner did not disconnect the source of power from the compressor conductors.
   B. The electrician failed to use a voltmeter to verify absence of voltage.
   C. The fish tape became energized from damage to the conductor inside the conduit.
   D. The electrician failed to install rubber mats on the floor in front of the equipment.

3. Which of the following facts of the case might have been a contributing factor to the incident?
   A. The nonconductive fish tape was inadequate.
   B. The voltage-rated gloves worn by the worker were underrated.
   C. The fuses should have been removed from the disconnecting means.
   D. The inadequate marking of the disconnecting means.

4. The electrician's body became a part of the motor circuit by being inserted in the circuit as described by which of the following?
   A. Between the energized conductor and the motor
   B. Between the energized conductor and the concrete floor
   C. Between the energized conductor and the fish tape
   D. Between two separate phase conductors

5. The contractor's electrician accepted the fact that the building maintenance man had removed the source of energy without questioning it. What do you think is the reason he did ask the maintenance man to demonstrate that the sources of energy had been removed?
   A. No requirement existed in the contractor's safety program to question a client's accuracy.
   B. Electricians are not qualified to create an electrically safe work condition.
   C. The culture in the contractor's organization was to accept all information and actions provided by a client. The electrician did not take responsibility for his own safety and that of his coworkers.
   D. The single-line diagram accurately illustrated the installation.

Ready for Review

- A culture among electrical workers deemphasizes safety by allowing and expecting them to perform work on energized equipment. Factors that contribute to this culture include the following:
  - Ignorance of laws
  - Ignorance of hazards
  - Lack of awareness of the true costs of energized work
  - Lack of awareness of the consequences associated with energized work
  - Pressure to perform the work regardless of the risks
  - Pressure to do the job quickly regardless of the risks
  - Fear that deenergizing the equipment/circuit will cost time and money
  - Preference to not wear PPE
  - Ego/nonchalant attitude toward risk
  - False sense of security
- The tendency to accept the risk of an electrical injury must change to recognize that risk of an electrical injury is unacceptable.
- The NIOSH FACE program is a research program designed to identify and study fatal occupational injuries. The program aims to formulate prevention strategies, but it does not enforce state or federal standards.
- NIOSH performs on-site investigations where fatalities have occurred. Data from investigations is used to write investigative reports. Surveillance and investigative reports are maintained by NIOSH in a database that researchers use for analysis.
- Pressures to perform work on energized equipment include the following:
  - Pressure from customers to work on energized equipment when a shutdown is necessary
  - Pressure from management to perform energized work when it is not justified
- In reality, workers who do not follow the rules expose themselves to injury or death. They also expose the contractor and their clients to undue risks such as increased insurance premiums and loss of production.
- Customers may not understand the total costs and risks associated with working on energized equipment; therefore, the contractor and employees must be part of ensuring a safe working environment.
- When a shutdown is avoided in the interest of time or money, it can become an unscheduled shutdown. Equipment that cannot be shut down for a few minutes ultimately could be shut down for days or weeks or more.
- A general disregard for safety stems from many sources, including a lack of knowledge in the past that allowed dangerous methods for testing voltage, such as the tasting method and the finger method.
- Electrical safety-related work practices, like all practices and procedures, require updating over time. Today's practices and procedures should be brought up-to-date with today's knowledge of hazards.
- An incomplete understanding of hazards contributes to a disregard for safety measures. The safest and only acceptable approach is to eliminate the hazard by creating an electrically safe work condition, unless the employer demonstrates a true need for work on energized equipment.
- The employer should ensure that a written hazard assessment is performed, hazards are engineered out, and that adequate protection against hazards are provided and used when the need for energized work is demonstrated.
- Electrical hazards include electrical shock, electrocution, arc flash burns, and injuries from the pressure wave of an arc blast.
- Workers can suffer injury or death if circuits and equipment are not worked on in an electrically safe work condition, which is defined specifically by *NFPA 70E*.
- Items such as GFCIs and PPE are helpful in protecting against electrical shock, but they are effective only when used and maintained properly. Such protective equipment does not eliminate the hazards.
- Survivors of electrical incidents have varied neuropsychiatric complaints that seem unrelated to the incident, but might be related. Complaints commonly described by electrical incident survivors include hearing loss, headache, memory changes, disorientation, slowing of mental processes, agita-

tion, confusion, irritability, affective disorders, and PTSD. Other complaints include ringing in their ears (tinnitus) and a loss of hearing, which can be attributed to blast effects.

- The potential for injury and death is related directly to the energy output from an electrical fault. However, the quantity and form of energy transfer and the affected individual's biologic characteristics are critical in predicting the extent of injury.
- Voltage testing is an example of a task that is recognized as infeasible to perform on deenergized equipment, but the vast majority of work performed on energized equipment does not qualify as work allowed by OSHA and *NFPA 70E*.
- Factors that need to be considered when making decisions regarding electrical safety include priorities, the importance of training, and how to ensure that training is effective.
- The costs of shutting down sometimes are considered without considering the costs of working on energized equipment. Costs of working on energized equipment include the cost of human life, equipment, loss of production, insurance premium increases, and much more. If performing a shutdown seems inconvenient, employers should consider whether a shutdown could occur at night or on a weekend.
- Electrical workers might feel a time pressure to finish a job as quickly as possible. However, safety overrides the unrealistic expectation that all work can be performed immediately. Ensuring that the worker understands the magnitude of the hazards present and has taken the necessary precautions to ensure safe work is more important.
- The employee performing the work must be qualified to perform the tasks to which he or she is assigned, and he or she must use appropriate PPE.
- The total cost of implementing an arc flash protective clothing program is far less than that of not wearing PPE. In fact, it has been shown that the least costly approach is to begin with a detailed arc flash hazard analysis.
- The ramifications of buying into an unsafe electrical culture must be considered. Injury and death are the consequences that have a huge toll on the worker and all those associated with him or her.

Vocabulary

arc blast‡ Dangerous condition caused by the release of energy in an electric arc, usually associated with electrical distribution equipment.

arc flash‡ Dangerous condition caused by the release of energy in an electric arc, usually associated with electrical distribution equipment.

arcing fault‡ Fault characterized by an electrical arc through the air.

electrically safe work condition* A state in which an electric conductor or circuit part has been disconnected from energized parts, locked/tagged in accordance with established standards, tested to ensure the absence of voltage, and grounded if determined necessary.

flame-resistant (FR)* The property of a material whereby combustion is prevented, terminated, or inhibited following the application of a flaming or nonflaming source of ignition, with or without subsequent removal of the ignition source.

> FPN: Flame resistance can be an inherent property of a material or it can be imparted by a specific treatment applied to the material.

ground-fault circuit interrupter (GFCI)* A device intended for the protection of personnel that functions to deenergize a circuit or portion thereof within an established period of time when a current to ground exceeds the values established for a Class A device.

> FPN: Class A ground-fault circuit-interrupters trip when the current to ground is 6 mA or higher and do not trip when the current to ground is less than 4 mA. For further information, see UL 943, *Standard for Ground-Fault Circuit Interrupters*.

*This is the *NFPA 70E* definition.
‡This definition is from the National Fire Protection Association's (NFPA's) *Pocket Dictionary of Electrical Terms*.

OPENING CASE

On March 17, 1986, an electrician's apprentice (the victim) and a Journeyman electrician were installing two new switch boxes on the walls of a room being renovated in an office building. The circuits in the room where the new switch boxes were being installed were deenergized, with the exception of the circuit to an existing metal switch box suspended by conduit from the ceiling. Prior to installing the new boxes, the Journeyman electrician momentarily left the room and told the victim that "they would figure out how to wire the boxes" when he returned. The victim was expected to wait for the return of the Journeyman electrician prior to performing any work; however, while the victim was alone, he elected to disassemble the energized switch box suspended from the ceiling.

Due to the relocation of walls during the renovation, this suspended switch box was to be replaced by a new box. The suspended switch box was energized by a 277-volt circuit from the adjacent room. A metal-sheathed cable provided electricity to this switch box; however, this cable was outside of the conduit and entered the switch box through the side of the box. Apparently, the victim thought the suspended box was deenergized, since it was in the same room as the new boxes being installed.

The victim reached into the suspended box and, using wire cutters, cut the conductors from each of the four terminal connections in the box. Then, with his left hand, the victim pulled the metal sheathed

Electrical Hazard Awareness CHAPTER 2

Introduction

Electricity has become such an integral part of our society that it is often taken for granted. Yet electricity remains a very dangerous hazard for people working on or near it. Many electrical circuits do not directly pose serious shock or burn hazards by themselves. However, many of these circuits are found adjacent to circuits with potentially lethal levels of energy. Even a minor shock can cause a worker to drop a tool into the circuit. Involuntary reaction to a shock might also result in bruises, bone fractures, and even death from collisions or falls.

The following are recognized as examples of some common electrical hazards that can cause injury and even death while a person works on or near electrical equipment and systems:

- Electric shock
- Electric burns from contact (current) and flash (radiant)
- Arc blast impact from expanding air and vaporized materials

Workplace Hazards

Contractors and their employees are well aware that they face health and safety challenges on a daily, if not on a minute-to-minute basis. Potential dangers include working on or near energized circuits and equipment. Perhaps employees do not fully understand the lockout/tagout program put in place by their employer. It could be that an employee has not fully retained his or her hazardous communication training, including how to read a Material Safety Data Sheet (MSDS), or he or she does not remember the location and

OPENING CASE, CONT.

conductor out of the switch box that he was holding in his right hand. Apparently, the bare conductors contacted the box and/or his left hand. The victim provided a path to ground and was electrocuted. Burn marks found on the victim's right hand were consistent with the shape of the box.

The victim was found 14 feet from the switch box. Emergency Medical Service (EMS) personnel responded and administered advance cardiac life support (ACLS) procedures. Attempts to resuscitate the victim were unsuccessful. The victim was pronounced dead on arrival at a nearby hospital.

Adapted from: Fatality Assessment and Control Evaluation (FACE) Program Case 87-34. Accessed 4/22/08 at http://www.cdc.gov/niosh/face/In-house/full8734.html.

1. When is it appropriate to have an apprentice work unattended while performing tasks related to demolition/renovation?

2. When is it appropriate to have a worker of any level of experience work alone while performing tasks related to demolition/renovation?

Chapter Outline

- ■ **Introduction**
- ■ **Workplace Hazards**
- ■ **Electrical Shock**
- ■ **Arcing Faults: Arc Flash and Arc Blast**
- ■ **Summary**
- ■ **Lessons Learned**
- ■ **Current Knowledge**

Learning Objectives

1. Identify electrical hazards.
2. Explain the effects of current on the human body.
3. Describe an arcing fault event.
4. List the effects an arcing fault can have on the human body.

References

1. *NFPA 70E®*, 2009 Edition
2. Cooper Bussmann® Safety BASICs™

OSHA Tip

Lockout is one of the main protective measures that can be taken to prevent workers from working on an energized circuit. It is a specific procedure that involves placing a lockout device on an energy-isolating device, making it physically impossible to operate the equipment until the lockout device is removed.

OSHA defines lockout as follows:

lockout The placement of a lockout device on an energy-isolating device, in accordance with an established procedure, ensuring that the energy-isolating device and the equipment being controlled cannot be operated until the lockout device is removed.

availability of the hazardous communication written program and the required list of hazardous chemicals. Hazards might include exposure to laser equipment used by the carpenter, paints and solvents used by the painter, an improperly built scaffold, or perhaps an excavation without properly designed sloping, benching, trench shields, or other protective systems. Unfortunately, not all of those exposed to these and other hazards fully appreciate the exposures they face, nor always know how best to avoid the many potential dangers that can cause injury and death.

While the workplace must be evaluated to identify and eliminate all hazards, this program will focus primarily on electrical hazards. These may include fire, falls and falling objects, electrical shock, and the hazards associated with arcing faults including arc flash and arc blast. However, this publication will primarily focus on electrical shock, arc flash, and arc blast.

OSHA Tip

Tagout is a protective measure in which a prominent warning device, such as a tag, is placed on the energy-isolating device, indicating that the equipment must not be operated. The tagout device warns that a worker might be injured if he or she operates the energy-isolating device.

OSHA defines tagout as follows:

tagout The placement of a tagout device on an energy-isolating device, in accordance with an established procedure, to indicate that the energy-isolating device and the equipment being controlled may not be operated until the tagout device is removed.

Electrical Shock

Based on data compiled by the U.S. Department of Labor's Bureau of Labor Statistics, between the years of 1992 and 2002, electricity was the cause of death in 3378 on-the-job fatalities. In addition, for the same 10-year period, close to 47,000 workers were nonfatally injured in electrical incidents. Approximately half of these nonfatal injuries took place in the manufacturing and construction industries alone. This works out to be almost 180 nonfatal electrical injuries in just these two industries every month, and this only includes accidents that are reported.

Most personnel are aware of the danger of electrical shock, including electrocution. It is the one electrical hazard around which many electrical safety standards have been built. However, few really understand just how little current is required to cause injury or even death. In fact, even the current drawn by a 7½ watt, 120-volt lamp, passing across the chest, from hand-to-hand or hand-to-foot, is enough to cause electrocution.

The effects of electric current on the human body depend on the following:

- Circuit characteristics (current, resistance, frequency, and voltage)
- Contact resistance and internal resistance of the body
- The current's pathway through the body, determined by contact location and internal body chemistry
- Duration of the contact
- Environmental conditions that affect the body's contact resistance

Skin Resistance

To understand the currents possible in the human body, it is important to understand the contact resistance of the skin. The skin's resistance can change as a function of the moisture present in its external and internal layers, with changes due to such factors as ambient temperatures, humidity, fright, and anxiety. The variance in these skin contact resistances is displayed in **Table 2.1**.

Body tissue, vital organs, blood vessels, and nerve (nonfat) tissue in the human body contain water and

 CAUTION

Even current as little as that drawn by a 7½ watt, 120-volt lamp can be enough to cause electrocution.

Table 2.1	Human Resistance Values for Skin-Contact Conditions	
	Resistance (ohms)	
Condition	**Dry**	**Wet**
Finger touch	40,000 to 1000	4000 to 15,000
Hand holding wire	15,000 to 50,000	3000 to 6000
Finger-thumb grasp	10,000 to 30,000	2000 to 5000
Hand holding pliers	5000 to 10,000	1000 to 3000
Palm touch	3000 to 8000	1000 to 2000
Hand around $1^1/_2$-inch pipe	1000 to 3000	500 to 1500
Two hands around $1^1/_2$-inch pipe	500 to 1500	250 to 750
Hand immersed	N/A	200 to 500
Foot immersed	N/A	100 to 300
Human body, internal, excluding skin	200 to 1000	

N/A, not applicable.
Data source: Kouwenhoven, W. B., and Milnor, W. R., "Field Treatment of Electric Shock Cases—1," AIEE Trans. Power Apparatus and Systems, Volume 76, pp. 82–84, April 1957; discussion pp. 84–87.

electrolytes and are highly conductive with limited resistance to alternating electric current. As the skin is broken down by electrical current, resistance drops, and current levels increase.

Consider an example of a person with hand-to-hand resistance of only 1000 ohms. The voltage determines the amount of current passing through the body.

While 1000 ohms might appear to be low, even lower levels can be approached by someone with sweat-soaked noninsulating gloves on both hands and a full-hand grasp of an energized conductor and a grounded pipe or conduit. Moreover, cuts, abrasions, or blisters on hands can negate skin resistance, leaving only internal body resistance to oppose current flow. A circuit in the range of 50 volts could be dangerous in this instance.

Using Ohm's law, we can calculate the current (I, in amperes) in a circuit based on the circuit voltage (V) and the resistance (R). **Ohm's law** states that current (amperes) equals voltage (volts) divided by resistance (ohms), in equation form:

$$I \text{ (amperes)} = \frac{V \text{ (volts)}}{R \text{ (ohms)}}$$

Example 1: $I = \dfrac{480}{1000} = 0.480 \text{ amps (480 mA)}$

Example 2: $I = \dfrac{120}{1000} = 0.120 \text{ amps (120 mA)}$

Electrical currents can cause muscles to lock up, resulting in the inability of a person to release his or her grip from the current source. This is known as the **let-go threshold** current. At 60 Hz, most people have a "let-go" limit of 10 to 40 milliamperes (mA) (**Table 2.2**).

Potential injury (current flow) also increases with time. A victim who cannot "let go" of a current source is much more likely to be electrocuted than someone whose reaction removes him or her from the circuit more quickly. A victim who is exposed for only a fraction of a second is less likely to sustain an injury.

Extent of Injury

The most damaging path for electrical current is through the chest cavity and head (**Figure 2.1**). In short, any prolonged exposure to 60 Hz current of 10 mA or more might be fatal. Fatal ventricular fibrillation of

Table 2.2	Effects of Electrical Shock (60 Hz AC)
Response	**60 Hz, AC Current (mA)**
Tingling sensation	0.5 to 3
Muscle contraction and pain	3 to 10
"Let-go" threshold	10 to 40
Respiratory paralysis	30 to 75
Heart fibrillation; might clamp tight	100 to 200
Tissue and organs burn	More than 1500

Data source: Kouwenhoven, W. B., and Milnor, W. R., "Field Treatment of Electric Shock Cases—1," AIEE Trans. Power Apparatus and Systems, Volume 76, pp. 82–84, April 1957; discussion pp. 84–87.

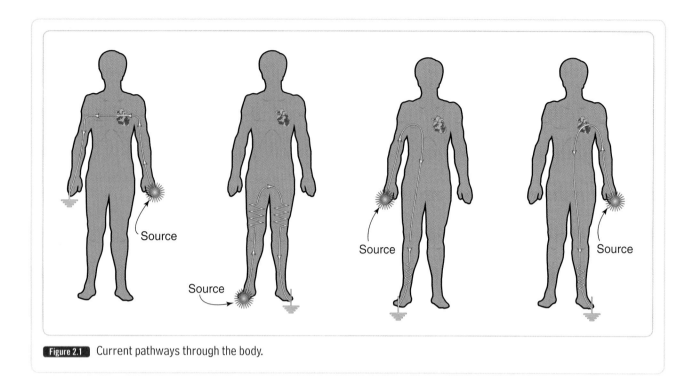

Figure 2.1   Current pathways through the body.

the heart (a state in which the heart does not contract, but instead twitches, resulting in stopping of rhythmic pumping action) can be initiated by a current flow of as little as several milliamperes. These injuries can cause fatalities resulting from either direct paralysis of the respiratory system, failure of the rhythmic heart pumping action, or immediate heart stoppage.

During fibrillation, the victim might become unconscious. On the other hand, he or she might be conscious, deny needing help, walk a few feet, and then collapse. Death could occur within a few minutes or take hours. Prompt medical attention is needed for anyone receiving electrical shock. Many of these people can be saved, provided they receive proper medical treatment, including cardiopulmonary resuscitation (CPR), quickly.

Think of electrical shock injuries as "icebergs," where most of the injury is unseen below the surface. Entrance and exit wounds are usually coagulated areas and might have some charring or these areas might be missing, having "exploded" away from the body due to the level of energy present. The smaller the area of contact, the greater the heat produced. For a given current, damage in the limbs might be the greatest, due to the higher current flux per unit of cross-sectional area.

Within the body, the current can burn internal body parts in its path. This type of injury might be difficult to diagnose, as the only initial signs of injury are the entry and exit wounds. Damage to the internal tissues, while not apparent immediately, might cause delayed internal tissue swelling and irritation. Prompt medical attention can minimize possible loss of blood circulation and the potential for amputation of the affected extremity can also prevent death.

Prevention

All electrocutions are preventable. A significant part of OSHA requirements are dedicated to electrical safety. OSHA compliance is considered a minimum requirement and seen as a very good place to start for improving the safety of the workplace. Current OSHA regulations were promulgated many years ago.

Several standards offer guidance regarding safe approach distances to minimize the possibility of shock from exposed electrical conductors of different voltage levels. One of the most recent, and perhaps the most authoritative guidance, is presented in *NFPA 70E*, *Standard for Electrical Safety in the Workplace*, in Section 130.2. *Electrical Safety in the Workplace* and NFPA-70E are registered trademarks of the National Fire Protection Association, Quincy, MA. The requirements related to approach boundaries will be covered in a subsequent chapter.

OSHA Tip

OSHA compliance is considered a minimum requirement.

Arcing Faults: Arc Flash and Arc Blast

Arcing Fault Basics

 Figure 2.2 is a graphical model of an **arcing fault** and the physical consequences that can occur. The unique aspect of an arcing fault is that the fault current flows through the air between conductors or a conductor(s) and a grounded part. The arc has an associated arc voltage, because of arc **impedance**. The product of the fault current and arc voltage in a concentrated area results in tremendous energy released in several forms.

The resulting energies can be in the form of radiant heat, intense light, and tremendous pressures. Intense radiant heat from the arcing source travels at the speed of light. The temperature of the arc terminals can reach approximately 35,000°F, or about four times as hot as the surface of the sun. The high arc temperature changes the state of conductors from solid to hot molten metal and to vapor. The immediate vaporization of the conductors is an explosive change in state from solid to vapor. Copper vapor expands to 67,000 times the volume of solid copper. A copper conductive component the size of a penny could expand as it vaporizes and disperse to the size of a refrigerator. Because of the expansive vaporization of conductive metal, a line-to-line or line-to-ground arcing fault can escalate into a 3-phase arcing fault in less than a thousandth of a second.

The release of intense thermal energy superheats the immediate surrounding air. The air also expands in an explosive manner. The rapid vaporization of conductors and superheating of air result in a pressure wave of air and gases and a conductive plasma cloud that can engulf a person. The thermal shock and pressures can violently destroy circuit components. The pressure waves hurl the destroyed, fragmented components like shrapnel at high velocity; shrapnel fragments can be expelled in excess of 700 miles per hour, or about the speed at which shotgun pellets leave the gun barrel. Molten metal droplets at high temperatures typically are blown out from the event due to the pressure waves.

Testing has proven that the arcing fault current magnitude and time duration are among the most critical variables in determining the energy released. It is important to note that the predictability of arcing faults and the energy released by an arcing fault are subject to significant variance. Some of the variables that affect the outcome include the following:

- Available bolted short-circuit current
- The time the fault is permitted to flow (speed of the **overcurrent protective device** [OCPD])
- Arc gap spacing
- Size of the enclosure (or no enclosure)
- Power factor of fault
- System voltage
- Whether the arcing fault can sustain itself
- Type of system grounding scheme
- Distance of the worker's body parts from the arc

Typically, engineering data that the industry provides about arcing faults is based on specific values of these variables. For instance, for 600 volts and less systems, much of the data has been gathered from testing on systems with an arc-gap spacing of 1.25 inches and **incident energy** determined at 18 inches from the point of the arcing fault.

Arc Flash

Arc flash is the hazard associated with the release of thermal energy during an arcing fault. The most common unit of measure used to quantify the arc flash hazard is calories per square centimeter (cal/cm^2). This unit of measurement is the amount of heat energy imposed on a surface area at a given distance from the arcing fault. Arc flash suits and flame-resistant (FR) clothing also have an **arc rating** expressed in cal/cm^2 and are designed to help protect workers from the arc flash hazard. *NFPA 70E* and the current edition of the Institute of Electrical and Electronics Engineers (IEEE) 1584, *Guide For Performing Arc Flash Hazard Calculations*, illustrate methods to estimate the amount of thermal energy (incident energy) available. The arc flash hazard

Molten metal
35,000°F

Radiant heat and UV at speed of light

Pressure waves

Sound waves

Shrapnel at 700 mph

Copper Vapor: Solid to Vapor Expands by 67,000 times

Hot air—rapid expansion

Intense light

Figure 2.2 Effects of an electrical arc. The two rods represent copper conductors.

70E Highlights

NFPA 70E Section 130.7(C)(10) Fine Print Note (FPN) No. 2:

> The PPE requirements of this section are intended to protect a person from arc flash and shock hazards. While some situations could result in burns to the skin, even with the protection selected, burn injury should be reduced and survivable. Due to the explosive effects of some arc events, physical trauma injuries could occur. The PPE requirements of Table 130.7(C)(10) do not address protection against physical trauma other than exposure to the thermal effects of an arc flash.

analysis allows a person to select the personal protective equipment (PPE) to protect against the thermal energy. Incident energy analysis is discussed in detail in subsequent chapters of this publication.

Arc Blast

Arc blast is associated with the release of tremendous pressure. The industry is trying to devise ways to quantify the risks associated with arc blast. A number of the potential hazards associated with arc blast were covered in the discussion of arcing fault basics above. However, there is little or no information on arc blast hazard risk assessment or on protecting workers from the arc blast hazard at this time. The arc flash suits and other arc-rated FR clothing used to protect people from arc flash hazards might not protect people from arc blast.

How Arcing Faults Can Affect Humans

Nearly everyone is aware that an electrical shock is a hazard that ultimately can lead to death. However, while many people have experienced minor shocks, few have realized any real consequences, making them somewhat complacent. In contrast, too few people are aware of the extreme nature of electrical arcing faults, the potential of severe burns from arc flash, and the potential injuries due to high pressures from arc blast. However, this is starting to change; people are learning that the affects of an arcing fault can be devastating.

In recent years, awareness of arc flash hazards has been increasing. Recent studies of reported electrical injuries have indicated that many documented injury cases were burns resulting from exposure to electrical arcs. Reports indicate that, each year, more than 2000 people are admitted to burn centers in the U.S. with severe electrical burns. Electrical burns are considered extremely hazardous for a number of reasons. Contact with the circuit is not necessary to incur a serious, even deadly, burn. Serious or fatal burns can occur at distances of more than 10 feet from the source of a flash.

Burns are the prevalent consequence of electrical incidents. These can be due to contact (shock hazard) or arc flash. Three basic types are mentioned below:

- *Electrical burns due to current flow*—tissue damage (whether skin deep or deeper) occurs because the body is unable to dissipate the heat from the current flow through the body. The damage to a person's tissue can be internal and initially not obvious from external examination. Typically, electrical burns are slow to heal and frequently result in amputation.
- *Arc burns by radiant or convective heat*—caused by electrical arcs. Temperatures generated by electric arcs can burn flesh and ignite clothing at distances of 10 feet or more.
- *Thermal contact burns (conductive heat)*—normally experienced from skin contact with the hot surfaces of overheated electric conductors or a person's clothing that ignites due to an arc flash.

Studies show that when the skin temperature is as low as 110°F, the body's temperature equilibrium begins to break down in about 6 hours. At 158°F, a one-second duration is sufficient to cause total cell destruction. Skin at temperatures of 205°F for more than one-tenth of one second can cause incurable, third-degree burns (Table 2.3).

In addition to burn injuries, victims of arcing faults can experience damage to their sight, hearing, lungs, skeletal system, respiratory system, muscular system, and nervous system. The speed of an arcing fault event can be so rapid that the human system cannot react quickly enough for a worker to take corrective measures. The radiant thermal waves, the high pressure waves, the spewing of hot molten metal, the intense light, the

CAUTION

It is not necessary to be in contact with the circuit to incur a serious burn. Serious or fatal burns can occur at distances of more than 10 feet from energized conductors.

CAUTION

Skin exposure to temperatures of 205°F for as little as one-tenth of one second can cause incurable, third-degree burns.

Table 2.3	Skin Temperature Tolerance Relationship	
Skin Temperature	**Duration**	**Damage Caused**
110°F	6.0 hours	Cell breakdown begins
158°F	1.0 second	Total cell destruction
176°F	0.1 second	Curable (second-degree) burn
205°F	0.1 second	Incurable (third-degree) burn

Source: Bussmann Safety BASICs Handbook, Courtesy of Cooper Bussmann, Inc. 2004.

Table 2.4	Key Thresholds for Injury from an Arcing Fault
Threshold	**Measurement**
Just curable burn threshold	80°C/176°F (0.1 second)
Incurable burn threshold	96°C/205°F (just under the temperature where water will boil) for 0.1 second
Eardrum rupture threshold	720 lbs/ft^2
Lung damage threshold	1728–2160 lbs/ft^2 (approximately the equivalent to having a compact car resting its weight on your chest)
OSHA required ear protection threshold	85 decibel (db) for sustained time period (Note: an increase of 3 db is equivalent to doubling the sound level.)

hurling shrapnel, and the hot conductive plasma cloud can be devastating in a small fraction of a second. The intense thermal energy released can cause severe burns or ignite flammable clothing. Molten metal blown out can burn skin or ignite flammable clothing and cause serious burns over much of the body. A person can gasp and inhale hot air and vaporized metal, sustaining severe injury to their respiratory system. The tremendous pressure blast from the vaporization of conducting materials and superheating of air can fracture ribs, collapse lungs, and knock workers off ladders or blow them across a room.

What is often difficult for people to comprehend is that the time in which the arcing fault event runs its course might be only a small fraction of a second. In a matter of only a thousandth of a second or so, a single-phase arcing fault can escalate to a 3-phase arcing fault. Tremendous energy can be released in a few hundredths of a second. Humans cannot detect, much less comprehend and react to, events in these time frames.

Sometimes a greater respect for arcing fault and shock hazards is afforded to medium and high-voltage systems. However, injury reports show serious accidents are occurring at an alarming rate on systems of 600 volts or less, in part because of the high fault currents that are possible. But also, designers, management, and workers mistakenly tend not to take the same necessary precautions taken when designing or working on medium- and high-voltage systems.

Staged Arc Flash Tests

An ad hoc electrical safety working group within the IEEE Petroleum and Chemical Industry Committee conducted staged arc flash tests to investigate arcing fault hazards. These tests and others are detailed in "Staged Tests Increase Awareness of Arc Fault Hazards in Electrical Equipment," IEEE Petroleum and Chemical Industry Conference Record, September 1997, pp. 313–322. One finding of this IEEE paper is that current-limiting OCPDs reduce damage and arc fault energy (provided the fault current is within the current-limiting range). To better assess the benefit of limiting the current of an arcing fault, it is important to note some key thresholds of injury for humans (Table 2.4). Results of the staged arc flash tests were recorded by sensors on mannequins and can be compared to these thresholds.

Test 4, Test 3, and Test 1

The results of three of those tests are reviewed here and identified as Test 4, Test 3, and Test 1. All three of these tests were conducted on the same electrical circuit set-up with an available bolted 3-phase, short-circuit current of 22,600 symmetrical root mean square (rms) amperes at 480 volts. In each case, an arcing fault was initiated in a size 1 combination motor controller enclosure with the door open, as if an electrician were working on the unit while energized or before it was placed in an electrically safe work condition. Test 4 and Test 3 were identical except for the OCPD protecting the circuit. In Test 4 (Figure 2.3), a 640-ampere circuit breaker with a short-time delay is protecting the circuit; the circuit was cleared in six cycles. In Test 3 (Figure 2.4), 601-ampere (KRP-C-601SP), current-limiting fuses (Class L) are protecting the circuit; they opened the fault current in less than one-half cycle and limited the current. In addition, the arcing fault was initiated on the line side of the branch circuit device in both Test 4 and Test 3 (the fault is on the feeder circuit, but within the controller enclosure). In Test 1

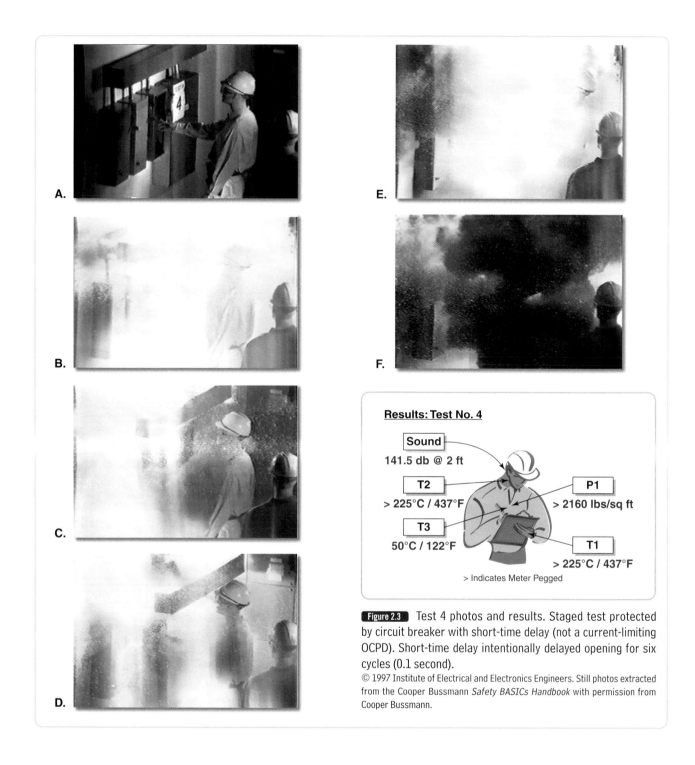

Results: Test No. 4

Sound
141.5 db @ 2 ft

T2
> 225°C / 437°F

P1
> 2160 lbs/sq ft

T3
50°C / 122°F

T1
> 225°C / 437°F

> Indicates Meter Pegged

Figure 2.3 Test 4 photos and results. Staged test protected by circuit breaker with short-time delay (not a current-limiting OCPD). Short-time delay intentionally delayed opening for six cycles (0.1 second).

© 1997 Institute of Electrical and Electronics Engineers. Still photos extracted from the Cooper Bussmann *Safety BASICs Handbook* with permission from Cooper Bussmann.

(**Figure 2.5**), the arcing fault is initiated on the load side of the 30-ampere branch-circuit OCPDs (LPS-RK 30SP) current-limiting fuses (Class RK1). These fuses limited this fault current to a much lower amount and cleared the circuit in approximately one-fourth cycle or less.

Following are the results recorded from the various sensors on the mannequin closest to the arcing fault. T1 and T2 recorded the temperature on the bare hand and neck, respectively. The hand with T1 sensor was very close to the arcing fault. T3 recorded the temperature on the chest under the cotton shirt. P1 recorded the pressure on the chest. Also, the sound level was measured at the ear. Some results "pegged the meter." That is, the specific measurements were unable to be recorded because the actual level exceeded the range of the sensor/recorder setting. These values are shown as >, which indicates that the actual value exceeded the value given,

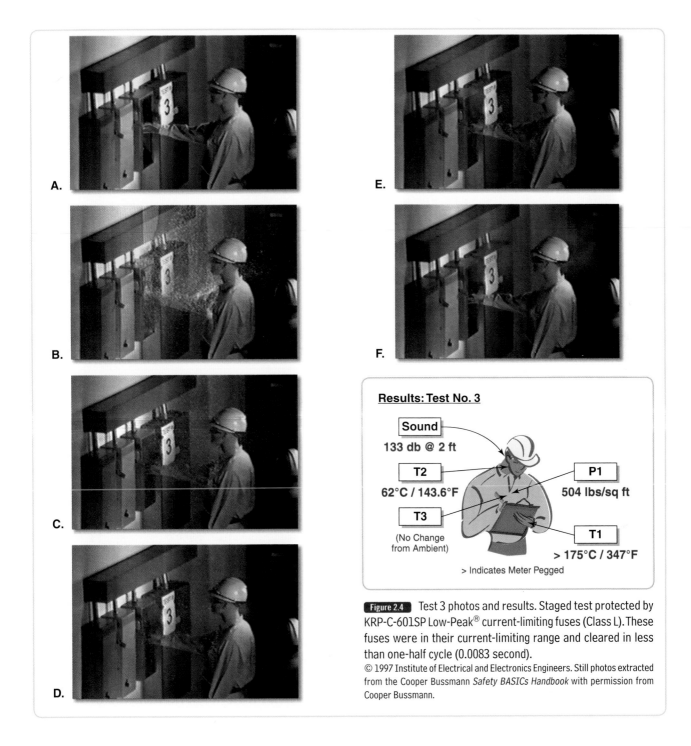

A.

B.

C.

D.

E.

F.

Results: Test No. 3

Sound
133 db @ 2 ft

T2
62°C / 143.6°F

T3
(No Change from Ambient)

P1
504 lbs/sq ft

T1
> 175°C / 347°F

> Indicates Meter Pegged

Figure 2.4 Test 3 photos and results. Staged test protected by KRP-C-601SP Low-Peak® current-limiting fuses (Class L). These fuses were in their current-limiting range and cleared in less than one-half cycle (0.0083 second).
© 1997 Institute of Electrical and Electronics Engineers. Still photos extracted from the Cooper Bussmann *Safety BASICs Handbook* with permission from Cooper Bussmann.

but it is unknown how high a level the actual value attained.

The Role of Overcurrent Protective Devices in Electrical Safety

If an arcing fault occurs while a worker is in close proximity to the fault, the survivability of the worker is mostly dependent upon the following:

- The characteristics of the OCPDs
- The arcing fault current
- Precautions the worker has taken prior to the event, such as wearing PPE appropriate for the hazard

The selection and performance of OCPDs play a significant role in electrical safety. Extensive tests and analysis by those in the industry have shown that the energy released during an arcing fault is related primarily

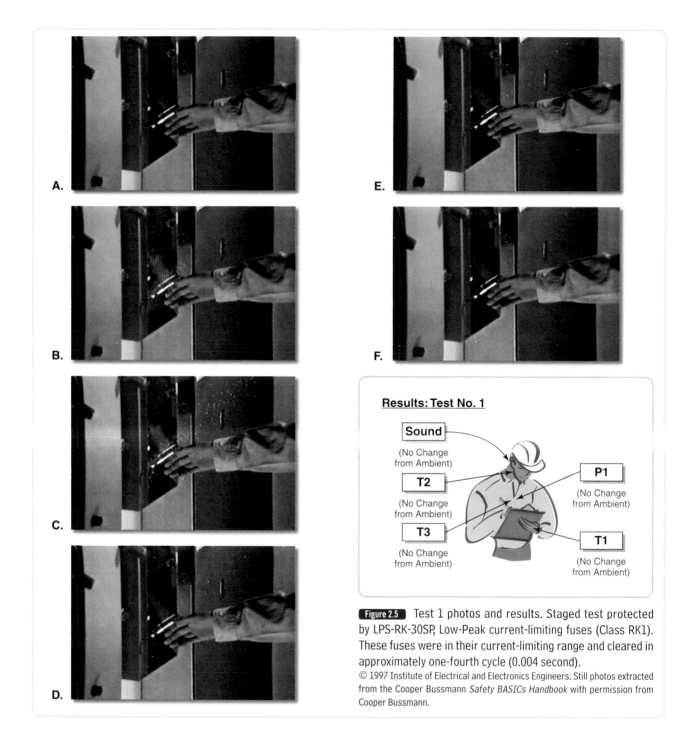

Results: Test No. 1

Sound
(No Change from Ambient)

T2
(No Change from Ambient)

T3
(No Change from Ambient)

P1
(No Change from Ambient)

T1
(No Change from Ambient)

Figure 2.5 Test 1 photos and results. Staged test protected by LPS-RK-30SP, Low-Peak current-limiting fuses (Class RK1). These fuses were in their current-limiting range and cleared in approximately one-fourth cycle (0.004 second).

© 1997 Institute of Electrical and Electronics Engineers. Still photos extracted from the Cooper Bussmann *Safety BASICs Handbook* with permission from Cooper Bussmann.

to two characteristics of the OCPD protecting the affected circuit:

- The time it takes the OCPD to open. The faster the fault is cleared by the OCPD, the lower the energy released.
- The amount of fault current the OCPD lets through. Current-limiting OCPDs may reduce the current let-through (when the fault current is within the current-limiting range of the OCPD) and can reduce the energy released.

The lower the energy released, the better for both worker safety and equipment protection. The photos and recording sensor readings from the staged tests illustrate this point very well.

Conclusions that can be drawn from the staged tests:

1. Arcing faults can release tremendous amounts of energy in many forms in a very short period of time as indicated by the measured values compared to key thresholds of injury for humans given in Table 2.4. Although the circuit in Test 4 was protected by a 640-ampere OCPD, it was a noncurrent limiting device and took six cycles (0.1 second) to open.

2. The OCPD's characteristic can have a significant impact on the outcome. A 601-ampere, current-limiting OCPD protected the circuit in Test 3. The current that flowed was reduced (limited) and the clearing time was one-half cycle or less. This was a significant reduction compared to Test 4 (▌Table 2.5▐). Compare the Test 3 and Test 4 results to the values in the Table 2.4, *Key Thresholds for Injury From an Arcing Fault*, and note the difference in exposure. In addition, note that the results of Test 1 are significantly less than those in Test 4 and even those in Test 3. The reason is that Test 1 utilized a much smaller (30-ampere) current-limiting device.

   Test 3 and Test 1 both show that there are benefits of using current-limiting OCPDs. Test 1 proves the point that the greater the current limitation, the more the arcing fault energy may be reduced. Both Test 3 and Test 1 utilized very current-limiting fuses, but the lower ampere-rated fuses limit the current more than the larger ampere-rated fuses. It is important to note that the fault current must be in the current-limiting range of the OCPD in order to receive the benefit of the lower current let-through. See ▌Figure 2.6▐ that

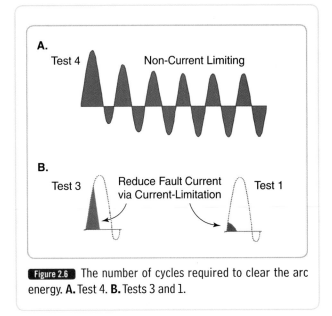

Figure 2.6 The number of cycles required to clear the arc energy. **A.** Test 4. **B.** Tests 3 and 1.

depicts the oscillographs of Tests 4, Test 3, and Test 1.

3. The cotton shirt reduced the thermal energy exposure on the chest (T3 measured temperature under the cotton shirt). This illustrates the benefit of workers wearing protective garments.

Summary

Recognizing the many hazards a worker might be exposed to and understanding the severe consequences of that exposure are important steps in convincing everyone involved that much must be done to provide a workplace free from recognized hazards. In summary, the hazards associated with electrical work could include the following:

- Fire
- Falls and falling objects
- Electrical shock
- Hazards associated with arcing faults including arc flash and arc blast

Developing a safety plan, training to the plan, evaluating the workplace for hazards, eliminating the hazards, understanding the consequences of noncompliance, and implementing safety-related work practices are examples of considerations that should go a long way towards allowing workers to go home the same way they come to work each day.

Table 2.5	Comparison of Staged Arc Flash Tests	
Test	**Protective Device Used**	**Clearing Time**
Test 4	640-ampere, noncurrent-limiting device	Six cycles
Test 3	KRP-C 601SP, 601-ampere, current-limiting fuses (Class L)	Less than one-half cycle
Test 1	LPS-RK 30SP, 30-ampere, current-limiting fuses (Class RK1)	One-fourth cycle

Electrician Dies from Explosion Injuries

On August 21, 1986, an electrician died as a result of injuries he received in an electrical explosion that occurred while he was wiring around a circuit breaker with blown fuses in a transformer room. Note: Although electrical discharge would more accurately describe what actually occurred, the more common term *electrical explosion* is used in this report.

The employer is an electrical contractor specializing in commercial and residential wiring who employs 13 electricians. The firm has neither a written safety program or a safety policy, nor management of the safety function. The only training provided is on-the-job training. The employer pays one-half the tuition for workers taking night school courses in the field of electricity or electronics.

On the day of the accident, the victim and his helper were to install branch circuit conductors to an existing 20-ampere, 277-volt circuit breaker in a 480/277 electrical panelbox. This circuit breaker was to control the power to an above-ceiling, water-cooled air-conditioning unit. The panelbox was located in the transformer room of a high-rise office building. As the victim was trying to pull the #12 copper wire from the air-conditioning unit into the top of the panelbox, the fishtape with which he was pulling the wire accidentally entered the rear of a meter base located at the top of the panelbox. This action damaged the meter base and created a dead short, causing the three fuses in the circuit breaker in an adjacent electrical panel to open. At this point the victim removed the meter from the meter socket and checked the continuity of the phases with a flashlight-type continuity tester. The continuity tester (rated at 3 volts) could not detect the dead short in the high-voltage (480 volts) phases. The victim then reinstalled the meter into the meter socket.

The circuit breaker that had opened controlled the ceiling lights in the building tenant's computer room. Since the victim did not have the correct replacement fuses, he went to a local electrical supply warehouse to obtain them. The warehouse did not have replacement fuses in stock, but they would have them delivered by the following morning. The victim then returned to the building, where he found the office staff collecting table lamps to illuminate the computer room. The victim told the staff that the lights wouldn't be necessary, that he would be able to bypass the circuit breaker with the blown fuses and supply power to the overhead lights. He said he was going to connect the three fuse terminals to the "live" bus terminals on an adjacent 480-volt circuit breaker bus using jumper wires. This would bypass the blown fuses and provide light to the computer room.

The victim connected a jumper wire to each of the three fuse terminals (referred to as Phases A, B, and C). The victim then connected the first jumper on Phase A to one of the energized bus terminals on the 480-volt circuit breaker. Phase A now was energized from the meter socket to the fuse terminals. However, because of the dead short in the meter socket, Phase B also was energized back to the fuse terminal and jumper. The victim then prepared to connect Phase B to the energized bus. The energized tip of Phase B touched the energized bus connection, and the 480-volt circuit breaker arced.

Several office workers standing in the hall outside the transformer room were burned by the explosion. One stated that shortly after the blast, the helper, who was holding a flashlight for the victim, staggered out of the transformer room with his clothes on fire. The office worker extinguished the helper's burning clothes, and EMS was summoned. Fifteen people were taken to the hospital for treatment (most for smoke inhalation); however, five were admitted to the hospital's burn unit. The victim died later that evening. The helper remains in the hospital in critical condition at the time of this report.

The cause of death was listed as massive electrical burns. The victim was burned over 95 percent of his body.

Recommendations/Discussion

Recommendation 1: Electrical safety devices should never be bypassed.

Discussion: Electrical safety devices (e.g., fuses, circuit breakers) are incorporated into an electrical system to alert the user of an existing problem. By bypassing safety devices, one eliminates safety features designed into an electrical system. This might result in electrical explosions, fires, or injuries. In this case, although an inconvenience was created by the outage of the overhead lights, the computer room could have and should have been lighted by an alternative means until the appropriate fuses could be obtained.

Recommendation 2: Electrical systems should be deenergized prior to any work being performed on them.

Discussion: The incoming power to the circuitry being worked on was not deenergized before the "repair work" was attempted. The circuitry was not deenergized, because it would have caused a power outage to portions of the office building. A job of this type should be scheduled at a time (a weekend or before or after hours) when the incoming power could be deenergized without disrupting operations. The employer has no written safety program or job procedures. Specific job procedures should be developed for tasks that are performed by employees, including deenergizing circuits before beginning to work on them. These procedures should detail the various safety hazards associated with these tasks. Once these specific procedures have been developed, the employer should ensure that they are implemented and enforced by a qualified person at each job site.

Recommendation 3: Employers should ensure that employees are trained in the use of appropriate electrical testing devices.

Discussion: The victim used the incorrect testing device when checking the phases in the meter; thus, the short went undetected. Had the victim used the appropriate testing device, the short would have been detected, and the accident could have been prevented. In addition, the victim could have tested the exposed ends of the jumper wires with a voltmeter before making the connections, which would have alerted the victim to the presence of electrical energy.

Adapted from: FACE Program Report No. 1986-52. Accessed 4/23/08 at http://www.cdc.gov/niosh/face/In-house/full8652.html.

Questions

1. Small businesses with few employees are required to provide which of the following information to employees?
   A. Copies of the single-line diagrams
   B. Copies of OSHA's electrical safety regulations, including the Preamble
   C. Electrical safety policies, programs, and procedures
   D. Electrical work licenses

2. Employers are required to provide which of the following training requirements for a qualified employee?
   A. Electrical safety hazard training, including shock, electrocution, arc flash, and arc blast
   B. How to read and interpret schematic and single-line diagrams
   C. Providing training during working hours or paying 100 percent of tuition at night school
   D. How to safely bypass overcurrent devices

3. After the short circuit occurred in the meter base, the worker should have tested for damage with which of the following instead of the flashlight-type continuity tester?
   A. A digital voltmeter with a Category IV transient voltage rating
   B. An analog multimeter
   C. Close visual inspection
   D. An insulation megger tester

4. Pulling wire into an energized panel should be done using which of the following?
   A. A metal fishtape
   B. A nylon fishtape with a metal bullnose
   C. A nonconductive fishtape
   D. Wire should not be pulled into an energized panel

5. Why was an electrically safe work condition not established after the fuses were blown in the circuit breaker?
   A. The owner could not locate enough desk lamps to illuminate the room.
   B. The contractor's electrician assumed that the owner required continuity of service for the lighting.
   C. The contractor's electrician was qualified to install jumpers when he believed it was necessary.
   D. The contractor's electrician had a signed energized work permit.

6. Which of the following precautions would have prevented the office workers standing outside the transformer room from being burned?
   A. Adequate electrical safety procedures
   B. Wearing protective clothing rated at 4 cal/cm^2
   C. Remaining outside the arc flash protection boundary (AFPB)
   D. The placement of an adequate label on the surface of the panelbox

Ready for Review

- Electrical circuits can be serious shock or burn hazards, either in and of themselves or because they are adjacent to a circuit with potentially lethal levels of energy.
- Common electrical hazards include electric shock, electric burns, and arc blast. These can cause injury and death.
- Working on or near energized circuits and equipment is dangerous. Workers must understand and follow their employers' lockout/tagout program.
- Very little current is required to cause injury or death. Even the current drawn by a 7½ watt, 120-volt lamp can be enough to cause electrocution.
- The effects of electric current on the human body depend on circuit characteristics, contact resistance, internal resistance of the body, the current's pathway through the body, duration of contact, and environmental conditions affecting the body's resistance.
- Skin resistance is the electrical resistance introduced by human skin in a circuit in which the body is a conducting pathway.
- Skin resistance can change with moisture, temperature changes, humidity, fright, and anxiety. Cuts, abrasions, or blisters can negate the skin's resistance.
- Ohm's law states that current equals voltage divided by resistance. This principle can be used to calculate the current in a circuit.
- Electrical currents can cause muscles to lock up, resulting in the inability of a person to release his or her grip from the current source. This is known as the let-go threshold current.
- The most damaging path for electrical current is through the chest cavity and head. The heart can experience ventricular fibrillation as the result of only a few milliamps.
- A person who has experienced ventricular fibrillation might remain conscious, only to collapse and die moments later. All workers who receive an electrical shock must receive immediate medical treatment.
- The damage from an electrical shock injury is largely internal and unseen. Damage in the limbs might be the greatest due to the higher current flux per unit of cross-sectional area. Prompt medical attention can minimize the possible loss of blood circulation, the potential for amputation, and death.
- With an arcing fault, the fault current flows through the air between conductors or between a conductor and a grounded part. The product of the fault current and the arc voltage in a concentrated area is tremendous energy.
- An arcing fault results in an intense release of heat (as high as 35,000°F), copper vapor, molten copper, shrapnel, intense light, rapid expansion (hot air), and intense pressure and sound waves. Obviously, these exposures can be fatal to anyone near the arcing fault.
- Arcing fault current magnitude and time duration are among the most critical variables in determining how much energy is released. Other factors include available bolted short-circuit current, the speed of the OCPD, arc gap spacing, the size of the enclosure, the power factor of the fault, the system's voltage, whether the arc can sustain itself, the type of grounding scheme, and the distance of the worker's body from the arc.
- Arc flash is the hazard associated with the energy release during an arcing fault. Arc flash is measured in cal/cm^2; protective clothing has an arc rating measured in these units as well.
- Arc blast is the hazard associated with the pressure released during an arcing fault.
- Arcing faults can be extremely hazardous. Contact with the circuit is not necessary for a serious or deadly burn to occur; such burns can occur at distances of more than 10 feet from energized conductors.
- The three types of burns that can result from electrical incidents include those due to current flow, burns due to incident energy exposure, and contact burns.
- Even if the burn is not fatal, serious injury to sight, hearing, lungs, the skeletal system, the respiratory system, the muscular system, and the nervous system can result.

- An arcing fault event can occur in a fraction of a second; a human cannot detect and react to it quickly enough.
- Serious accidents can occur on systems of 600 volts or less. Do not assume that a "low voltage" means low risk. The same precautions are needed when working on these systems as when working on higher voltage systems.
- If an arcing fault occurs, the survivability of the worker depends mostly on the characteristics of the OCPDs, the arc fault current, and the precautions the worker has taken prior to the event.
- The two characteristics of an OCPD that relate to the amount of energy released during an arcing fault include the time it takes the device to open and the amount of fault current the device lets through. The lower the energy released, the better.
- Staged tests conducted by the IEEE Petroleum and Chemical Industry Committee investigated fault hazards. The primary conclusions of these tests were as follows:
  - Arcing faults can release tremendous amounts of energy in many forms in a very short period of time.
  - The OCPD's characteristics can have significant impacts on the outcome.
  - Wearing PPE is extremely important and beneficial.

Vocabulary

arc blast‡ Dangerous condition caused by the release of energy in an electric arc, usually associated with electrical distribution equipment.

arc flash‡ Dangerous condition caused by the release of energy in an electric arc, usually associated with electrical distribution equipment.

arc rating* The value attributed to materials that describes their performance to exposure to an electrical arc discharge The arc rating is expressed in cal/cm^2 and is derived from the determined value of the arc thermal performance value (ATPV) or energy of breakopen threshold (E_{BT}) (should a material system exhibit a breakopen response below the ATPV value) derived from the determined value of ATPV or E_{BT}.

FPN: *Breakopen* is a material response evidenced by the formation of one or more holes in the innermost layer of flame-resistant material that would allow flame to pass through the material.

arcing fault‡ Fault characterized by an electrical arc through the air.

impedance‡ Ratio of the voltage drop across a circuit element to the current flowing through the same circuit element, measured in ohms (Ω).

incident energy* The amount of energy impressed on a surface, a certain distance from the source, generated during an electrical arc event. One of the units used to measure incident energy is calories per centimeter squared (cal/cm^2).

let-go threshold The electrical current level at which the brain's electrical signals to muscles can no longer overcome the signals introduced by an external electrical system. Because these external signals lock muscles in the contracted position, the body may not be able to let go when the brain is telling it to do so.

lockout§ The placement of a lockout device on an energy isolating device, in accordance with an established procedure, ensuring that the energy isolating device and the equipment being controlled cannot be operated until the lockout device is removed.

Ohm's law‡ Mathematical relationship between voltage, current, and resistance in an electric circuit. Ohm's law states that the current flowing in a circuit is proportional to electromotive force (voltage) and inversely proportional to resistance: I = E/R.

overcurrent protective device (OCPD)‡ General term that includes both fuses and circuit breakers.

tagout§ The placement of a tagout device on an energy isolating device, in accordance with an established procedure, to indicate that the energy isolating device and the equipment being controlled may not be operated until the tagout device is removed.

*This is the *NFPA 70E* definition.
‡This definition is from the National Fire Protection Association's (NFPA's) *Pocket Dictionary of Electrical Terms*.
§This is the OSHA definition.

OPENING CASE

A 38-year old employee was severely burned while attempting to test an electrical power circuit. On the day of the incident, the electrician was asked by a coworker to determine why an 85-horsepower (hp) electric motor was not functioning. The deceased used a multimeter rated at 600 volts to check out the power supply source to the motor. The motor was connected to the electrical circuit inside a 14,400-volt, 600-ampere metal electrical switch box. When he attempted to check the fuse on this circuit, an electrical arc ionized the air in the cabinet, and a flash fire occurred. The injured worker was burned over 50 percent of his body and died 18 days later.

Investigation of this work-related fatality was prompted by a report from the OSHA Area Office. The investigation included interviews with the company safety director and coworkers. The incident site and equipment were photographed, reports were obtained from the county coroner and the responding ambulance team, and medical records were obtained from the treating hospital. Representatives of an independent engineering laboratory contracted by the insurance company to analyze the multimeter used were also interviewed.

OSHA Requirements

CHAPTER 3

Introduction

The **Occupational Safety and Health Administration (OSHA)** was established some four decades ago when the U.S. Congress passed the *Occupational Safety and Health Act* (OSH Act) of 1970. Its purpose, in part, is "to assure so far as possible every working man and woman in the nation safe and healthful working conditions and to preserve our human resources." President Richard M. Nixon signed this legislation into law on December 29, 1970. Requirements were put into place that, when followed, go a long way towards avoiding the hazards of working on or near exposed energized parts. Implementing and following all of OSHA's requirements including, but not limited to, those for training, selection and use of work practices, and the use of equipment and safeguards for personnel protection will help ensure that employees are provided with a workplace free from recognized hazards.

OSHA: History and Purpose

OSHA publication 2056-07R, entitled *All About OSHA* and published in 2003, includes the following paragraphs that describe OSHA's history and purpose.

The U.S. Congress created OSHA when it passed the *Occupational Safety and Health Act of 1970* (**Figure 3.1**). Its sole intent was to provide a safe work environment for workers.

Prior to 1970, no uniform or comprehensive provisions existed to protect against workplace safety and health hazards. At that time, job-related accidents accounted for more than 14,000 worker deaths, with nearly 2.5 million workers disabled by workplace

OPENING CASE, CONT.

The company employs 200 people. The company had a full-time safety officer and a written safety program. The company had been in business for 90 years. The deceased had worked at the company 18 months and was not authorized to work on voltage over 440 volts. The company conducted on-the-job training, but effects of the training were not measured.

The cause of death, as determined by autopsy and listed on the death certificate, was thermal burns as a consequence of an electrical flash fire.

Adapted from: Colorado Fatality and Assessment and Control Evaluation (FACE) Investigation 92CO039. Accessed 5/12/08 at http://www.cdc.gov/niosh/face/stateface/co/92co039.html.

1. Is it possible for a safety procedure to prevent this from happening?

2. Who is responsible for generating the procedure, if such a procedure is possible?

3. What was the worker's role in the incident?

Chapter Outline

- **Introduction**
- **OSHA: History and Purpose**
- **Summary**
- **Lessons Learned**
- **Current Knowledge**

Learning Objectives

1. Describe the Section 5 of the OSH Act and the 29 Code of Federal Regulations (CFR) Part 1926 requirements presented, including the related subparts where those 29 CFR Part 1926 requirements are located.
2. Explain the 29 CFR Part 1910 requirements presented and the various subparts where those requirements are located.
3. Identify the performance- and prescriptive-based requirements presented and how performance-based requirements differ from prescriptive-based requirements.

References

1. *NFPA 70E*®, 2009 Edition
2. OSHA 29 CFR Part 1926
3. OSHA 29 CFR Part 1910
4. *National Electrical Code*®

Figure 3.1 OSHA was created when Congress passed the OSH Act of 1970.

accidents and injuries, while estimated new cases of occupational disease totaled 300,000. Even with the many successes that OSHA has enjoyed, significant hazards and unsafe conditions still exist in the workplace today.

OSHA Requirements

OSHA requirements are not recommendations. The requirements set forth in the OSHA standards are law. By passing the *Occupational Safety and Health Act of 1970*, Congress authorized enforcement of the standards developed under the Act.

Several OSHA requirements address working on or near exposed energized parts for both construction and maintenance work. The information discussed in this chapter is not intended to be all-inclusive or the basis for a safety program. Instead, the intent is to point out important requirements that must be considered as the OSHA requirements, which apply to any work being performed. OSHA's requirements generally are written in **performance language**. Requirements defined in performance language define requirements without spelling out how to achieve the result.

A good place to begin is in Section 5, Duties, of the OSH Act of 1970. The performance language found in Section 5(a) reads as follows:

5. Duties (a) Each employer

(1) shall furnish to each of his employees employment and a place of employment which are free from recognized hazards that are causing or are likely to cause death or serious physical harm to his employees;

(2) shall comply with occupational safety and health standards promulgated under this Act.

The broad language in Section 5 could cover virtually any hazard to which a worker might be exposed. OSHA might, and often does, cite Section 5(a)(1) of the Occupational Safety and Health Act of 1970 (the Act) when a more specific requirement does not exist.

Sections 5(a)(1) and Section 5(a)(2) of the Act outline the responsibilities of the employer, but what about *employee* responsibility? Each employee is required to comply with the requirements defined by the employer. Employee responsibility is addressed in Section 5(b) of the Act, and reads as follows:

5. Duties (b) Each employee shall comply with occupational safety and health standards and all rules, regulations, and orders issued pursuant to this Act, which are applicable to his own actions and conduct.

Several requirements found in **29 CFR Part 1926** for Construction and in **29 CFR Part 1910** for General Industry might be applicable to any work that is performed. Since most workers and contractors are engaged in work covered by both of these standards, this chapter discusses some requirements that are applicable to electrical work from Parts 1926 (Construction) and 1910 (General Industry). This chapter begins with a look at a number of the Part 1926 requirements.

29 CFR Part 1926

The importance of safety training and education cannot be overstated. OSHA addresses training in Section 1926.21 of 29 CFR, Part 1926, entitled "employer responsibility." Employers are required to instruct employees to recognize and avoid unsafe conditions, in general, and are additionally required to instruct employees about hazards related to confined or enclosed space work as indicated in the following requirements (**Figure 3.2**):

Figure 3.2 OSHA requires that employees working in confined spaces receive instruction on the related hazards and precautions.

1926.21 (b)(2) The employer shall instruct each employee in the recognition and avoidance of unsafe conditions and the regulations applicable to his work environment to control or eliminate any hazards or other exposure to illness or injury.

1926.21(b)(6)(i) All employees required to enter into confined or enclosed spaces shall be instructed as to the nature of the hazards involved, the necessary precautions to be taken, and in the use of protective and emergency equipment required. The employer shall comply with any specific regulations that apply to work in dangerous or potentially dangerous areas.

The above requirements mandate that each employer train employees to recognize and avoid unsafe conditions and are an example of performance language. The requirements apply generally (rather than to one particular hazard), since they are located in 29 CFR Part 1926, Subpart C, General Safety and Health Provisions. Another example of requirements that apply generally can be found in a number of the performance-based requirements contained in 29 CFR Part 1926, Subpart E, Criteria for Personal Protective Equipment. The requirements shown below require, among other things, that personal protective equipment (PPE) be provided and used whenever it is necessary and that the employer be responsible for the adequacy of the PPE (even if employees provide their own).

1926.95(a) Protective equipment, including personal protective equipment for eyes, face, head, and extremities, protective clothing, respiratory devices, and protective shields and barriers, shall be provided, used, and maintained in a sanitary and reliable condition wherever it is necessary by reason of hazards of processes or environment, chemical hazards, radiological hazards, or mechanical irritants encountered in a manner capable of causing injury or impairment in the function of any part of the body through absorption, inhalation or physical contact.

1926.95(b) Where employees provide their own protective equipment, the employer shall be responsible to assure its adequacy, including proper maintenance, and sanitation of such equipment.

Ground-Fault Circuit Protection

Subpart K in 29 CFR Part 1926 addresses electrical safety requirements that are necessary for the practical safeguarding of employees involved in construction work. Section 1926.404 addresses wiring design and protection. Requirements for branch circuit ground-fault protection are contained in 1926.404(b).

It is important to note that OSHA's requirements for ground-fault protection differ from those contained in the *National Electrical Code (NEC®)*. The *NEC* generally requires ground-fault protection to be accomplished by **ground-fault circuit interrupters (GFCIs)** (**Figure 3.3**) that provide ground-fault protection for 125-volt, single-phase, 15-, 20-, and 30-ampere receptacle outlets that are not part of the permanent wiring. OSHA's 29 CFR Part 1926 requirements generally permit ground-fault protection for 125-volt, single-phase, 15- and 20-ampere receptacle outlets to be accomplished by either GFCIs or by an **assured equipment grounding conductor program** as indicated in 1926.404(b)(1)(i), while 1926.404(b)(1)(ii) contains the requirements related to GFCI.

1926.404(b)(1)(i) General. The employer shall use either ground-fault circuit interrupters as specified in paragraph (b)(1)(ii) of this section or an assured equipment grounding conductor program as specified in paragraph (b)(1)(iii) of this section to protect employees on construction sites. These requirements are in addition to any other requirements for equipment grounding conductors.

1926.404(b)(1)(ii) Ground-fault circuit interrupters. All 120-volt, single-phase 15- and 20-ampere receptacle outlets on construction sites, which are not a part of the permanent wiring of the building or structure and which are in use by employees, shall have approved ground-fault circuit interrupters for personnel protection. Receptacles on a 2-wire, single-phase portable or vehicle-mounted generator rated not more than 5 kW, where the circuit conductors of the generator are insulated from the generator frame and all other grounded surfaces, need not be protected with ground-fault circuit interrupters.

The requirements for the assured equipment grounding conductor program are addressed in 1926.404(b)(1)(iii). The employer is required to develop and implement an assured equipment grounding conductor program on construction sites as follows:

1926.404(b)(1)(iii) Assured equipment grounding conductor program. The employer shall establish and implement an assured equipment grounding conductor program on construction sites covering all cord sets, receptacles which are not a part of the building or structure, and equipment connected by cord and plug which are available for use or used by employees. This program shall comply with the following minimum requirements:

One note before we list those minimum requirements. Again, OSHA's 29 CFR Part 1926 requirements generally allow ground-fault protection for 125-volt, single-phase, 15- and 20-ampere receptacle outlets to be accomplished by *either* GFCIs *or* by an assured equipment

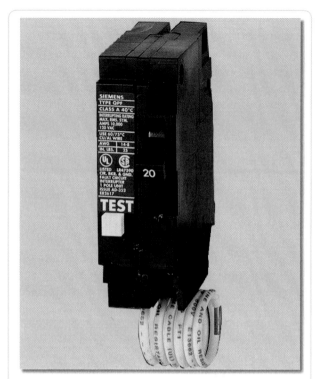

A. A GFCI circuit breaker.

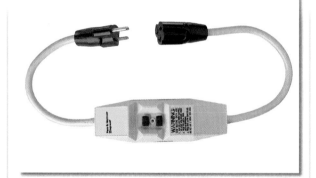

B. A raintight GFCI with open neutral protection.

C. A temporary power outlet unit.

Figure 3.3 GFCI is a common form of ground-fault protection for employees on construction sites.

grounding conductor program. The assured equipment grounding conductor program on construction sites must be in writing and be implemented by a competent person in accordance with the prescriptive-based requirements of 1926.404(b)(1)(iii) (A) through (G) as indicated below:

1926.404(b)(1)(iii)(A) A written description of the program, including the specific procedures adopted by the employer, shall be available at the jobsite for inspection and copying by the Assistant Secretary and any affected employee.

1926.404(b)(1)(iii)(B) The employer shall designate one or more competent persons (as defined in 1926.32(f)) to implement the program.

1926.404(b)(1)(iii)(C) Each cord set, attachment cap, plug and receptacle of cord sets, and any equipment connected by cord and plug, except cord sets and receptacles that are fixed and not exposed to damage, shall be visually inspected before each day's use for external defects, such as deformed or missing pins or insulation damage, and for indications of possible internal damage. Equipment found damaged or defective shall not be used until repaired.

1926.404(b)(1)(iii)(D) The following tests shall be performed on all cord sets, receptacles that are not a part of the permanent wiring of the building or structure, and cord- and plug-connected equipment required to be grounded:

1926.404(b)(1)(iii)(D)(1) All equipment grounding conductors shall be tested for continuity and shall be electrically continuous.

1926.404(b)(1)(iii)(D)(2) Each receptacle and attachment cap or plug shall be tested for correct attachment of the equipment grounding conductor. The equipment grounding conductor shall be connected to its proper terminal.

1926.404(b)(1)(iii)(E) All required tests shall be performed:

1926.404(b)(1)(iii)(E)(1) Before first use;

1926.404(b)(1)(iii)(E)(2) Before equipment is returned to service following any repairs;

1926.404(b)(1)(iii)(E)(3) Before equipment is used after any incident which can be reasonably suspected to have caused damage (for example, when a cord set is run over); and

1926.404(b)(1)(iii)(E)(4) At intervals not to exceed 3 months, except that cord sets and receptacles that are fixed and not exposed to damage shall be tested at intervals not exceeding 6 months.

1926.404(b)(1)(iii)(F) The employer shall not make available or permit the use by employees of any equipment that has not met the requirements of this paragraph (b)(1)(iii) of this section.

1926.404(b)(1)(iii)(G) Tests performed as required in this paragraph shall be recorded. This test record shall identify each receptacle, cord set, and cord- and plug-connected equipment that passed the test and shall indicate the last date it was tested or the interval for which it was tested. This record shall be kept by means of logs, color coding, or other effective means and shall be maintained until replaced by a more current record. The record shall be made available on the job site for inspection by the Assistant Secretary and any affected employee.

Each cord set (**Figure 3.4**), attachment cap, and plug and receptacle of cord sets must be visually inspected before each day's use.

Figure 3.4 Inspect each cord set, attachment cap, and plug and receptacle of cord sets before each day's use.

The assured equipment grounding conductor program requirements are in prescriptive language, since the requirements define what must be done to comply. The requirements are quite detailed and are one reason why an employer might choose to use GFCIs for ground-fault protection. However, OSHA requires the employer to establish and implement an assured equipment grounding conductor program on construction sites covering all cord sets, receptacles that are not a part of the building or structure, and equipment connected by cord and plug that are available for use or used by employees unless they are protected by GFCIs.

Employee Protection from Hazards

Now examine a number of performance-based safety-related work practices contained in Sections 1926.416 and 1926.417. In addition to covering the hazards arising from the use of electricity at job sites, sections 1926.416 and 1926.417 cover the hazards arising from the accidental contact by employees with all energized lines. Some requirements are discussed in the following text and require, among other things, that employers inform employees of hazards and any necessary protective measures. Equipment must be deenergized, grounded, and have tags attached. Deenergized circuits must be rendered inoperative and have tags attached (**Figure 3.5**).

1926.416(a) Protection of employees.

(1) No employer shall permit an employee to work in such proximity to any part of an electric power circuit that the employee could contact the electric power circuit in the course of work, unless the employee is protected against electric shock by deenergizing the circuit and grounding it or by guarding it effectively by insulation or other means.

(3) Before work is begun the employer shall ascertain by inquiry or direct observation, or by instruments, whether any part of an energized electric power circuit, exposed or concealed, is so located that the performance of the work may bring any person, tool, or machine into physical or electrical contact with the electric power circuit. The employer shall post and maintain proper warning signs where such a circuit exists. The employer shall advise employees of the location of such lines, the hazards involved, and the protective measures to be taken.

1926.417(b) Equipment and Circuits.

Equipment or circuits that are deenergized shall be rendered inoperative and shall have tags attached at all points where such equipment or circuits can be energized.

As our discussion of 29 CFR Part 1926 (Construction) requirements concludes, each student must note how performance requirements differ from prescriptive requirements. Performance requirements include safety-related work practice requirements in Sections 1926.416 and 1926.417. These sections illustrate requirements without spelling out how to comply. Examples of prescriptive requirements include an assured equipment grounding conducted program as previ-

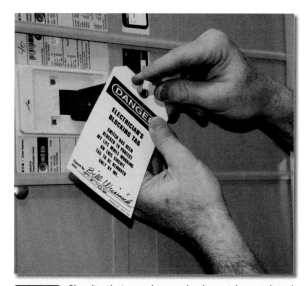

Figure 3.5 Circuits that are deenergized must be rendered inoperative and shall have tags attached.
Photo courtesy of National Electrical Contractors Association (NECA).

ously discussed and include details of what must be done to comply.

29 CFR Part 1910

As mentioned earlier in this chapter, many important requirements are also contained in 29 CFR Part 1910 (General Industry). Again, this discussion is not intended to be all-inclusive, but rather to point out a number of important requirements.

A hazard assessment and training are fundamental parts of identifying hazards to which workers might be exposed and determining the PPE necessary to protect those workers. Training is intended to ensure that workers understand the use and limitations of PPE. OSHA requires that a hazard assessment be conducted and that the hazard assessment be in writing.

Personal Protective Equipment

The following requirements are from Subpart I, PPE. Subpart I requires, among other things, that PPE be provided and used. It also requires that the employer ensure adequacy of PPE (including PPE provided by the employee). The hazard assessment must be in writing to facilitate the selection and use of PPE.

1910.132(a) Application. Protective equipment, including personal protective equipment for eyes, face, head, and extremities, protective clothing, respiratory devices, and protective shields and barriers, shall be provided, used, and maintained in a sanitary and reliable condition wherever it is necessary by reason of hazards of processes or environment, chemical hazards, radiological hazards, or mechanical irritants encountered in a manner capable of causing injury or impairment in the function of any part of the body through absorption, inhalation, or physical contact.

1910.132(b) Employee-owned equipment. Where employees provide their own protective equipment, the employer shall be responsible to assure its adequacy, including proper maintenance and sanitation of such equipment.

1910.132(c) Design. All personal protective equipment shall be of safe design and construction for the work to be performed.

1910.132(d) Hazard assessment and equipment selection.

(1) The employer shall assess the workplace to determine if hazards are present, or are likely to be present, which necessitate the use of personal protective equipment (PPE). If such hazards are present, or likely to be present, the employer shall:

(i)   Select, and have each affected employee use, the types of PPE that will protect the affected employee from the hazards identified in the hazard assessment;

(ii) Communicate selection decisions to each affected employee; and,

(iii) Select PPE that properly fits each affected employee. Note: Nonmandatory Appendix B contains an example of procedures that would comply with the requirement for a hazard assessment.

(2) The employer shall verify that the required workplace hazard assessment has been performed through a written certification that identifies the workplace evaluated; the person certifying that the evaluation has been performed; the date(s) of the hazard assessment; and, which identifies the document as a certification of hazard assessment.

The employer must determine and require each affected employee use PPE, which will protect the affected employee from the hazards identified in the hazard assessment (**Figure 3.6**).

Training

The following requirements, also from Subpart I, address a number of requirements related to training and retraining. Among other things, the employer must provide and document training about selecting and using PPE. The employer must have the employee demonstrate an understanding of the training (as indicated in the following requirements):

1910.132(f) Training.

1910.132(f)(1) The employer shall provide training to each employee who is required by this section to use PPE. Each such employee shall be trained to know at least the following:

1910.132(f)(1)(i) When PPE is necessary;

1910.132(f)(1)(ii) What PPE is necessary;

1910.132(f)(1)(iii) How to properly don, doff, adjust, and wear PPE;

Figure 3.6   An example of PPE appropriate for the hazard.

1910.132(f)(1)(iv) The limitations of the PPE; and,

1910.132(f)(1)(v) The proper care, maintenance, useful life and disposal of the PPE.

1910.132(f)(2) Each affected employee shall demonstrate an understanding of the training specified in paragraph (f)(1) of this section, and the ability to use PPE properly, before being allowed to perform work requiring the use of PPE.

1910.132(f)(3) When the employer has reason to believe that any affected employee who has already been trained does not have the understanding and skill required by paragraph (f)(2) of this section, the employer shall retrain each such employee. Circumstances where retraining is required include, but are not limited to, situations where:

1910.132(f)(3)(i) Changes in the workplace render previous training obsolete; or

1910.132(f)(3)(ii) Changes in the types of PPE to be used render previous training obsolete; or

1910.132(f)(3)(iii) Inadequacies in an affected employee's knowledge or use of assigned PPE indicate that the employee has not retained the requisite understanding or skill.

1910.132(f)(4) The employer shall verify that each affected employee has received and understood the required training through a written certification that contains the name of each employee trained, the date(s) of training, and that identifies the subject of the certification.

Section 1910.137 of Subpart I addresses electrical protective equipment. These requirements address, but are not limited to, in-service care and use of electrical protective equipment such as rubber insulating gloves, blankets, and sleeves including daily inspection and periodic electrical tests. Both performance and prescriptive requirements are contained in Section 1910.137, electrical protective equipment.

1910.137 Electrical Protective Devices

1910.137(b) "In-service care and use."

1910.137(b)(1) Electrical protective equipment shall be maintained in a safe, reliable condition.

1910.137(b)(2) The following specific requirements apply to insulating blankets, covers, line hose, gloves, and sleeves made of rubber:

1910.137(b)(2)(i) Maximum use voltages shall conform to those listed in Table I-5.

1910.137(b)(2)(ii) Insulating equipment shall be inspected for damage before each day's use and immediately following any incident that can reasonably be suspected of having caused damage. Insulating

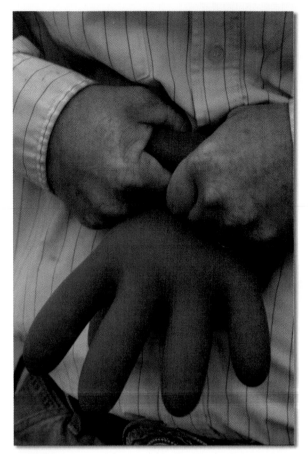

Figure 3.7 Insulating gloves being given an air test prior to use.

gloves shall be given an air test, along with the inspection (see Figure 3-5).

1910.137(b)(2)(viii) Electrical protective equipment shall be subjected to periodic electrical tests. Test voltages and the maximum intervals between tests shall be in accordance with Table I-5 and Table I-6.

Figure 3.7 shows a step of an air test being done on insulating gloves being prior to use. **Table 3.1** and **Table 3.2** show the above-referenced OSHA Tables I-5 and I-6, respectively.

Many of these requirements covered were performance-based. The regulations require, for example, that PPE be selected and used based on the written assessment but do not indicate what the hazard assessment must look like or what it must include. Subpart S, Electrical, also contains performance requirements in many cases.

The importance of safety training and education cannot be overstated. The requirements from Subpart S are a few more examples of important considerations

Table 3.1	OSHA 1910.137 Table I-5: Rubber Insulating Equipment Voltage Requirements		
Class of Equipment	Maximum Use Voltage* a-c-rms	Retest Voltage† a-c-rms	Retest Voltage† d-c-avg
0	1000	5000	20,000
1	7500	10,000	40,000
2	17,000	20,000	50,000
3	26,500	30,000	60,000
4	36,000	40,000	70,000

*The maximum use voltage is the a-c voltage (rms) classification of the protective equipment that designates the maximum nominal design voltage of the energized system that may be safely worked. The nominal design voltage is equal to the phase-to-phase voltage on multiphase circuits. However, the phase-to-ground potential is considered to be the nominal design voltage:
1. If there is no multiphase exposure in a system area and if the voltage exposure is limited to the phase-to-ground potential, or
2. If the electrical equipment and devices are insulated or isolated or both so that the multiphase exposure on a grounded wye circuit is removed.
†The proof-test voltage shall be applied continuously for at least 1 minute, but no more than 3 minutes.

Table 3.2	OSHA 1910.137 Table I-6: Rubber Insulating Equipment Test Intervals
Type of Equipment	When to Test
Rubber insulating line hose	Upon indication that insulating value is suspect.
Rubber insulating covers	Upon indication that insulating value is suspect.
Rubber insulating blankets	Before first issue and every 12 months thereafter.*
Rubber insulating gloves	Before first issue and every 6 months thereafter.*
Rubber insulating sleeves	Before first issue and every 12 months thereafter.*

*If the insulating equipment has been electrically tested but not issued for service, it may not be placed into service unless it has been electrically tested within the previous 12 months.

that are contained in 29 CFR 1910. It is important to note that only a select number of the requirements were included. All of OSHA's requirements must be reviewed and implemented in their entirety, as applicable.

Following are a few of the requirements that address training and include, among other things, requirements that a qualified person must be trained to understand and to be familiar with. Also note that, while the required training can be either on the job or in the classroom, the degree of the training is dependent on the risk to the employee.

1910.332—Training.

1910.332(b) Content of training.

1910.332(b)(1) Practices addressed in this standard. Employees shall be trained in and familiar with the safety-related work practices required by 1910.331 through 1910.335 that pertain to their respective job assignments.

1910.332(b)(2) Additional requirements for unqualified persons. Employees who are covered by paragraph (a) of this section but who are not qualified persons shall also be trained in and familiar with any electrically related safety practices not specifically addressed by 1910.331 through 1910.335 but which are necessary for their safety.

1910.332(b)(3) Additional requirements for qualified persons. Qualified persons (i.e., those permitted to work on or near exposed energized parts) shall, at a minimum, be trained in and familiar with the following:

1910.332(b)(3)(i) The skills and techniques necessary to distinguish exposed live parts from other parts of electric equipment.

1910.332(b)(3)(ii) The skills and techniques necessary to determine the nominal voltage of exposed live parts, and

1910.332(b)(3)(iii) The clearance distances specified in 1910.333(c) and the corresponding voltages to which the qualified person will be exposed.

Note 1: For the purposes of 1910.331 through 1910.335, a person must have the training required by paragraph (b)(3) of this section in order to be considered a qualified person.

Note 2: Qualified persons whose work on energized equipment involves either direct contact or contact by means of tools or materials must also have the training needed to meet 1910.333(C)(2).

1910.332(c) Type of training. The training required by this section shall be of the classroom or on-the-job type. The degree of training provided shall be determined by the risk to the employee.

Preventing Electric Shock and Other Injuries

Several requirements related to safety-related work practices must be used to prevent electric shock or other injuries associated with energized work. Section 1910.333(a) identifies general requirements related to the selection and use of work practices, 1910.333(b) defines requirements related to working on or near exposed *deenergized* parts, and 1910.333(c) addresses requirements related to working on or near exposed *energized* parts.

Among other things, OSHA requires safety-related work practices to be implemented to protect from electrical hazards. In 1910.333(a), (a)(1), and (a)(2), OSHA indicates that deenergizing must be the primary safety-related work practice. An important part of understanding and implementing safety-related work practices is recognizing that live parts to which an employee might be exposed must be deenergized before the employee works on or near them, unless the employer can demonstrate that deenergizing introduces additional or increased hazards or is infeasible due to equipment design or operational limitations. This requirement is illustrated in the following extract from 29 CFR Part 1910, Subpart S. Examples of increased or additional hazards and examples of work that might be performed on or near energized circuit parts because of infeasibility due to equipment design or operational limitations are suggested in the following extracted information.

1910.333—Selection and use of work practices.

1910.333(a) "General." Safety-related work practices shall be employed to prevent electric shock or other injuries resulting from either direct or indirect electrical contacts, when work is performed near or on equipment or circuits which are or may be energized. The specific safety-related work practices shall be consistent with the nature and extent of the associated electrical hazards.

1910.333(a)(1) "Deenergized parts." Live parts to which an employee may be exposed shall be deenergized before the employee works on or near them, unless the employer can demonstrate that deenergizing introduces additional or increased hazards or is infeasible due to equipment design or operational limitations. Live parts that operate at less than 50 volts to ground need not be deenergized if there will

OSHA Tip

The two main OSHA rules that pertain to electrical safety are 29 CFR 1910.147 and 29 CFR 1910.333. OSHA 29 CFR 1910.147 defines requirements to control hazardous energy. OSHA CFR 1910.333, on the other hand, defines requirements applicable to work on exposed deenergized parts or near enough to them to expose the employee to any electrical hazard they present.

be no increased exposure to electrical burns or to explosion due to electric arcs.

Note 1: Examples of increased or additional hazards include interruption of life support equipment, deactivation of emergency alarm systems, shutdown of hazardous location ventilation equipment, or removal of illumination for an area.

Note 2: Examples of work that may be performed on or near energized circuit parts because of infeasibility due to equipment design or operational limitations include testing of electric circuits that can only be performed with the circuit energized and work on circuits that form an integral part of a continuous industrial process in a chemical plant that would otherwise need to be completely shut down in order to permit work on one circuit or piece of equipment.

Note 3: Work on or near deenergized parts is covered by paragraph (b) of this section.

1910.333(a)(2) "Energized parts." If the exposed live parts are not deenergized (i.e., for reasons of increased or additional hazards or infeasibility), other safety-related work practices shall be used to protect employees who may be exposed to the electrical hazards involved. Such work practices shall protect employees against contact with energized circuit parts directly with any part of their body or indirectly through some other conductive object. The work practices that are used shall be suitable for the conditions under which the work is to be performed and for the voltage level of the exposed electric conductors or circuit parts. Specific work practice requirements are detailed in paragraph (c) of this section.

Among other things, OSHA requires lockout or tagout (or both) of circuits that energize parts of fixed electric equipment where employees are exposed to contact in Section 1910.333(b). These requirements are detailed, much like Section 1910.147, control of hazardous energy (lockout/tagout), located in Subpart J. In Section 1910.333(b)(1), OSHA outlines the application, and 1910.333(b)(2) defines the requirements that must be followed for locking and tagging.

Section 1910.333(b)(2) is further broken down into five categories as 1910.333(b)(2)(i) through 1910.333(b)(2)(v), covering (i) Procedures (in writing), (ii) Deenergizing equipment, (iii) Application of locks and tags (Figure 3.8), (iv) Verification of deenergized condition, and (v) Reenergizing equipment.

1910.333(b) "Working on or near exposed deenergized parts."

1910.333(b)(1) "Application." This paragraph applies to work on exposed deenergized parts or near enough to them to expose the employee to any electrical hazard they present. Conductors and parts of electric equipment that have been deenergized but have not been locked out or

Figure 3.8 Applying a lock and tag.

tagged in accordance with paragraph (b) of this section shall be treated as energized parts, and paragraph (c) of this section applies to work on or near them.

1910.333(b)(2) also addresses the application of locks and tags.

1910.333(b)(2) "Lockout and Tagging." While any employee is exposed to contact with parts of fixed electric equipment or circuits which have been deenergized, the circuits energizing the parts shall be locked out or tagged or both in accordance with the requirements of this paragraph. The requirements shall be followed in the order in which they are presented [i.e., paragraph (b)(2)(i) first, then paragraph (b)(2)(ii), etc.].

Note 1: As used in this section, fixed equipment refers to equipment fastened in place or connected by permanent wiring methods.

Note 2: Lockout and tagging procedures that comply with paragraphs (c) through (f) of 1910.147 will also be deemed to comply with paragraph (b)(2) of this section provided that:

[1] The procedures address the electrical safety hazards covered by this Subpart; and

[2] The procedures also incorporate the requirements of paragraphs (b)(2)(iii)(D) and (b)(2)(iv)(B) of this section.

1910.333(b)(2)(i) "Procedures." The employer shall maintain a written copy of the procedures outlined in paragraph (b)(2) and shall make it available for inspection by employees and by the Assistant Secretary of Labor and his or her authorized representatives.

Note: The written procedures may be in the form of a copy of paragraph (b) of this section.

1910.333(b)(2)(ii) "Deenergizing equipment."

1910.333(b)(2)(ii)(A) Safe procedures for deenergizing circuits and equipment shall be determined before circuits or equipment are deenergized.

1910.333(b)(2)(ii)(B) The circuits and equipment to be worked on shall be disconnected from all electric energy sources. Control circuit devices, such as push buttons, selector switches, and interlocks, may not be used as the sole means for deenergizing circuits or equipment. Interlocks for electric equipment may not be used as a substitute for lockout and tagging procedures.

1910.333(b)(2)(ii)(C) Stored electric energy which might endanger personnel shall be released. Capacitors shall be discharged and high capacitance elements shall be short-circuited and grounded, if the stored electric energy might endanger personnel.

Note: If the capacitors or associated equipment are handled in meeting this requirement, they shall be treated as energized.

1910.333(b)(2)(ii)(D) Stored nonelectrical energy in devices that could reenergize electric circuit parts shall be blocked or relieved to the extent that the circuit parts could not be accidentally energized by the device.

1910.333(b)(2)(iii) "Application of locks and tags."

1910.333(b)(2)(iii)(A) A lock and a tag shall be placed on each disconnecting means used to deenergize circuits and equipment on which work is to be performed, except as provided in paragraphs (b)(2)(iii)(C) and (b)(2)(iii)(E) of this section. The lock shall be attached so as to prevent persons from operating the disconnecting means unless they resort to undue force or the use of tools.

1910.333(b)(2)(iii)(B) Each tag shall contain a statement prohibiting unauthorized operation of the disconnecting means and removal of the tag.

1910.333(b)(2)(iii)(C) If a lock cannot be applied, or if the employer can demonstrate that tagging procedures will provide a level of safety equivalent to that obtained by the use of a lock, a tag may be used without a lock.

1910.333(b)(2)(iii)(D) A tag used without a lock, as permitted by paragraph (b)(2)(iii)(C) of this section, shall be supplemented by at least one additional safety measure that provides a level of safety equivalent to that obtained by use of a lock. Examples of additional safety measures include the removal of an isolating circuit element, blocking of a controlling switch, or opening of an extra disconnecting device.

1910.333(b)(2)(iii)(E) A lock may be placed without a tag only under the following conditions:

1910.333(b)(2)(iii)(E)(1) Only one circuit or piece of equipment is deenergized, and

1910.333(b)(2)(iii)(E)(2) The lockout period does not extend beyond the work shift, and

Figure 3.9 A lock may be placed without a tag under specific conditions only.

1910.333(b)(2)(iii)(E)(3) Employees exposed to the hazards associated with reenergizing the circuit or equipment are familiar with this procedure.

Figure 3.9 shows a lock placed without a tag. A lock without a tag may be used only under specific conditions (see the above discussion).

1910.333(b)(2)(iv) Verification of deenergized condition. The requirements of this paragraph shall be met before any circuits or equipment can be considered and worked as deenergized.

1910.333(b)(2)(iv)(A) A qualified person shall operate the equipment operating controls or otherwise verify that the equipment cannot be restarted.

1910.333(b)(2)(iv)(B) A qualified person shall use test equipment to test the circuit elements and electrical parts of equipment to which employees will be exposed and shall verify that the circuit elements and equipment parts are deenergized. The test shall also determine if any energized condition exists as a result of inadvertently induced voltage or unrelated voltage backfeed even though specific parts of the circuit have been deenergized and presumed to be safe. If the circuit to be tested is over 600 volts, nominal, the test equipment shall be checked for proper operation immediately after this test.

1910.333(b)(2)(v) "Reenergizing equipment." These requirements shall be met, in the order given, before circuits or equipment are reenergized, even temporarily.

1910.333(b)(2)(v)(A) A qualified person shall conduct tests and visual inspections, as necessary, to verify that all tools, electrical jumpers, shorts, grounds, and other such devices have been removed, so that the circuits and equipment can be safely energized.

1910.333(b)(2)(v)(B) Employees exposed to the hazards associated with reenergizing the circuit or equipment shall be warned to stay clear of circuits and equipment.

1910.333(b)(2)(v)(C) Each lock and tag shall be removed by the employee who applied it or under his or her direct supervision. However, if this employee is absent from the workplace, then the lock or tag may be removed by a qualified person designated to perform this task provided that:

1910.333(b)(2)(v)(C)(1) The employer ensures that the employee who applied the lock or tag is not available at the workplace, and

1910.333(b)(2)(v)(C)(2) The employer ensures that the employee is aware that the lock or tag has been removed before he or she resumes work at that workplace.

1910.333(b)(2)(v)(D) There shall be a visual determination that all employees are clear of the circuits and equipment.

In 1910.333 (c)(2) and (5), OSHA permits only qualified persons to work on energized electric circuit parts or equipment. The same section indicates that qualified persons must be capable of working safely on energized circuits and be familiar with the proper use of special precautionary techniques, PPE, insulating and shielding materials, and insulated tools.

1910.333(c) "Working on or near exposed energized parts."

1910.333(c)(2) "Work on energized equipment." Only qualified persons may work on electric circuit parts or equipment that have not been deenergized under the procedures of paragraph (b) of this section. Such persons shall be capable of working safely on energized circuits and shall be familiar with the proper use of special precautionary techniques, personal protective equipment, insulating and shielding materials, and insulated tools.

1910.333(c)(5) "Confined or enclosed work spaces." When an employee works in a confined or enclosed space (such as a manhole or vault) that contains exposed energized parts, the employer shall provide, and the employee shall use, protective shields, protective barriers, or insulating materials as necessary to avoid inadvertent contact with these parts. Doors, hinged panels, and the like shall be secured to prevent their swinging into an employee and causing the employee to contact exposed energized parts.

In section 1910.334 of 1910, Subpart S, OSHA addresses the use of equipment. Paragraph (b)(2) covers reclosing circuits after protective device operation and test instruments and equipment in paragraph (c)(2). Generally, repetitive manual reclosing of circuit breakers or reenergizing circuits by replacing fuses is prohibited as indicated below:

1910.334—Use of equipment.

1910.334(b) "Electric power and lighting circuits."

1910.334(b)(2) "Reclosing circuits after protective device operation." After a circuit is deenergized by a circuit protective device, the circuit protective device, the circuit may not be manually reenergized until it has been determined that the equipment and circuit can be safely energized. The repetitive manual reclosing of circuit breakers or reenergizing circuits through replaced fuses is prohibited.

Note: When it can be determined from the design of the circuit and the overcurrent devices involved that the automatic operation of a device was caused by an overload rather than a fault condition, no examination of the circuit or connected equipment is needed before the circuit is reenergized.

1910.334(c) "Test instruments and equipment."

1910.334(c)(1) "Use." Only qualified persons may perform testing work on electric circuits or equipment.

1910.334(c)(3) "Rating of equipment." Test instruments and equipment and their accessories shall be rated for the circuits and equipment to which they will be connected and shall be designed for the environment in which they will be used.

Recall that the primary safety-related work practice is deenergizing with lockout and tagout. The requirements from 29 CFR 1910.333(b)(2) indicate that lockout and tagout procedures must be in writing and must be followed in the order they are presented. This and all OSHA requirements must be understood and implemented in their entirety.

Only after the employer demonstrates that deenergizing introduces additional or increased hazards or is infeasible due to equipment design or operational limitations does OSHA permit an employee to work on or near live parts that have not been deenergized. Other safety-related work practices, such as PPE, are required once the employer demonstrates that it is infeasible or a greater hazard to deenergize. If the need to work energized is demonstrated by the employer, it is important to remember that other safety-related work practices must be used and might include electrical protective equipment, insulated tools, and protective shields, barriers or insulating materials. Safeguards for personnel protection are discussed in the following (and several other) requirements:

1910.335—Safeguards for personnel protection.

1910.335(a) Use of protective equipment.

1910.335(a)(1) Personal protective equipment.

1910.335(a)(1)(i) Employees working in areas where there are potential electrical hazards shall be provided with, and shall use, electrical protective equipment that is appropriate for the specific parts of the body to be protected and for the work to be performed.

Employers must provide electrical protective equipment, and employees must use the equipment that is appropriate for the specific parts of the body to be protected (**Figure 3.10**).

1910.335(a)(2)(i) When working near exposed energized conductors or circuit parts, each employee shall use insulated tools or handling equipment if the tools or handling equipment might make contact with such conductors or parts (**Figure 3.11**). If the insulating capability of insulated tools or handling equipment is subject to damage, the insulating material shall be protected.

1910.335(a)(2)(ii) Protective shields, protective barriers, or insulating materials shall be used to protect each employee from shock, burns, or other electrically related injuries while that employee is working near exposed energized parts which might be accidentally contacted or where dangerous electric heating or arcing might occur (see Figure 3-10). When normally enclosed live parts are exposed for maintenance or repair, they shall be guarded to protect unqualified persons from contact with the live parts.

Figure 3.10 Employers must provide and employees must use electrical protective equipment.

Figure 3.11 Insulated tools are tested to 10,000 volts and have a maximum use rating of 1000 volts.

Insulating materials must be used to protect each employee while working near exposed energized parts that might be touched accidentally (**Figure 3.12**).

Summary

Each day workers are exposed to hazards associated with working on or close enough to energized parts where they could be exposed to electrical hazards. When appropriate requirements are followed, incidents, injuries, and fatalities can be reduced. Many of these requirements were discussed in this chapter. The requirements covered here are only a few of the issues that should be considered when assessing workplace hazards and developing an effective safety program. Implementing and following the requirements for training, selection and use of work practices, the use of equipment and safeguards for personnel protection will help ensure that the workplace is free from recognized hazards. A fundamental part of providing a workplace free

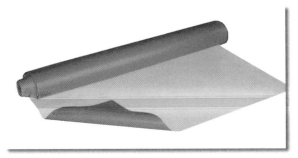

Figure 3.12 Insulating materials must be used to protect each employee while working near exposed energized parts that might be touched accidentally.

from recognized hazards is deenergizing and lockout/tagout meeting as required by Sections 1910.147 and 1910.333(b)(2). A written hazard assessment plays a critical role for an employer to demonstrate that deenergizing introduces additional or increased hazards or is infeasible due to equipment design or operational limitations. PPE must be selected and used based on the level of the hazards identified in the written hazard assessment.

Lessons *learned*

Licensed Electrician Electrocuted

On July 4, 2003, a contract crew of five electricians (job foreman, decedent who was also the lead electrician, and three other electricians) was working at a manufacturing facility that was shutdown for the holiday. Besides a plant employee working in the office, they were the only workers on site and had complete control of the facility. The contractor had already installed wiring in the newly constructed addition to the manufacturing facility and was connecting service for two air-conditioning units (3-phase, 480 volts, 30-ampere service and 35-ampere service) and service for a lighting panel (3-phase, 277/480 volts with 200 amperes). Each of the three services had its own breaker located approximately 130 ft away on the wall, near the ground. A 4 ft × 4 ft × 12 in. deep metal junction box, 20 ft in the air, rested on two metal tracks suspended from the ceiling, housed the junction of wiring from the new addition and the main building.

The work crew began working at 7:00 a.m. At 9:00 a.m., the owner of the contracting company stopped by and checked on the workers. Work was progressing nicely, and the crew had approximately 1 hour left of work to finish so they could go home and enjoy the holiday. As the owner left the work site, the outside temperature was a hot and humid 85°F, with a dew point of 71°F. It is unknown what the temperature was inside where the men were working. The area where the men were working was not air-conditioned, nor did they have fans running to cool the work environment.

At approximately 9:45 a.m., the job foreman and another employee were on the ground gathering up tools and putting them away while waiting for instructions from the lead electrician (decedent) and another worker who were in the 4 × 4 ft junction box. A fifth employee was assisting the two workers in the junction box from an 8 × 6 ft scissor lift located adjacent to the junction box. The workers in the junction box were pulling three sets of wiring service (two sets for air-conditioning units; one set for lighting) from the breaker box in the main plant and connecting it to new electrical service in a new addition to the facility. Each set of wiring had its own breaker on a breaker panel, which the foreman, but not the lead electrician, had locked out, modifying the normally followed lockout/tagout procedure. Normally, the employee (in this instance the lead electrician) performing the work would place their lockout/tagout equipment on the necessary breakers and then remove the lockout/tagout equipment when work was completed and the breakers could be reenergized.

The decedent was hot, sweaty, and not wearing a shirt. After running wires and completing connections for the lighting service from the main building breaker panel to the new addition, the two electricians in the junction box began pulling wires for the two air-conditioning units. The lead electrician was pulling the wires for the air-conditioning service from the breaker box in the main building and was getting ready to connect the wires to the new wiring in the addition. As he pulled the wires into the junction box, he routed the wires under his legs, tapping the ends of the wires into his right hand to make them even. He was not wearing insulated gloves as he handled the wires and made the connections. At the same time, the decedent was ready for the breaker to the lighting service panel to be turned on and instructed the foreman to throw the breaker to the "on" position. The foreman, thinking he should throw all three breakers to the "on" position, walked over to the breaker panel, removed his lockout/ tagout on all three breakers. He proceeded to throw all three to the "on" position. This sent electricity through the wires into the lead electrician's hand, killing him.

Reportedly, the victim looked at his coworker in the junction box and said "help me," then collapsed. The worker on the lift and the other worker in the junction box called down to the foreman to contact emergency services, which he immediately did. As emergency services were en route, the other employee in the junction box and employee in the scissor lift placed the victim in the scissor lift and performed cardio-pulmonary resuscitation (CPR) until paramedics arrived. Paramedics took the victim to a nearby hospital, where a physician notified the coroner, who arrived at 10:25 a.m. and declared the victim dead.

The cause of death was electrocution. At the time of the incident, the five-man crew had 10 more minutes of work to complete the day's agenda.

Recommendations/Discussion

Recommendation 1: Employees should always follow company lockout/tagout procedures.

Discussion: The company had lockout/tagout procedures in place that were to be used in these types of situations. All employees were equipped with lockout/tagout equipment, but because they were the only workers at the facility, only the foreman used his. Normally, all employees who could be adversely affected by the electrical equipment would have locked out the breaker panel. The occupational safety and health regulation 1926.416(a)(1) states that "No employer shall permit an employee to work in such proximity to any part of an electric power circuit that the employee could contact the electric power circuit in the course of work, unless the employee is protected against electric shock by deenergizing the circuit and grounding it or by guarding it effectively by insulation or other means." The job foreman had used his lockout/tagout, but the decedent and the other worker in the junction box had not put their lockout/tagouts on the breaker. This was done to save time. All three men should have removed their lockout/tagouts from the breaker. The two men in the junction box needed to lower themselves to the ground, walk to the breakers and, with the foreman, all three electricians then remove their lockout/tagouts so that the correct breaker could be energized. Gloves were not required in this instance since the electrical circuit had been deenergized. However, if the circuit had been hot, gloves would have been required.

Recommendation 2: Communication between workers should be clear and precise.

Discussion: Communication between workers should be clear and precise when instructing other workers on what they want the other workers to do. "Throw the breaker" could mean anything. The men involved in this incident had worked together for several years and knew each other but still miscommunicated. Only the person performing the work should throw the breaker when the work is completed. In this case, the lead electrician should have been the one to throw the breaker in the "on" position after completing the work.

Adapted from: FACE Investigation 03KY115. Accessed 5/12/08 at http://www.cdc.gov/niosh/face/stateface/ky/03KY115.html.

Questions

1. Which of the following was the basic cause of the electrocution?
   - **A.** The decedent should have installed locks and tags on all of the circuit breakers.
   - **B.** The decedent should have used a scaffold with an insulating link for access to the junction box.
   - **C.** The electrician should have tested for voltage with a Cat III meter.
   - **D.** The lead electrician's helper did not warn the worker to keep clear of the service conductors.

2. The "normal" lockout/tagout procedure called for each worker who might be exposed to electrocution to install locks. Why did the crew choose only for the foreman to install locks and tags?
   - **A.** The procedure required the foreman to lock out for his or her crew.
   - **B.** The plant permitted only one person to install locks and tags on plant equipment such as the circuit breakers.
   - **C.** With the exception of the two plant employees, only one crew of workers was on site at the time, and they would not expose their coworkers to shock or electrocution.
   - **D.** The OSHA regulations required locks and tags be installed by plant employees.

3. The work was associated with modifying an existing installation. Therefore, which of the following regulations covered lockout/tagout associated with the work?
   - **A.** 29 CFR 1910.333, covering lockout/tagout for general industry
   - **B.** 29 CFR 1926.417, covering lockout/tagout for construction
   - **C.** 29 CFR 1910.137, covering electrical PPE
   - **D.** 29 CFR 1926.404, covering wiring design and protection

4. The temperature at the location of the junction box was high, and workers were perspiring. What role did the temperature in the building play in the incident?

A. The decedent was hot and perspiring. Therefore, the worker's skin was highly conductive.

B. The temperature was elevated, so the decedent was in a hurry to complete the installation and get to a cooler place.

C. The helpers were hot, so they were relaxing in the coolest place they could find.

D. The temperature at the location had no bearing on the incident or the fatality.

5. Both the lead electrician and the foreman made an assumption. Which of the following potential assumptions is most likely?

A. The foreman clearly understood the request made by the lead electrician, and the lead electrician assumed that the foreman knew he was talking about the lighting circuit.

B. The lead electrician assumed that the workers on the ground were following the completeness of the installation.

C. The foreman assumed that the appropriate OSHA rule was 29 CFR 1926.417, which gave him authority to remove the lock at any time.

D. The lead electrician assumed that the foreman could see that the area lighting would be the first service that needed to be restored.

Ready for Review

- The mission of OSHA, established in 1970, is to provide worker safety and health protection to workers in the U.S.
- Prior to 1970, there were no uniform or comprehensive provisions to protect against workplace hazards, resulting in countless deaths and injuries on the job.
- Even with OSHA requirements in place, significant hazards and unsafe conditions still exist in the workplace.
- OSHA's requirements include specific requirements pertaining to electrical work and must be followed. Electrical workers must be intimately familiar with these requirements.
- Both employers and employees are responsible for following OSHA requirements.
- Many of OSHA's requirements are written in performance language, meaning they specify what needs to be done but not how to do it.
- Employers are responsible for instructing employees to recognize and avoid unsafe conditions and also to instruct employees on the hazards of confined or enclosed space work.
- 29 CFR 1926 is the part of the OSH Act of 1970 that covers safety and health regulations for construction. Part 1926 indicates that the employer is responsible for the adequacy of the PPE.
- OSHA's requirements for ground-fault protection differ from those in the *NEC*. OSHA allows ground-fault protection for 125-volt, single-phase, 15- and 20-ampere receptacle outlets to be accomplished by either GFCIs or by an assured equipment grounding conductor program.
- The assured equipment grounding conductor program on construction sites must be in writing and implemented by a competent person.
- Employers must inform employees of hazards and the protective measures necessary.
- Equipment is generally required to be worked on deenergized and grounded with tags attached.
- 29 CFR Part 1910 is the part of the OSH Act of 1970 that covers occupational safety and health standards for general industry.
- OSHA includes requirements for employee training, which must cover certain items and be documented. The degree of training is dependent on the risk to the employee.
- Certain requirements regarding in-service care and use of electrical protective devices must be followed, for example, periodic tests must be performed.
- OSHA includes specific requirements related to safety-related work practices that must be used to prevent electric shock or other injuries associated with energized work.
- Deenergizing is a primary safety-related work practice. Live parts must be recognized and deenergized before the employee works on them, unless the employer can demonstrate that deenergizing introduces additional hazards or is infeasible due to equipment design or operational limitations.
- Other safety-related work practices, such as PPE, are required once the employer demonstrates that it is infeasible or a greater hazard to deenergize.
- OSHA requires that circuits energizing parts of fixed electric equipment or circuits that have been deenergized be locked out, tagged, or both where employees are exposed to contact.
- Only qualified persons may work on energized electric circuit parts or equipment. Qualified persons must be capable of working safely on energized circuits and be familiar with safety measures and protection.
- Lockout/tagout procedures must be in writing and must be followed in the order they are presented.

Vocabulary

29 CFR Part 1910 The part of the Code of Federal Regulations that covers occupational safety and health standards for general industry.

29 CFR Part 1926 The part of the Code of Federal Regulations that covers safety and health regulations for construction.

assured equipment grounding conductor program One of the methods recognized by OSHA to provide ground-fault protection to protect employees.

ground-fault circuit interrupter (GFCI)* A device intended for the protection of personnel that functions to deenergize a circuit or portion thereof within an established period of time when a current to ground exceeds the values established for a Class A device.

FPN: Class A ground-fault circuit-interrupters trip when the current to ground is 6 mA or higher and do not trip when the current to ground is less than 4 mA. For further information, see UL 943, *Standard for Ground-Fault Circuit Interrupters.*

Occupational Safety and Health Administration (OSHA) The U.S. government agency that sets standards for worker safety and health protection, including outlining requirements for electrical safety for employers and employees.

performance language Wording that indicates that something must be done, but not how to do it. For example, "the authorized employee shall have knowledge of the type and magnitude of the energy, the hazards of the energy to be controlled, and the method or means to control the energy."

*This is the *NFPA 70E* definition.

OPENING CASE

On July 13, 1983, an electrician with about 10 years experience was working at a coal-fired power plant. At about noon, he was electrocuted while replacing a limit switch on a coal sampler. The victim had been on vacation up to and including the day before the incident.

By 11:30 a.m. that day, the victim received a briefing on the need to replace a limit switch. The victim walked to the building containing the coal sampler and supposedly took a normal lunch break from 11:45 a.m. to 12:15 p.m. At about 12:25 p.m., three workers saw the victim lying face-up underneath a conveyor belt. On the ground next to the victim were a pack of cigarettes and two folded one-dollar bills.

The body had no signs associated with a fall and was not in a position that would result from a fall. Examination of the body showed electrical burns on the right hand, aspiration of foodstuff into the secondary and tertiary bronchi, contusion of the left mastoid scalp, and abrasion of the right mid-tibial area. The medical examiner concluded that the cause of death was electrocution.

The probable sequence of events follows: The victim was standing on the conveyor belt guard installing the new limit switch. Two of the three wires were connected, and the victim was connecting the last wire, the 220-volt energized wire. The wire coming out of the conduit was too short to reach the switch. The victim probably grabbed the wire with his right hand and attempted to pull it further out of the conduit. As he did, the bottom of part of his hand contacted the limit switch. When the wire hit the upper part of his palm, a circuit was completed. Due to the relatively low voltage, the victim was not killed instantly;

Design Considerations

Introduction

Engineering and system design have a significant impact on worker safety. While work practices ensure that electrical work is performed in a safe manner, by minimizing exposure to hazards, the initial opportunity to eliminate or minimize a worker's exposure to an electrical hazard is with the design of the system, before a worker ever has a chance to become exposed to the hazard(s).

Design Considerations

Several design-engineering techniques can enhance the overall safety of the electrical system. The following design practices enhance personnel safety:

- Choosing overcurrent protective devices (OCPDs) for maximum current limitation to limit the amount of energy released during arcing faults
- Choosing OCPDs for consistent operation over many years to minimize unexpected arc flash exposure
- Selecting rejection style (Class J, T, and L) fuses to ensure that continued, controlled arc flash protection levels are designed into the system as future replacements are installed

OPENING CASE, CONT.

his heart probably went into arrhythmia. He probably felt uncomfortable, decided to get down and have a smoke, and died seconds later.

The major etiologic factor in this fatal incident was the failure of the victim to follow standard procedures for locking out electrical power. The failure apparently resulted from inattention by the victim rather than the difficulty of the procedure or a lack of time to do the job. An explanation for the failure could be a somewhat cavalier attitude of the victim toward relatively small voltages. The research team observed such an attitude in other electricians at the plant.

Adapted from: Fatal Accident Circumstances and Epidemiology (FACE) Investigation #83PA08. Accessed 5/7/08 at http://www.cdc.gov/niosh/face/In-house/full8308.html.

1. What is low voltage? What distinguishes low voltage from high voltage?

2. What might cause a worker to become inattentive?

3. What specific procedures should be followed before performing this type of work?

Chapter Outline

- **Introduction**
- **Design Considerations**
- **Current Limitation of Overcurrent Protective Devices**
- **Devices and Measures That Enhance Electrical Safety**
- **Summary**
- **Lessons Learned**
- **Current Knowledge**

Learning Objectives

1. Understand the important role that system design can play in eliminating and reducing hazards.
2. Understand the relationships among the OCPD used, incident energy, and short-circuit (fault) current.
3. Identify engineering and design practices that can enhance electrical safety.

References

1. *NFPA 70E*®, 2009 Edition
2. Cooper Bussmann® Safety BASICs™

- Selecting a shunt-trip option for switches 800 amperes and larger to reduce arc flash hazard levels
- Choosing Type 2 protection for motor controllers to limit arc flash hazard exposure levels
- Installing finger-safe products, covers, and insulating barriers to reduce exposure to live parts
- Installing disconnecting means within sight to encourage following lockout/tagout procedures
- Installing selectively coordinated OCPDs to reduce exposure to arc flash hazards
- Selecting impedance-grounded systems to limit the amount of energy released during arcing faults
- Installing main overcurrent protection to reduce arc flash hazard levels
- Breaking larger loads into smaller ones to reduce arc flash hazard levels
- Utilizing adjustable instantaneous trip settings to reduce arc flash hazard exposure levels
- Eliminating short-time delay to reduce arc flash hazard exposure levels
- Installing zone-selective interlocking to reduce arc flash hazard exposure levels
- Installing remote monitoring to reduce the number of exposures to electrical hazards
- Selecting arc-resistant switchgear to divert arc energy away from workers, reducing exposure to arc flash hazards
- Installing remote racking of power circuit breakers to reduce exposure to arc flash hazards
- Installing remote opening and closing of switches and circuit breakers to reduce exposure to arc flash hazards

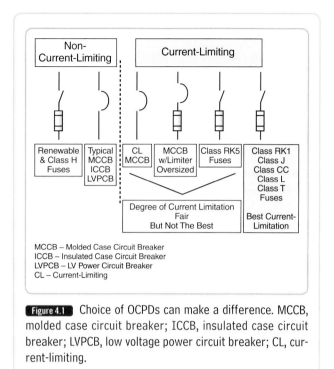

Figure 4.1 Choice of OCPDs can make a difference. MCCB, molded case circuit breaker; ICCB, insulated case circuit breaker; LVPCB, low voltage power circuit breaker; CL, current-limiting.

Non-Current-Limiting Overcurrent Protective Devices

Renewable and Class H fuses are outdated types of fuses that are not current-limiting. These type fuses are not recommended for use.

The typical molded case circuit breakers, insulated case circuit breakers, and low-voltage power circuit breakers are not **listed** as current-limiting. These devices do not significantly reduce the level of short-circuit currents, and they take longer to open. These devices can permit large amounts of energy to be released during an arcing fault. For instance, even if the short-circuit current is in the instantaneous setting range of a circuit breaker, the higher the short-circuit current, the more energy that is released. Circuit breakers require periodic maintenance and/or replacement to better ensure they will operate as intended. If not maintained properly, circuit breaker clearing times can extend beyond their specifications, translating into arcing fault energy that is significantly greater than expected.

Current-Limiting Overcurrent Protective Devices

The four types of devices depicted on the right in Figure 4.1 are all current-limiting. **Current-limiting overcurrent protective devices** provide the benefit of reducing the arcing fault energy in higher short-circuit currents by reducing both the current magnitude and time duration (when the arcing short-circuit current is within their current-limiting range).

Current Limitation of Overcurrent Protective Devices

Choosing and installing current-limiting OCPDs is an important safety design feature. The magnitude of the arcing fault current and the length of time the current flows are directly related to the energy released by an arcing fault. OCPDs limit the magnitude of the short-circuit current that flows and reduce the time duration of the short-circuit current. The current-limiting overcurrent devices are designed to reduce these characteristics, which reduces the energy released in an arcing fault. OCPDs can be either noncurrent-limiting or current-limiting. **Figure 4.1** illustrates this choice.

However, different degrees of current-limitation exist. Different devices become current-limiting at different levels of short-circuit current. If the arcing short-circuit current is in the current-limiting range of current-limiting fuses, the energy released during an arcing fault typically does not increase as the short-circuit current increases. This is a very important point.

Current-limiting molded case circuit breakers are a better choice than standard molded case circuit breakers. The degree of current-limitation is generally moderate but can vary significantly. Underwriters Laboratories (UL) 489, the Molded Case Circuit Breaker Standard, does not establish different short-circuit let-through values for peak current (Ip) and ampere² seconds (I^2t) for various ampere-rated circuit breakers. (UL tests products and certifies those that meet specific safety standards.) I^2t is proportional to the amount of thermal damage to the circuit under short-circuit conditions. I^2t is a measurable value that is used to evaluate fault protection performance of OCPDs. The lower the I^2t that an OCPD lets-through, the lower the thermal energy released into the circuit. Ip is the peak value of current let-through by the OCPD during a short-circuit. Magnetic damage to the circuit under short-circuit conditions is proportional to the square of the peak current, or Ip^2. The lower the square of the Ip, the lower the magnetic damage to the circuit. Periodic maintenance and testing is necessary for all circuit breakers to help ensure that they will operate as intended.

Standard circuit breakers that incorporate fuses as limiters are another current-limiting alternative. The limiter is intended only to provide current-limiting short-circuit protection. However, the fuse limiters normally are oversized to permit the circuit breaker to operate for lower level short-circuit currents. Therefore, these fuse limiters give less protection than current-limiting fuses sized to the load, such as when the circuit is a fusible switch system.

The result with the circuit breaker/limiter alternative is typically higher arcing fault energy releases. For instance, the circuit breaker limiter might be sized at 2 to 10 times the equivalent current-limiting fuses that would be used instead of a circuit breaker. As an example, a 600-ampere circuit breaker with fuse limiters may have limiters that are equivalent to 1600 amperes or greater fuses. Six hundred-ampere Class RK1 or Class J fuses would typically provide much lower arc flash incident energy than a limiter that is equivalent to a 1600-ampere fuse. Properly sized Class RK1 and Class J fuses enter the current-limiting range sooner and let-through less fault energy than a 1600-ampere limiter.

Class RK5 fuses provide current-limiting protection. The level of current-limiting ability is good. A better choice for applications using Class R fuse clips are Class RK1 fuses, because these fuses are more current-limiting and enter their current-limiting range at lower fault levels.

Class J, Class RK1, Class CC, Class L, and Class T fuses offer the best practical current-limiting protection. They have a significantly better degree of current limitation than the other alternatives discussed. They also typically enter their current-limiting range at lower currents than the other fuses or limiter alternatives. These types of fuses provide the greatest current limitation for general protection and motor circuit protection.

Table 4.1 illustrates the potential benefits of using fuses that have greater current-limiting ability. In evaluating arc flash protection, the overcurrent protective device's I^2t let-through is a direct indicator of the arc flash energy that would be released. Table 4.1 compares the UL 248 Fuse Standards and UL 489 Molded Case Circuit Breaker Standard maximum permitted I^2t let-through limits. These values shown are the maximum limits. Commercially available products have values lower than shown in this table.

The UL 248 Fuse Standards set short-circuit I^2t let-through limits for current-limiting fuse types such as

Table 4.1	**UL Standard Limits for OCPDs**				

UL Standard Maximum I^2t (Ampere² Seconds) Let-Through Limits for 50,000-Ampere Short-Circuit Test

Device Ampere Rating	Class J 600 Volts	Class RK1 600 Volts	Class RK5 600 Volts	Current-Limiting Molded Case Circuit Breaker	Standard Molded Case Circuit Breaker
600 ampere	2,500,000	3,000,000	10,000,000	20,750,000	No limit
400 ampere	1,000,000	1,200,000	5,200,000	20,750,000	No limit
200 ampere	200,000	400,000	1,600,000	20,750,000	No limit

Classes J, RK1, and RK5. Different limits are set for each fuse major case size such as 30, 60, 100, 200, 400, and 600. Fuses that are tested and listed as current-limiting are marked "current-limiting."

UL 489 molded case circuit breakers do not have I²t let-through limits for circuit breakers that are not tested and listed as current-limiting. These circuit breakers will not be marked "current-limiting." Circuit breakers that are marked "current-limiting" have I²t let-through limits, which is the lower of either what the manufacturer claims or the symmetrical short-circuit calibration wave for a half cycle without the circuit breaker in the circuit. UL 489 does not require current-limiting circuit breaker I²t let-through limits to apply when the circuit breakers are tested under "busbar" test conditions. UL 489 does not set different I²t let-through limits for different circuit breaker ampere ratings or frame sizes.

Figure 4.2 and **Figure 4.3** illustrate the importance of using OCPDs that have better current-limiting ability. The dotted line represents the asymmetrical short-circuit current that could flow with 50,000 symmetrical amperes available—the Ip could reach 115,000 amperes. The UL Ip limit for a 400-ampere Class RK5 fuse is 50,000 amperes; for a 400-ampere, RK1 fuse the Ip limit is 33,000 amperes; and for a 400-ampere Class J fuse, it is 25,000 amperes.

Figure 4.4 through Figure 4.10 illustrate the relationship between the available bolted short-circuit current, the arcing current, and the incident energy, when various types of circuit breakers are used.

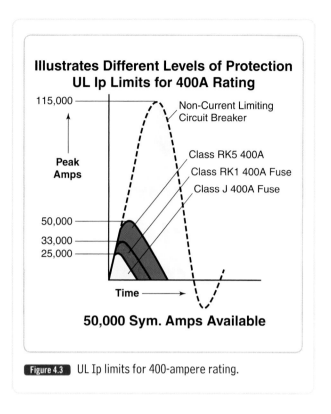

Figure 4.3   UL Ip limits for 400-ampere rating.

Figure 4.4 illustrates the incident energy curve for typical types of properly maintained circuit breakers. The current-limiting circuit breaker has the lowest incident energy, followed by standard noncurrent-limiting circuit breakers with instantaneous trips, followed by circuit breakers with short-time delay. Note that for low-level faults, the incident energy can be quite high. This occurs when the arcing current is less than the

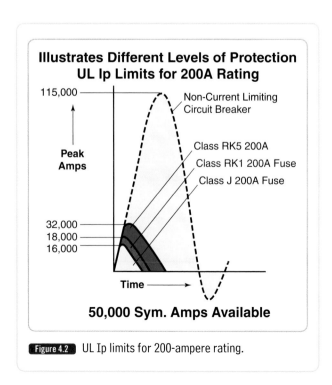

Figure 4.2   UL Ip limits for 200-ampere rating.

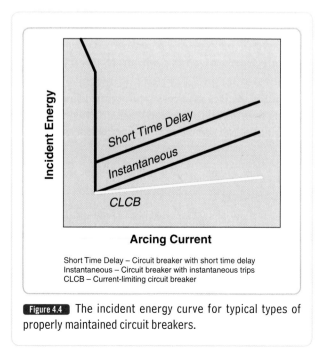

Figure 4.4   The incident energy curve for typical types of properly maintained circuit breakers.

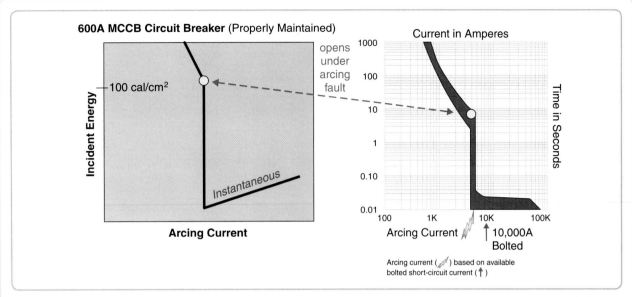

Figure 4.5 Example showing incident energy when using a properly maintained circuit breaker, when arcing current is below the instantaneous trip of the circuit breaker.

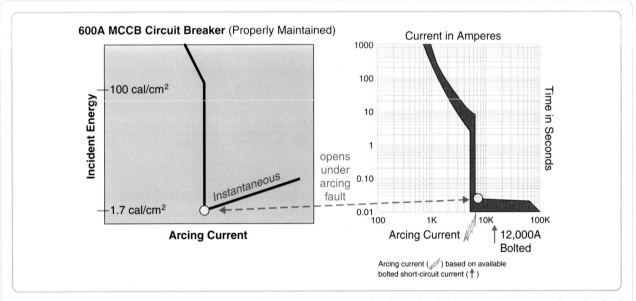

Figure 4.6 Example of an arcing current that exceeds the instantaneous trip of the circuit breaker, using a properly maintained 600-ampere molded case circuit breaker (MCCB).

current-limiting threshold, or instantaneous trip of the circuit breaker.

Figure 4.5 illustrates that arcing currents that are less than the instantaneous trip of the circuit breaker, where opening time can take seconds, translate to very high incident energies, often at 100 cal/cm² or more.

Figure 4.6 illustrates that once the arcing current exceeds the instantaneous trip of the circuit breaker, the incident energy will be greatly reduced.

Figure 4.7 illustrates that as the arcing current in-

creases above the instantaneous trip, the incident energy also increases.

Figure 4.8 illustrates that the curve for a current-limiting circuit breaker is not only below that of a standard circuit breaker, but that the curve has a lesser slope, meaning that there is a proportionally lower increase as the arcing current increases.

Figure 4.9 illustrates that the intentional delay significantly increases the arc flash energy above that of standard circuit breakers.

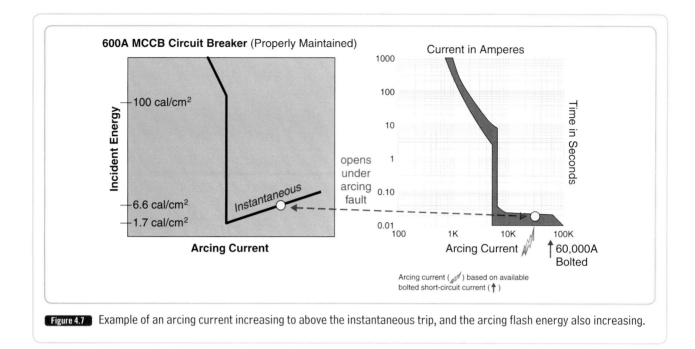

Figure 4.7   Example of an arcing current increasing to above the instantaneous trip, and the arcing flash energy also increasing.

Figure 4.10 compares incident energy for all three types of circuit breakers and corresponding time-current characteristics in a color-coded format.

Figure 4.11 compares the arc flash level of Class RK5 fuses with that of Class RK1 fuses.

Figure 4.12 through Figure 4.14 show examples with a 600-ampere Class RK1 fuse.

Figure 4.12 shows a very high arc flash energy for a low-level arcing current, one that takes seconds to open.

Figure 4.13 shows that as the arcing current increases, the arc flash incident energy decreases. This shows that the arcing current is still below the current-limiting threshold of the fuse. Once the arcing current is large enough to reach the current-limiting threshold of the fuse, the incident energy will fall to 0.25 cal/cm^2 and remain at that level as the arcing current continues to increase.

Figure 4.14 shows the arcing current beyond the current-limiting threshold of the fuse. In general, arcing currents beyond this value do not show an increased arc flash incident energy level.

Finally, **Figure 4.15** compares incident energy for both types of fuses and corresponding time-current characteristics in a color-coded format.

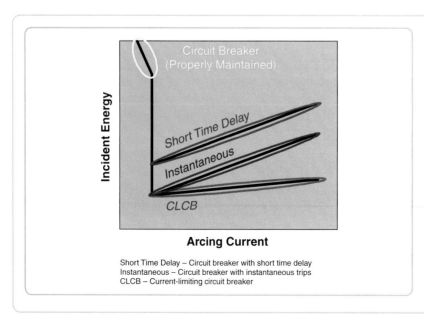

Short Time Delay – Circuit breaker with short time delay
Instantaneous – Circuit breaker with instantaneous trips
CLCB – Current-limiting circuit breaker

Figure 4.8   When a current-limiting circuit breaker is used, incident energy will not increase as much with increases in short-circuit current (as short-circuit current increases, incident energy increases, but not as much as with other types of circuit breakers).

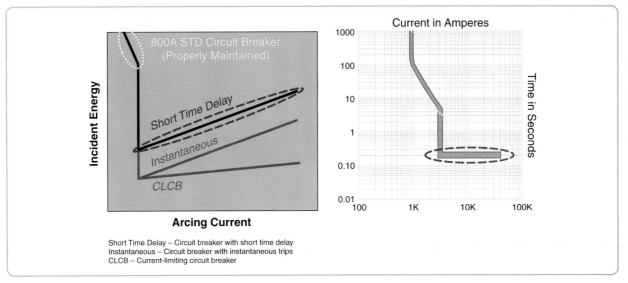

Figure 4.9 An intentional delay significantly increases the arc flash energy. This example uses a properly maintained 800-ampere short-time delay circuit breaker.

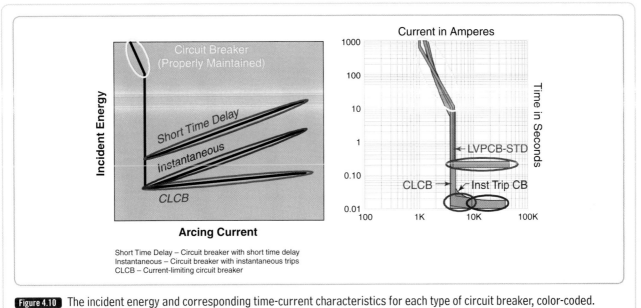

Figure 4.10 The incident energy and corresponding time-current characteristics for each type of circuit breaker, color-coded.

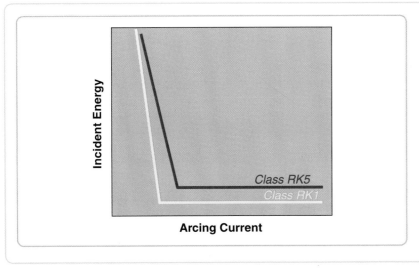

Figure 4.11 The arc flash level of Class RK5 and Class RK1 fuses, showing different degrees of current limitation.

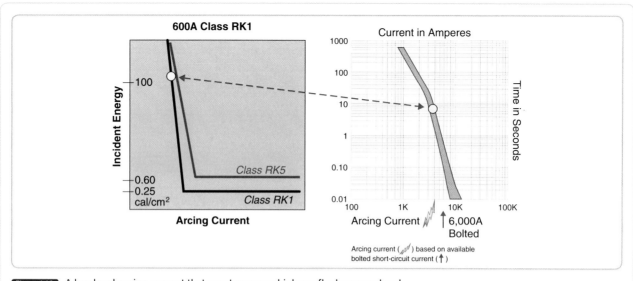

Figure 4.12 A low level arcing current that creates a very high arc flash energy level.

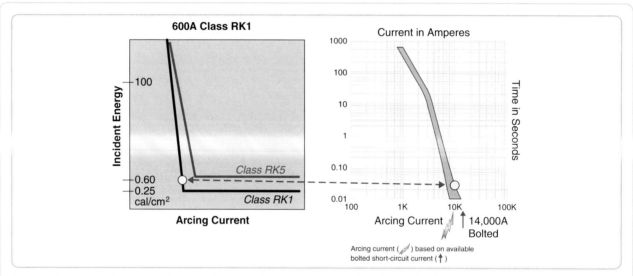

Figure 4.13 An example showing that as arcing current increases, the arc flash hazard level decreases.

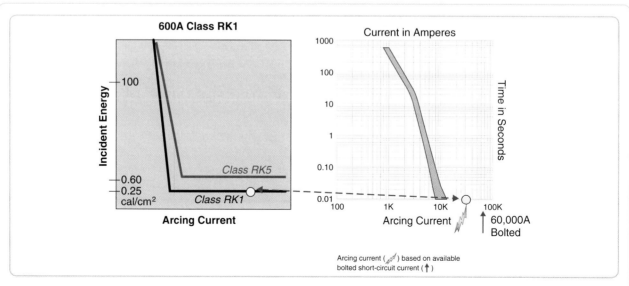

Figure 4.14 An example where the arcing current is beyond the current-limiting threshold of the fuse resulting in no increase of incident energy.

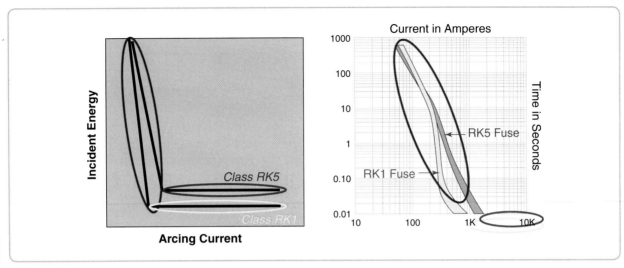

Figure 4.15 A color-coded comparison of incident energy and corresponding time-current characteristics for Class RK5 and Class RK1 fuses.

Devices and Measures That Enhance Electrical Safety

Consistency Over Time of Overcurrent Protective Devices

New information related to design, opening time, and condition of maintenance was added into *NFPA 70E* for the 2009 edition. The following are excerpts from various related sections of *NFPA 70E*:

130.3 Arc flash Hazard Analysis. The arc flash hazard analysis shall take into consideration the design of the overcurrent protective device and its opening time, including its condition of maintenance.

Fine Print Note (FPN) No. 1: Improper or inadequate maintenance can result in increased opening time of the overcurrent protective device, thus increasing the incident energy.

FPN No. 2: For additional direction for performing maintenance on overcurrent protective devices see Chapter 2, Safety-Related Maintenance Requirements.

210.5 Protective Devices. Protective devices shall be maintained to adequately withstand or interrupt available fault current.

FPN: Failure to properly maintain protective devices can have an adverse effect on the arc flash hazard analysis incident energy values.

225.1 Fuses. Fuses shall be maintained free of breaks or cracks in fuse cases, ferrules, and insulators. Fuse clips shall be maintained to provide adequate contact with fuses.

Fuseholders for current-limiting fuses shall not be modified to allow the insertion of fuses that are not current-limiting.

225.2 Molded-Case Circuit Breakers. Molded-case circuit breakers shall be maintained free of cracks in cases and cracked or broken operating handles.

225.3 Circuit Breaker Testing. Circuit breakers that interrupt faults approaching their interrupting ratings shall be inspected and tested in accordance with the manufacturer's instructions.

The reliability of OCPDs directly impacts arc flash hazards. The opening time of OCPDs is critical in the resultant arc flash energy released when an arcing fault occurs. The longer an OCPD takes to clear a given arcing short-circuit current, the greater the arc flash hazard. When an arcing fault occurs, or for that matter, when any short-circuit current occurs, the OCPD must be able to operate as intended. Therefore, the reliability of OCPDs is critical—they need to open as originally specified, otherwise the flash hazard can escalate to higher levels than expected.

Two different types of overcurrent protection technology provide different choices in reliability and maintenance requirements and might impact the flash hazard:

1. OCPDs that are reliable and do not require maintenance

2. OCPDs that require periodic maintenance according to the manufacturer's instructions and industry standards

Figure 4.16 Modern current-limiting fuses.

Current-Limiting Fuses

Modern fuses are reliable and retain their ability to open as originally designed under overcurrent conditions (**Figure 4.16**). When a fuse is replaced, a new factory-calibrated fuse is put into service, and the circuit has reliable protection with performance equal to the original specification. Modern current-limiting fuses do not require maintenance other than visual examination and ensuring that there is no damage from external thermal conditions or liquids.

Circuit Breakers

Circuit breakers are mechanical OCPDs that require periodic exercise, maintenance, testing, and possible replacement (**Figure 4.17**). A circuit breaker's reliability and operating speed are dependent upon its origi-

nal specification and its condition. A specific circuit breaker's condition is dependent on the following variables, some of which are not typically recorded and saved:

- Length of service
- Number of manual operations under load
- Number of operations due to overloads
- Number of fault interruptions
- Humidity
- Condensation
- Corrosive substances in the air
- Vibrations
- Invasion by foreign materials or liquids
- Thermal damage caused by loose connections
- Erosion of contacts
- Erosion of arc chutes

To help keep a circuit breaker within original specifications, a circuit breaker manufacturer's instructions for maintenance must be followed.

Failure to perform periodic maintenance or maintenance after interrupting a fault may result in the circuit breaker requiring longer clearing time or the inability to interrupt overcurrents. Either of these results can drastically affect the potential arc flash energy that is released.

Protective Device Maintenance as It Applies to the Arc flash Hazard, a technical paper by Dennis Neitzel of AVO Training Institute, is a good resource on this topic. The following excerpts from this paper highlight the importance of circuit breaker maintenance:

> "Where proper maintenance and testing (on circuit breakers) are not performed, extended clearing times could occur creating an unintentional time delay that will effect the results of flash hazard analysis . . ."
>
> "Fuses, although they are protective devices, do not have operating mechanisms that would require periodic maintenance: Therefore, this article will not address them . . ."
>
> "Circuit breakers installed in a system are often forgotten. Even though the breakers have been sitting in place supplying power to a circuit for years, there are several things that can go wrong. The circuit breaker can fail to open due to a burned out trip coil or because the mechanism is frozen due to dirt, dried lubricant, or corrosion. The overcurrent device can fail due to inactivity or a burned out electronic component. Many problems can occur when proper maintenance is not performed and the breaker fails to open under fault conditions. This combination of events can result in fires, damage to equipment or injuries to personnel." Courtesy of Dennis K. Neitzel.

Rejection Style Class J, T, R, and L Fuses

It is important to ensure continued arc flash protection levels as a facility ages. Class J, T, R, and L fuses provide an advantage in that these fuse classes are physically size

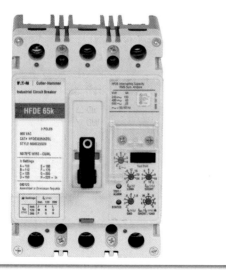

Figure 4.17 A circuit breaker.

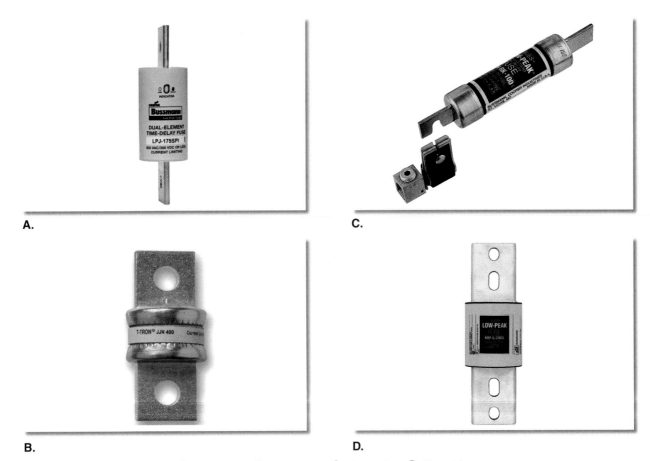

A.

B.

C.

D.

Figure 4.18 Rejection style fuses. **A.** Class J fuse. **B.** Class T fuse. **C.** Class R fuse. **D.** Class L fuse.

rejecting (**Figure 4.18**). No other class fuse can be inserted into mountings designed for these classes. This means that fuses with lower voltage ratings, lower interrupting ratings, or lower current limitation cannot be accidentally put into service. The designed-in level of arc flash protection will not change in the future as replacement devices are installed. With circuit breakers, replacement breakers with lower voltage ratings, lower interrupting ratings, and lower current limitation capabilities can be put into service, thereby reducing the designed-in level of arc flash protection.

Switches for Circuits 800 Amperes and Above

Switches with shunt-trip to open all three poles of the switch when the first fuse opens reduce arc flash energy levels (**Figure 4.19**). This option can be included on new switches or can be retrofitted on some existing switches. Tests have shown that this option can reduce the arc flash hazard levels on circuits with higher ampere ratings. A shunt trip is an electromechanical option that might require maintenance after an operation.

Figure 4.19 A switch with a shunt-trip.

"No Damage" Protection for Motor Controllers

Motor starters are susceptible to damage from short-circuit currents. If a worker needs to work within an energized motor starter enclosure, he or she will be exposed to a serious safety hazard. Specifying Type 2 motor starter protection can reduce the risk, because the level of current limitation typically required to obtain Type 2 protection also provides for excellent arc flash hazard reduction.

A choice of motor starter protection is available: Both UL 508E (outline of investigation) and International Electrotechnical Commission (IEC) 60947-4-1 differentiate between two types of protection for motor circuits. (IEC is an organization that sets international electrical standards.) The engineer or person with the responsibility to specify or choose the type of equipment can choose the level of motor starter protection desired: Type 1 or Type 2. The OCPD makes the difference.

Type 1 Protection

IEC 60947-4-1, **Type 1 protection**, requires that, under short-circuit conditions, the contactor or starter shall cause no danger to persons (with enclosure door closed) or installation and might not be suitable for further service without repair and replacement of parts. Note that damage is allowed, requiring partial or complete component replacement. It is possible for the overload devices to vaporize and the contacts to weld. Short-circuit protective devices interrupt the short-circuit current but are not required to prevent component damage. The requirements for Type 1 protection are similar to the requirements for listing to UL 508 (**Figure 4.20 A–C**). If a worker has any unprotected body parts near such an event, he/she might be injured.

Type 2 Protection

IEC 60947-4-1 and UL 508E (Outline of Investigation), **Type 2 protection**, requires that, under short-circuit conditions, the contactor or starter shall cause no danger to persons (with enclosure door closed) or installation and shall be suitable for further use. No damage is allowed to either the contactor or overload relay. Light contact welding is permitted, but contacts must be easily separable. "No damage" protection for the National Electrical Manufacturers Association (NEMA) and IEC motor starters can be provided only by a device that is able to limit the magnitude and the duration of short-circuit current (**Figure 4.21D**, **Figure 4.21E**, and **Figure 4.21F**).

Motor starter manufacturers test many but not all their combinations of contactors and overload relays with branch-circuit protection to verify that they meet Type 2 requirements. Tests are performed on both IEC and NEMA type devices. The tests include a low-level and a high-level (typically 100,000 amperes) short-circuit current. Overload relays are tested before and after both short-circuit tests to ensure that they remain calibrated. Dielectric (voltage withstand) tests also are conducted to prove insulation integrity after both the high- and low-level short-circuit tests.

Fuses that typically meet the requirements for Type 2 "no damage" protection that are the result of the controller manufacturers testing are Class J, Class CC, and Class RK1 fuses. As discussed in the two previous sections, these fuses are very current-limiting, which can protect the sensitive controller components.

A.   **B.**   **C.**

Figure 4.20 Type 1 Protection: Photos A, B, and C taken before, during, and after testing a motor circuit protector (MCP) intended to provide motor branch-circuit protection for 10 horsepower (hp), IEC Starter with 22,000 amps available at 480 volts. The heater elements vaporized, and the contacts were severely welded. This could be a hazard if the door is open and a worker is near.

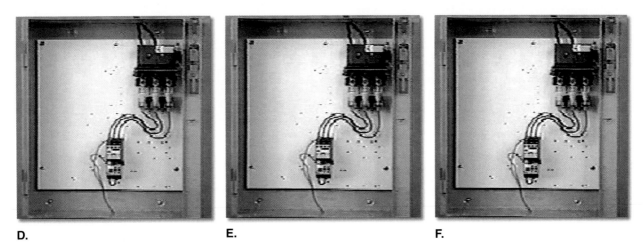

D.   **E.**   **F.**

Figure 4.21 Type 2 Protection: Photos D, E, and F taken before, during, and after same test circuit and same type starter during short-circuit interruption as in Figure 4.20 A, B, and C. The difference is that Cooper Bussmann LPJ_SP Class J current-limiting fuses provide the motor branch-circuit protection. This level of protection reduces the risk for workers.

Finger-Safe Products and Terminal Covers

Although most electrical workers and others are aware of the electrical shock hazard, it still is a prevalent cause of injury and death. One of the best ways to help minimize exposure to an electrical shock hazard is to use finger-safe products and nonconductive covers or barriers. Finger-safe products and covers reduce the chance of causing a shock or initiating an arcing fault. If all the electrical components are finger-safe or covered, a worker has a much lower chance of coming in contact with an energized conductor (shock hazard). The risk of any conductive part falling across bare, energized conductive parts and creating an arcing fault is greatly reduced (arc flash hazard).

Figure 4.22 shows several Cooper Bussmann items that can help minimize shock hazards and minimize

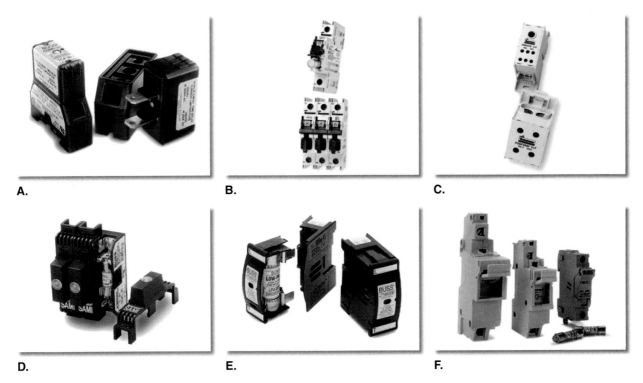

A.   **B.**   **C.**

D.   **E.**   **F.**

Figure 4.22 Examples of devices to help minimize hazards. **A.** CUBE Fuses. **B.** Compact Circuit Protector (CCP). **C.** PDBFS. **D.** SAMI fuse covers. **E.** Safety J fuse holders. **F.** CH fuse holders.

the initiation of an arcing fault: CUBEFuses™ (1 to 100 amperes) that are IP20 finger-safe and very current-limiting protective devices; Compact Circuit Protector (CCP), a fully rated, 600 Vac, UL 98 disconnect switch; PDBFS, a UL 1953 listed power distribution block with high short-circuit current rating; SAMI™ fuse covers for covering fuses; Safety J fuse holders for Class J fuses; CH fuse holders available for a variety of fuses All these devices can reduce the chance that a worker, tool, or other conductive item will come in contact with a live part.

Isolate the Circuit: Install "In-sight" Fusible Disconnect for Each Motor

Electrical systems should be designed to support maintenance, with easy, safe access to the equipment. This design should provide for isolating equipment for repair with a disconnecting means for implementation of lockout/tagout procedures. A sound design provides a disconnecting means at all motor loads in addition to the disconnecting means required within sight of the controller location that can be locked in the open position. Disconnecting means at the motor provide improved isolation and safety for maintenance and for use in case of an emergency.

Install horsepower rated fusible disconnects *within sight* (visible and within 50 ft) of every motor or driven machine (**Figure 4.23**). The provision for locking or adding a lock to the disconnecting means shall be installed on or at the switch or circuit breaker used as the disconnecting means and shall remain in place with or without the lock installed. Note: The *NEC®* defines "in-sight from" as follows:

> Where this *Code* specifies that one equipment shall be "in-sight from," "within sight from," or "within sight of," and so forth, another equipment, the specified equipment is to be visible and not more than 15 m (50 ft) distant from the other.

This measure fosters safer work practices and can be used for an emergency disconnect in case of an incident. An in-sight motor disconnect is more likely to be used by a worker for the lockout procedure to put equipment in an electrically safe work condition prior to doing work on the equipment.

An in-sight motor disconnect generally is required even if the disconnect within sight of the controller can be locked out. Some exceptions exist for specific industrial applications.

Selective Coordination

Today, more than ever, one of the most important parts of any installation is the electrical distribution system. Nothing can stop all activity, paralyze production, in-

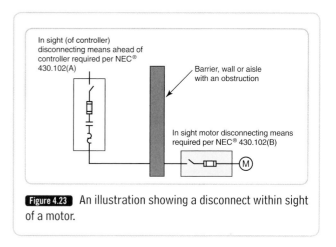

Figure 4.23 An illustration showing a disconnect within sight of a motor.

convenience and disconcert people, and possibly cause a panic more effectively than a major power failure.

Selective coordination is considered the act of isolating a faulted circuit from the remainder of the electrical system, thereby eliminating unnecessary power outages. The faulted circuit is isolated only by the selective operation of that OCPD closest to the overcurrent condition.

Isolation of a faulted circuit from the remainder of the installation is imperative in today's modern electrical systems. Power blackouts cannot be tolerated. An adequately engineered system enables only the protective device nearest the fault to open, leaving the remainder of the system undisturbed and preserving continuity of service.

Personnel safety is enhanced in a selectively coordinated system because the electrical worker is not unnecessarily exposed to arc flash hazards at upstream panels or switchboards. In a selectively coordinated system, the worker is exposed only at the level of the problem circuit. In a nonselectively coordinated system, the worker, while troubleshooting the open circuit, is required to work in upstream equipment, where arc flash energies are often significantly higher.

To illustrate the point, **Figure 4.24** shows a single-line diagram in which the arc flash energy at a branch-circuit panel is 1.6 cal/cm². If an overcurrent condition on the branch circuit opens only the overcurrent device in the branch-circuit panel, the worker is exposed to 1.6 cal/cm² while troubleshooting the circuit. If the feeder overcurrent device also opens, the worker is forced to work within the feeder panel and is unnecessarily exposed to a higher level of 6.7 cal/cm². If the main also unnecessarily opens, the worker is forced to work within the main panel, thereby exposing the worker to 12.3 cal/cm². Therefore, in the selectively coordinated system, the worker is exposed to only 1.6 cal/cm², but in the nonselectively coordinated system, the worker is exposed to 12.3 cal/cm².

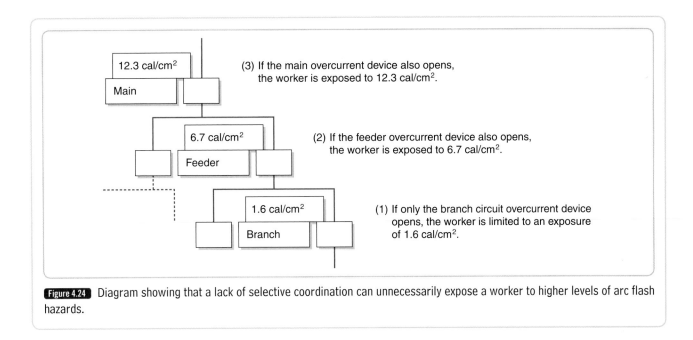

12.3 cal/cm²

Main

(3) If the main overcurrent device also opens, the worker is exposed to 12.3 cal/cm².

6.7 cal/cm²

Feeder

(2) If the feeder overcurrent device also opens, the worker is exposed to 6.7 cal/cm².

1.6 cal/cm²

Branch

(1) If only the branch circuit overcurrent device opens, the worker is limited to an exposure of 1.6 cal/cm².

Figure 4.24 Diagram showing that a lack of selective coordination can unnecessarily expose a worker to higher levels of arc flash hazards.

Install Impedance-Grounded Systems

Electrical systems are being designed with impedance-grounded wye systems. In this type of installation, an impedance is intentionally inserted between the center point of the wye of the transformer and ground, so that ground short-circuit current is limited by the inserted impedance. This type system can help reduce the probability that dangerous or destructive arcing faults can occur. For example, if a worker's screwdriver slips, simultaneously touching an energized bare phase termination and the enclosure with high-impedance-grounded wye systems, a high energy arc-fault would not be initiated, because the current is limited to only a few amperes by the impedance intentionally inserted in the ground return path (**Figure 4.25**). However, this type of system does not eliminate the hazard. If the worker's screwdriver simultaneously touches the energized bare terminations of two phases, a high-level arcing fault might occur. If a designer is planning high impedance-grounded wye systems or retrofitting an existing solidly grounded wye system, he or she must consider the single-pole interrupting capabilities of any circuit breakers and self-protected starters to be installed or already installed. Slash voltage-rated circuit breakers and self-protected starters cannot be used on any system other than solidly grounded wye systems.

Specify a Main on Each Service

The six-disconnect rule for service entrances permitted in *National Electrical Code*® Section 230.71 should not be used in lieu of a single main disconnect. Some designers use the six-disconnect rule to reduce the cost of the service equipment, but this choice might increase worker exposure to hazards. Without a main OCPD, the main bus and line terminals of the feeders are unprotected.

For instance, if a worker must work in the enclosure of one of the feeders, the compartment should be placed in an electrically safe work condition. To achieve that end, a main disconnect should be locked out. Afterwards, no energized conductors remain in the feeder device compartment.

Figure 4.26 illustrates that a system represented by the single-line diagram on the left has a much lower arc flash hazard level than a system represented by the one-line diagram to the right, which has no main overcurrent protection.

If a worker is performing a task within a feeder compartment with an exposed energized conductor, a main OCPD helps protect against arcing faults on the feeder device line terminals and the equipment main bus. The arcing fault hazard associated with the main device must be assessed since large ampere-rated OCPDs might permit high arc flash incident energies. However, in most cases, the main OCPD can provide better protection than the utility OCPD, which is located on the transformer primary.

70E Highlights

The 2009 edition of *NFPA 70E* contains a new annex, Annex O, that is helpful for generating designs that minimize exposure of workers to electrical hazards. The title of the annex is Safety-Related Design Requirements.

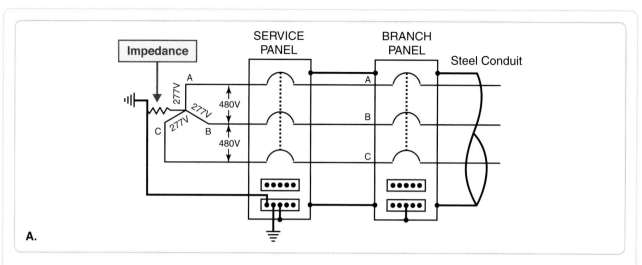

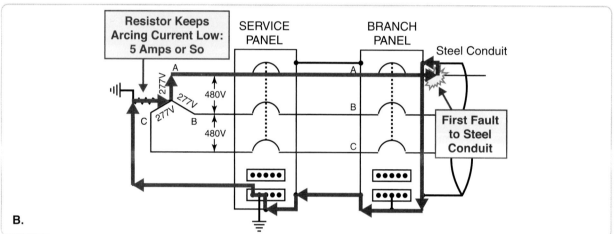

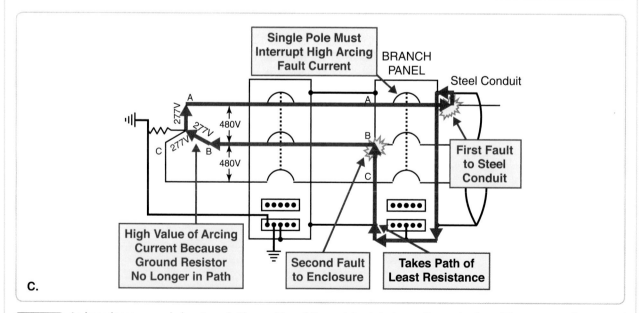

Figure 4.25 An impedance-grounded system. **A.** The position of the resistor is between the center tap of the wye transformer and ground. **B.** Because of the resistance, a fault occurring from phase to ground does not cause the OCPD to open. **C.** The first fault to ground *must* be removed before a second phase goes to ground, or a significant short-circuit current can occur across one pole of the branch-circuit device.

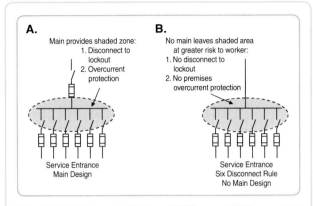

Figure 4.26 Single-line diagrams illustrating systems with and without overcurrent protection and how each design can impact the arc flash hazard level. **A.** With main specified. **B.** Without main specified.

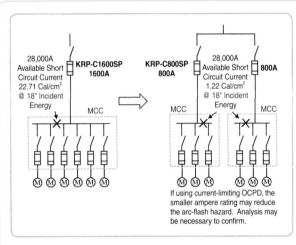

Figure 4.27 Example of splitting large feeders. Lower ampere-rated devices are typically more current limiting and thus can better reduce the arc flash hazard.

Break Up Large Circuits into Smaller Circuits

In the design stage, multiple, smaller circuits should be used instead of one higher ampere-rated circuit. In many cases, very large ampere-rated fuses and circuit breakers let through too much energy for a practical personal protective equipment (PPE) **arc rating**.

As an example, a 1600-ampere circuit could be broken into two 800-ampere circuits. An analysis of the incident energy available in each generally would indicate that two 800-ampere circuits would be better than one 1600-ampere circuit. For specific situations, an arc flash analysis should be completed, as variables can affect the outcome. This is especially beneficial when using current-limiting protective devices, since the lower ampere-rated devices are typically more current-limiting and thus can better reduce the arc flash hazard (**Figure 4.27**).

Adjustable Instantaneous Trip Setting

Some circuit breakers have an adjustable instantaneous trip setting (**Figure 4.28**). This setting allows adjustment to the current at which the circuit breaker can operate in its instantaneous mode. Like short-time delay settings, the adjustable trip setting is used to improve selective coordination. However, if the instantaneous trip is set too high, it might not operate in its instantaneous mode during an arcing fault. The result is a longer opening time, and a much higher level of energy release is possible during an arcing fault. The instantaneous trip settings should be adjusted as low as possible for the best arc flash energy reduction.

Eliminating Short-Time Delay

Some circuit breakers are equipped with a short-time delay setting, which is intended to delay operation of the circuit breaker under fault conditions (**Figure 4.29**). Short-

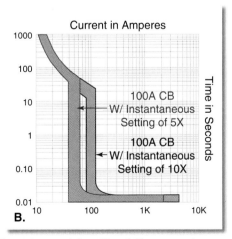

Figure 4.28 **A.** A circuit breaker with an adjustable instantaneous trip setting. **B.** Time-current curve showing the instantaneous trip setting at 5X = 500A and 10X = 1000A.

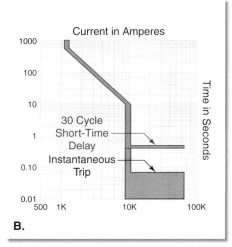

Figure 4.29 **A.** A circuit breaker with an adjustable short-time delay setting. **B.** Time-current curve showing the short-time setting at 30 cycles compared to an instantaneous trip setting.

time delay breakers are used on feeders and mains so that downstream molded case breakers can clear a fault without tripping the larger upstream circuit breaker. In many cases, a circuit breaker with a short-time delay setting does not have an instantaneous setting. Therefore, a fault is permitted to flow for an extended time. Under fault conditions, a short-time delay sensor intentionally delays signaling the circuit breaker to open for the time duration setting of the short-time delay. For example, a low voltage power circuit breaker with a short-time delay, and without instantaneous trip, permits a fault to flow for the length of time of the short-time delay setting, which might be 6, 12, 18, 24, or 30 cycles.

An adverse result is associated with using circuit breakers with short-time delay settings. If an arcing fault occurs on the circuit protected by a device with a short-time delay setting, a tremendous amount of fault energy might be released while the system waits for the circuit breaker short-time delay to time out. The longer an OCPD takes to open, the greater the flash hazard due to arcing faults. Experience shows that the arc flash hazard increases with the length of time the current is permitted to flow.

System designers and users should understand that using circuit breakers with short-time delay settings could greatly increase the arc flash energy. If an incident occurs when a worker is at or near the arc flash, the worker might be subjected to considerably more arc flash energy than if an instantaneous trip circuit breaker, a current-limiting circuit breaker, or current-limiting fuses were protecting the circuit.

Zone-Selective Interlocking

As previously discussed, short-time delay settings can permit extremely hazardous incident energy levels, but they might be necessary to achieve selective coordination. A possible solution to this dilemma is to use **zone-selective interlocking** (**Figure 4.30**). In this scheme, the circuit breakers with this option have communication wiring between the circuit breakers and the circuit breakers' sensing elements. For instance, the main and feeder circuit breakers might be equipped with zone-selective interlocking. For faults on the load side of the feeder circuit breaker, the main circuit breaker, if signaled by the feeder circuit breaker, might be set to have a short-time delay of 24 cycles. This allows for the main circuit breaker to wait for the feeder to open for faults on the feeder circuit. However, if the fault location is between the main and feeder circuit breakers, then the main circuit breaker cannot receive a signal from a feeder circuit breaker, and the main circuit breaker opens without an intentional delay.

Arc Flash Reducing Maintenance Switching

Specification of arc flash reducing maintenance switching, where short-time delay is used, reduces exposure levels by turning off the short-time delay function while working in the arc flash protection boundary (AFPB) (**Figure 4.31** and **Figure 4.32**). In this way, if an arc flash is accidentally started, the circuit breaker will trip without intentional delay, providing the fastest possible opening time for the circuit breaker. When the work is completed, the switch is set back to the delay mode to provide the required selective coordination. Some systems

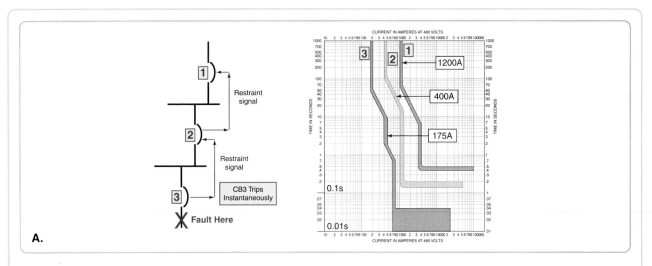

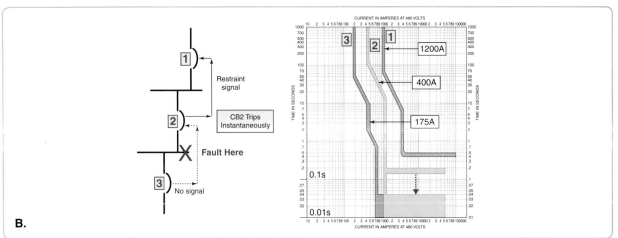

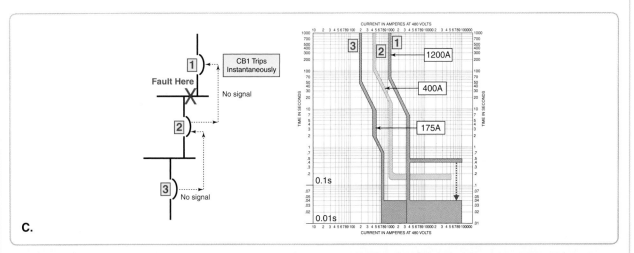

Figure 4.30 Zone-selective interlocking helps reduce arc flash hazard. **A.** Circuit breaker 3 sends a restraint signal to circuit breakers 1 and 2 for all faults on its load side. For faults beyond its instantaneous trip setting, on its load side, circuit breaker 3 opens without any intentional delay. **B.** For faults between circuit breaker 2 and circuit breaker 3, circuit breaker 3 sends no restraint signal to circuit breaker 2. Circuit breaker 2 sends a restraint signal to circuit breaker 1, so that circuit breaker 1 is restrained to selectively coordinate with circuit breaker 2 for this fault. Circuit breaker 2 will open without any intentional delay. Circuit breaker 2 then opens as quickly as possible, reducing the arc flash hazard. **C.** For faults between circuit breaker 1 and circuit breaker 2, neither circuit breaker 3 nor circuit breaker 2 sends a restraint signal to circuit breaker 1. Because circuit breaker 1 does not receive a restraint signal, circuit breaker 1 will open without any intentional delay. Circuit breaker 1 then opens as quickly as possible, reducing the arc flash hazard.

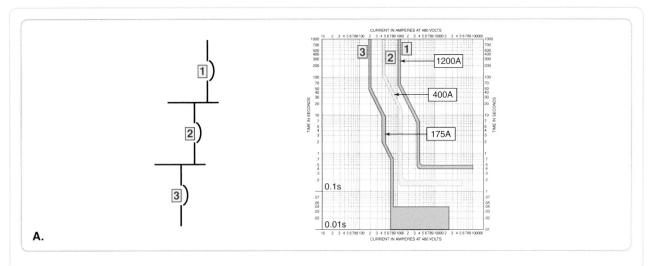

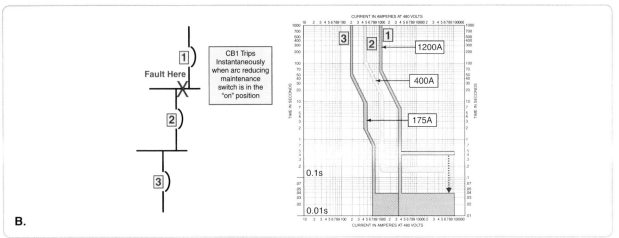

Figure 4.31 Arc flash reducing maintenance switching. **A.** A selectively coordinated system that has been achieved through the usage of short-time delay. **B.** Here, an arc reducing maintenance switch is utilized to provide an instantaneous trip for circuit breaker 1 when a worker is exposed to a possible arc flash hazard. Circuit breaker 1 then opens as quickly as possible, reducing the arc flash hazard. After the worker is no longer exposed to a potential arc flash hazard, the arc reducing maintenance switch turns off the instantaneous trip, returning the system to a selectively coordinated state.

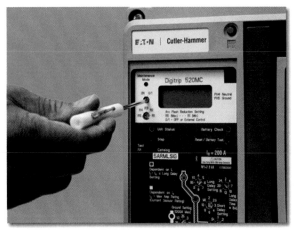

Figure 4.32 Arc flash reducing maintenance switching.

use a manual switch that the worker uses to turn the bypass on and off. Other systems use automatic means to sense whether the worker is "in" the equipment that automatically turns the bypass on and off.

Remote Monitoring

Specifying remote monitoring of voltage and current levels reduces exposure to electrical hazards by transferring the potentially hazardous troubleshooting activity from the actual live equipment to a remote computer screen with the equipment doors closed and latched (**Figure 4.33**). (The troubleshooting activity is performed remotely on a computer screen, so that the worker is not physically working close to the equipment.) These designs reduce the associated electrical hazards and reduce the number of times that required PPE must be worn.

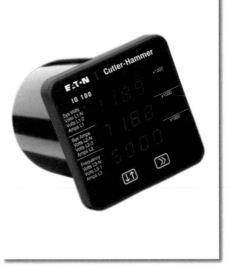

Figure 4.33 Remote monitoring devices for electrical system information (voltage, ampacity, etc.).

Arc-Resistant (Arc Diverting) Medium Voltage Switchgear

Arc-resistant switchgear can be installed to withstand internal arcing faults. Arc-resistant equipment typically is designed with stronger door hinges and latches, better door gaskets, and hinged enclosure top panels. The concept is to divert the resultant explosive hot gases and pressures from an internal arcing fault via the hinged enclosure top panels. If the switchgear is installed indoors, then a means of exhausting the hot gases to the outside of the building is required.

Arc-resistant equipment is rated to withstand specific levels of internal arcing faults with all the doors closed and latched. The rating does not apply with any door opened or any cover removed. Therefore, arc-resistant equipment does not protect a worker performing a task with an open door or panel. The term "arc-resistant" is a bit misleading. The internal switchgear must withstand an internal arcing fault and, therefore, the sheet metal and other components of the equipment must resist or withstand a specified arcing fault. However, a major feature of this equipment is diverting the arcing fault byproducts (hot ionized gases and blast) via the enclosure top panels. This feature helps to prevent the arcing fault from blowing open the doors or side panels and venting the arcing fault byproducts where a worker might be standing. **Figure 4.34** shows an example of arc-resistant switchgear.

Figure 4.34 Arc-resistant switchgear.

Remote Racking

Experience shows that racking circuit breakers into energized equipment is especially hazardous. Workers can help avoid the hazards associated with racking activities by installing remote racking equipment (**Figure 4.35**). Workers are at a safe distance from the actual racking operation when the action is completed. If an arc flash event occurs, workers should be outside the AFPB and be uninjured.

Figure 4.35 Remote racking.

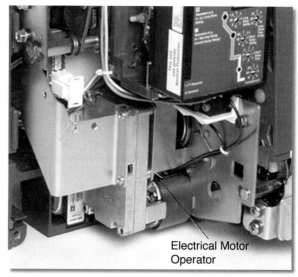

Electrical Motor
Operator

Figure 4.36 Motor operator for remote opening and closing of a circuit breaker.

Remote Opening and Closing

Opening and closing large switches and circuit breakers have caused serious arc flash accidents also. By opening and closing large switches and circuit breakers remotely, a worker might control the operation from a safe distance. If an arc flash incident should occur, the worker is not exposed to the hazard (**Figure 4.36**).

Summary

Electrical system design has a significant impact on personnel safety. System design including device and equipment selection, plays an important role in reducing or eliminating exposure to electrical hazards. Designers must carefully select equipment and circuit designs that provide maximum protection for workers.

Lessons *learned*

Journeyman Electrician Electrocuted When Lockout Attempt Fails

After deenergizing one controlled-access area, a 39-year-old Journeyman electrician died when he mistakenly entered an identical area that was energized. The employer is a large pulp and paper mill that has been in operation since 1961. The company employs approximately 500 individuals, including 19 electricians. The company has a full-time safety and health officer, a written safety policy, and detailed safety procedures. The victim had been employed by the company for the past 12 years as a Journeyman electrician.

On the day of the incident, the paper mill was in the middle of a "scheduled shutdown," an annual event that occurs when the mill ceases one of its two paper manufacturing operations, thereby reducing production by 50 percent. To return the idle unit to production as quickly as possible, maintenance crews work 12-hour shifts, 7 days a week, until the task is completed. The victim had been working these extended shifts for 3 weeks prior to the incident.

At the time of the incident, the victim and a coworker were engaged in routine maintenance and inspection of an electrostatic precipitator in a large recovery boiler. This inspection can only be done when the boiler has been shut down for an extended period of time due to the high temperatures in the area during boiler operation.

The transformers controlling the precipitators are located inside two 25-ft by 37-ft by 7-ft-high "precipitator penthouses." The only access to these penthouses is through separate 24-in-diameter access hatches located on the roof. The south penthouse (the unit to be inspected) was one of two identical units in the area that shared a common roof. Six transformers were located on the roof between the hatches leading to the north and south penthouses. Transformers 1, 3, and 5 controlled power to the south penthouse; transformers 2, 4, and 6 controlled power to the north penthouse.

The company utilizes a sophisticated "captive key" lockout system to control access to the penthouses. This procedure calls for a complex series of functions to be performed in exact sequence to obtain access. The normal sequence of events required to enter the penthouse is as follows:

1. Shut down the main breaker, located on ground floor.

2. Open the key control cabinet (ground floor) and obtain keys to power supply transformers (in this case, transformers 1, 3, and 5).

3. Go to the eighth floor and, using the keys from the control cabinet, shut down transformers 1, 3, and 5. As each transformer is locked out, the key used to lock out the transformer is retained and another key is released.

4. Take the three keys obtained from locking out the transformers back to the key control cabinet on the ground floor.

5. Insert the three keys into the control cabinet and turn them, thereby releasing one single key.

6. Take the single key obtained in step 5 back to the eighth floor. This key unlocks the hatch and allows access to the interior of the penthouse.

A few minutes after 1:00 a.m., the victim had completed the first three steps, locking out transformers 1, 3, and 5. As a result, the south transformer was deenergized. At this point the victim had the three keys needed to obtain the key to the penthouse from the key control panel. To obtain this key, the victim would have had to travel down the steps from the eighth floor to the first floor, obtain the required key from the key control cabinet, and return to the eighth floor to open the hatch to the penthouse.

The victim commented to his coworker that the lock to the penthouse hatch was broken and that they could save themselves a trip downstairs by entering the hatch. The victim then went over to the hatch of the (energized) north penthouse and pointed out the broken lock to his coworker. The coworker twice asked the victim if the area was secure and safe to enter. The victim replied "Yes," and proceeded to open the hatch and enter the penthouse while the coworker waited outside the hatch. Shortly after the victim entered the penthouse, the coworker heard a "pop" and saw a flash. The coworker called to the victim but received no reply.

The coworker then went to the nearest phone and called the utilities supervisor. The supervisor immediately sounded an alarm, summoning the plant emergency organization. The supervisor, four workers trained in first-aid, and a plant paramedic responded to the scene. Actual entry to the penthouse was delayed for several minutes, as these individuals were required to go through the entire lockout procedure described above to deenergize the north penthouse.

Company rescue personnel entered the penthouse approximately 10 minutes after the incident. The victim was observed lying on the floor, with small third-degree burns on his left arm and extensive burns on his right hand and forearm. Although the incident itself was not witnessed, it appeared that the victim had picked up a static ground cable with the intention of attaching it to a metal brace (standard procedure) when he made contact with the energized 50,000-volt transformer and was electrocuted.

Company personnel immediately began cardiopulmonary resuscitation (CPR) and continued it until the local ambulance service arrived on the scene approximately 20 minutes after the incident. The emergency medical personnel performed defibrillation twice, and continued CPR. They then discontinued CPR for approximately 1 minute while they removed the victim from the penthouse and placed him on a stretcher. The victim was transported to the local hospital, where he was pronounced dead on arrival approximately 1 hour and 15 minutes after the incident. The medical examiner gave the cause of death as cardiac arrest as a result of electrocution.

Recommendations/Discussion

Recommendation 1: Periodic safety inspections should be made to ensure that unsafe conditions are identified and corrected. Conditions likely to result in serious injury or death should be assigned a high priority.

Discussion: In this case, the company had an extremely sophisticated lockout system in place; however, the failure of one component of this system (the lock on the north hatch) permitted the victim to enter an energized area, which resulted in his electrocution. The broken lock had been reported 7 days prior to the incident, and a work order to repair it had been submitted. This order, assigned to a contractor working on the site, had been returned, marked completed; however, no one had followed up to verify that the lock had been repaired. Proper follow-up of this work order would have detected the unrepaired lock and could have prevented this fatality by correcting the flaw in the lockout system.

Recommendation 2: Employers must continue to stress the importance of following established safety procedures to all employees.

Discussion: In this case, the worker disregarded two company safety policies.

- The worker failed to complete the standard lockout procedure for entering the penthouse. If he had followed company policy and obtained the key, he might have realized that he was preparing to enter the wrong area.
- The worker disregarded company confined space entry procedures that require an entry permit to be obtained and that the air within the confined space be tested by a company technician. If this confined space procedure had been followed, at least one of the workers involved (the victim, coworker, or technician) probably would have realized that the victim was planning to enter the wrong hatch.

Recommendation 3: Auxiliary work lights should be available in all areas where maintenance work is routinely performed after dark.

Discussion: The site where the incident occurred had no lighting installed, and the victim entered the penthouse with only a flashlight for illumination. The area where the hatches are located is outdoors and the area normally would be dark. Auxiliary lighting in this area might have helped the victim realize he was entering the energized north hatch.

Recommendation 4: The employer should use more highly visible means of identifying similar accessways or work areas.

Discussion: The north and south hatches are identical in appearance. Working in the dark, after many days of long shifts, the employee became confused as to which hatch he was entering. An easily recognized visible marking, such as a color coding or the letters S or N painted on the hatch covers, might have alerted the worker that he was entering the wrong hatch.

Recommendation 5: Permanently installed safety equipment should be designed to minimize the possibility of accidental damage.

Discussion: In this case, a complex and sophisticated lockout system was defeated by the failure of one small component (the lock on the hatch). This lock was subject to damage whenever the hatch was opened. A simple metal guard, installed to prevent the lock from striking other objects when the hatch was opened, might have kept the lock from becoming damaged and prevented this fatality.

Recommendation 6: Electric switch panels, key control panels, and similar units with numerous identical or similar controls should have permanent, highly legible identification labels installed.

Discussion: The key control panel in this case contained numerous visually identical locks without any readily apparent means of differentiating between them. A simple, legible labeling system would help ensure proper identification of these units by involved workers.

Recommendation 7: Consideration should be given to the installation of an indicator light system on the hatches. Such a system would have a "green light" that would illuminate when the power to the penthouse was deenergized.

Discussion: An indicator light located in the hatch area would provide a ready check as to the state of the equipment inside the penthouse. Any failure to obtain a "green light" would serve to provide a visible warning to employees that the unit remained energized. Such a system possibly could have prevented this fatality.

Adapted from: FACE 89-18. Accessed 7/16/08 at http://www.cdc.gov/niosh/face/In-house/full8918.html.

Questions

1. Installing a disconnecting device within sight of the precipitator entrance would have which of the following effects?
   A. Enable the worker to efficiently install a lockout device
   B. Prevent a worker from entering the precipitator if the disconnect were horse-power rated
   C. Change the basic requirement to install lockout devices, since the worker could keep watch on the position of the disconnecting device
   D. Violate the OSHA regulation to install lockout devices

2. When was the trap set for the worker?
   A. When the procedure was written
   B. When the foreman assigned the worker to the task
   C. When the hatch entrance was not repaired after being broken
   D. When the safety man inspected the work before it began

3. The purpose of the static discharge conductor inside the precipitator was which of the following?
   A. To enable a worker to discharge the static voltage that permitted the precipitator to function.
   B. To enable a worker to install a safety ground set on the precipitator conductors.
   C. To enable a worker to verify that no voltage was present inside the precipitator enclosure before beginning work.
   D. To provide a ground-fault path for electrically powered hand tools that might be required within the precipitator enclosure.

4. Which of the following statements about the sequence of events necessary to ensure that the precipitator is deenergized and locked is true?

A. The sequence of events could be reorganized to minimize the number of times a worker must climb and descend eight flights of steps.

B. As written, each step in the sequence of events is necessary and must be performed in the order it is written.

C. The sequence of events could be shortened by eliminating one or more of the required steps.

D. As written, the sequence of events illustrates the scope and authority of the safety man.

5. The inside of the precipitator enclosure was a confined space. Which of the following requirements was related to working in the vicinity of the precipitator electrodes?

A. An authorized energize electrical work permit

B. An authorized confined space permit

C. An authorized excavation permit

D. A license as a person qualified to work in the vicinity of high-voltage static energy.

6. Which section of the OSHA rules applies to lockout/tagout of the electrical energy associated with the precipitator?

A. 29 CFR 1910.333

B. 29 CFR 1926.417

C. 29 CFR 1910.269

D. 29 CFR 1910.228

Ready for Review

- The relationship between short-circuit current, arcing current, and incident energy depends on several factors.
- When comparing circuit breakers, the current-limiting circuit breaker has the lowest incident energy, followed by standard non-current-limiting circuit breakers with instantaneous trips, followed by circuit breakers with short-time delay.
- Arc flash energy can be quite high in low-level arcing faults. Arc current that is less than the instantaneous trip of the circuit breaker translates to high arc flash energies.
- Once the arcing current reaches the instantaneous trip of the circuit breaker, the arc flash energy is greatly reduced.
- As arcing current increases above the instantaneous trip, arc flash energy again begins to increase.
- The incident energy curve for a current-limiting circuit breaker is below that of a standard circuit breaker, and also has less slope, meaning that the arc flash energy increase is proportionally lower.
- Short-time delay significantly increases the arc flash energy above that of standard circuit breakers.
- In general, arcing currents beyond the current-limiting threshold of the fuse do not show an increased arc flash level.
- One of the most important decisions impacting the arc flash hazard is the current-limiting ability of the OCPDs. OCPDs limit the magnitude of the short-circuit current and reduce the time duration of the short-circuit current, thereby reducing the arc flash hazard.
- Renewable and Class H fuses are outdated, are not current-limiting, and must not be used.
- Molded case circuit breakers, insulated case circuit breakers, and low-voltage power circuit breakers that are not listed as current-limiting, do not significantly reduce the level of short-circuit currents, and take longer to open.
- Class RK5, Class RK1, Class J, Class CC, and Class L fuses are all current-limiting and reduce the arcing fault energy released when the arcing current is within their current-limiting range.
- Current-limiting devices provide different degrees of protection based on the level of short-circuit current.
- The reliability of OCPDs is critical. The protective devices must open as originally specified, otherwise the flash hazard can escalate to higher levels than expected. Some OCPDs require maintenance.
- Current-limiting fuses do not require maintenance other than visual examination and ensuring that they have not been damaged.
- Circuit breakers are mechanical OCPDs that require periodic exercise, maintenance, testing, and possible replacement. Periodic maintenance and testing is necessary for all circuit breakers to help ensure that they will operate as intended.
- Failing to perform maintenance can increase clearing times and increase the arc flash hazard.
- Rejection style Class J, T, R, and L fuses are physically size-rejecting, meaning that fuses with lower voltage ratings, lower interrupting ratings, or lower current limiting ability cannot be accidentally put into service because they won't fit.
- Switches that shunt trip to open all three poles when the first fuse opens can reduce arc flash energy levels.

- Specifying Type 2 motor starter protection can reduce the arc flash hazard level.
- One way to help minimize the electrical shock hazard is to use finger-safe products and non-conductive covers or barriers.
- Horsepower (hp)-rated fusible disconnects should be installed within sight of every motor or driven machine. An in-sight motor disconnect is more likely to be used by a worker for the lockout procedure to put equipment in an electrically safe work condition prior to doing work on the equipment.
- Selective coordination is considered the act of isolating a faulted circuit from the remainder of the electrical system, thereby eliminating unnecessary power outages. Personnel safety is enhanced in a selectively coordinated system because the electrical worker is exposed only at the level of the problem circuit.
- Impedance-grounded wye systems can reduce the probability that dangerous or destructive arcing faults to ground will occur.
- Designers should select and install a main on each service. The six-disconnect rule for service entrances in lieu of a single main disconnect should not be used. Unless an OCPD is provided, the main bus and line terminals of the feeders are unprotected.
- Large circuits should be broken into smaller circuits.
- Some circuit breakers have an adjustable instantaneous trip setting, which allows adjustment to the current at which the circuit breaker will operate in its instantaneous mode. The setting on the trip might cause the circuit breaker not to operate during an arcing fault if it is set too high. The instantaneous trip settings should be adjusted as low as possible for the best arc flash energy reduction.

- Short-time delay mechanisms should not be used, since they allow a fault to flow for an extended period of time. (The arc flash hazard increases with the length of time the current is permitted to flow.)
- Zone-selective interlocking provides a solution to the high incident energy levels allowed by short-time delay settings. With this feature, the circuit breakers' sensing elements communicate, which allows the upstream circuit breaker to open instantaneously when a fault occurs between the upstream and downstream circuit breakers.
- Arc flash reducing maintenance switching, where short-time delay is utilized, reduces exposure levels by turning off the short-time delay function while working in the flash protection boundary. This option allows the circuit breaker to trip without intentional delay if an arc flash is accidentally started.
- Remote monitoring of voltage and current levels reduces exposure to electrical hazards. Troubleshooting is performed remotely on a computer screen so that the worker is not working physically close to the equipment.
- Arc-resistant switchgear might be installed. This type of equipment withstands internal arcing faults. Arc-resistant equipment is designed with features that divert hot gases and pressures from an internal arcing fault.
- Workers can avoid the hazards associated with racking circuit breakers by using equipment with remote racking capabilities. Workers can remain a safe distance from the actual racking operation.
- Remote opening and closing motors on large switches and circuit breakers permits workers to control the operation from a safe distance.

Vocabulary

arc rating[*] The value attributed to materials that describes their performance to exposure to an electrical arc discharge. The arc rating is expressed in cal/cm^2 and is derived from the determined value of the arc thermal performance value (ATPV) or energy of breakopen threshold (E_{BT}) should a material system exhibit a breakopen response below the ATPV value) derived from the determined value of ATPV or E_{BT}.

> FPN: *Breakopen* is a material response evidenced by the formation of one of more holes in the innermost layer of flame-resistant material that would allow flame to pass through the material.

current-limiting overcurrent protective device[†] A device that, when interrupting currents in its current-limiting range, reduces the current flowing in the faulted circuit to a magnitude substantially less than that obtainable in the same circuit if the device were replaced with a solid conductor having comparable impedance.

listed[†] Equipment, materials, or services included in a list published by an organization that is acceptable to the authority having jurisdiction (AHJ) and concerned with evaluation of products or services, that maintains periodic inspection of production of listed equipment or materials or periodic evaluation of services, and whose listing states that the equipment, material, or services either meets appropriate designated standards or has been tested and found suitable for a specified purpose.

selective coordination† Localization of an overcurrent condition to restrict outages to the circuit or equipment affected, accomplished by the choice of overcurrent protective devices and their ratings or settings.

Type 1 protection# A form of motor controller protection that requires that, under short-circuit conditions, the contactor or starter shall cause no danger to persons or installation and might not be suitable for further service without repair and replacement of parts.

Type 2 protection# A form of motor controller protection that requires that, under short-circuit conditions, the contactor or starter shall cause no danger to persons or installation and shall be suitable for further use. The risk of contact welding is recognized, in which case the manufacturer shall indicate the measures to be taken regarding the maintenance of the equipment.

zone-selective interlocking A system of communications between line-side and load-side circuit breakers allowing the line-side circuit breaker to open without delay for short-circuits located between the circuit breakers and, for purposes of obtaining selective coordination, allowing the line-side circuit breaker to delay operation for short-circuits on the load side of the load-side circuit breaker.

*This is the *NFPA 70E* definition.

†This is the *National Electrical Code®* (*NEC®*) definition.

#This definition is from International Electrotechnical Commission (IEC) 60947-4-1.

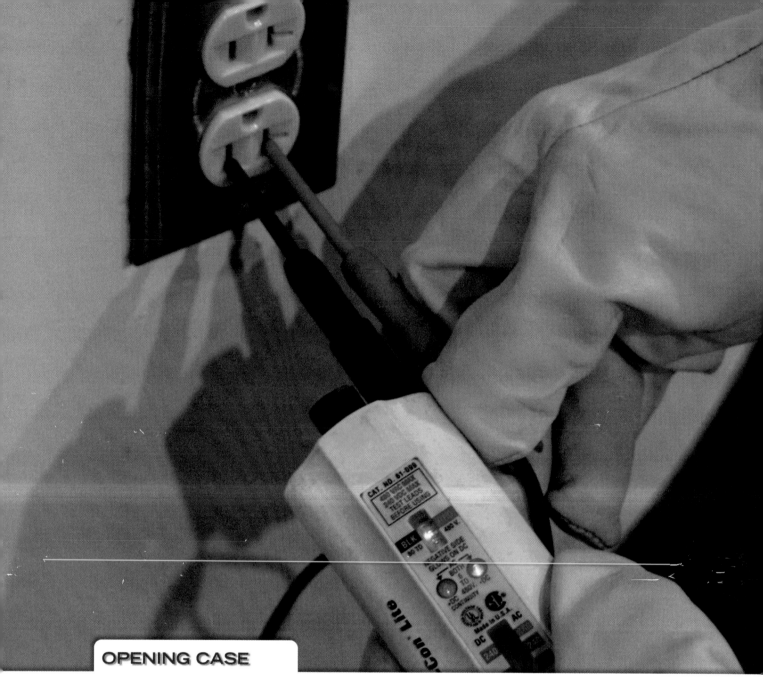

On June 23, 1987, a Journeyman electrician (the victim) and an electrician's helper were repairing several electrical malfunctions in a single-family residence. The workers had started at 7:30 a.m. and had worked at the house all day, except for approximately 2 hours when they went to check on another job. At 4:25 p.m. the victim called his office and notified his supervisor that they would not be able to finish the job by their normal quitting time of 4:30 p.m.

The electrician's helper had been trying to repair a 110-volt, 4-ft fluorescent light over a stainless steel sink in the kitchen. He had replaced the ballast; however, he could not get the light to operate properly. The electrician's helper asked the Journeyman electrician if he would try to repair the light. The electrician was sitting on the sink when he apparently contacted an energized wire on the load side of the ballast. Neither the electrician nor his helper had deenergized the circuit at the panel box or at the single-pole switch on the wall beside the sink.

At 4:35 p.m. the owner of the residence discovered the victim slumped over the sink. The owner pulled the victim away from the light and asked if he was okay. The victim responded that he was all

History, Evolution, Layout, Style, and Scope of *NFPA 70E®*

CHAPTER 5

Introduction

The **National Electrical Code®** is primarily an installation document. Its purpose is the practical safeguarding of persons and property from the hazards arising from the use of electricity. The **Occupational Safety and Health Administration (OSHA)** recognized the need for a consensus document comprised of electrical safety requirements to protect individuals working close enough to electrical equipment to be subject to electrical hazards. *NFPA 70E* is an approved **American National Standard** that provides requirements addressing electrical safety-related work practices and maintenance of electrical systems. *NFPA 70E* utilizes a layout and numbering style similar to the NEC and has become an invaluable tool to help employers comply with the requirements of OSHA and provide a workplace free from recognized hazards where the work is covered by the scope of *NFPA 70E*.

History of *NFPA 70E*

The National Fire Protection Association (NFPA) Standards Council established the Committee on Electrical Safety Requirements for Employee Workplaces in 1976 at the request of OSHA. This NFPA Technical Committee was charged with the responsibility of developing electrical safety standards to help OSHA reduce electrically-related injuries and fatalities.

The stated purpose of the *NEC®* is the practical safeguarding of persons and property from the hazards arising from the use of electricity. It is important to note that, prior to the development of *NFPA 70E*, OSHA's electrical standards were primarily based on

Chapter Outline

- ■ Introduction
- ■ History of *NFPA 70E*
- ■ Evolution of *NFPA 70E*
- ■ Layout of *NFPA 70E*
- ■ Numbering Style
- ■ *NFPA 70E* Today
- ■ Summary
- ■ Lessons Learned
- ■ Current Knowledge

Learning Objectives

1. Become familiar with the unique history and evolution of *NFPA 70E*.
2. Understand the format and numbering style used in *NFPA 70E*.
3. Become familiar with the scope of *NFPA 70E*.

References

1. *NFPA 70E*, 2009 Edition
2. Appendix A: OSHA Citation Examples

OPENING CASE, CONT.

right, but, clearly, he was not. The electrician's helper and the home owner contacted the local fire department and attempted to make the victim comfortable. Fire department personnel arrived at 4:50 p.m. The victim was transported to a nearby hospital emergency room where he died a short time later. (The reason the fluorescent light would not operate properly was later determined to be the result of a burned-out lamp.)

The medical examiner determined that the cause of death was cardiac arrhythmia due to electrocution.

Adapted from: Fatality Assessment and Control Evaluation (FACE) 87-55. Accessed 5/9/08 at http://www.cdc.gov/niosh/face/In-house/full8755.html.

1. Why didn't either the electrician or the helper first deenergize the circuit?

2. Under what conditions is working on an energized circuit necessary?

Figure 5.1 The *National Electrical Code* is primarily used by those designing, installing, and inspecting electrical installations.

the *NEC*. The *NEC* was and still is a document primarily used by those designing, installing, and inspecting electrical installations (**Figure 5.1**). Since OSHA's electrical requirements additionally address employees and employers in the workplace, the need existed for an electrical safety-related work practices consensus document addressing electrical safety-related work practices and maintenance of electrical systems.

The First Version

In the mid 1970s, OSHA determined that the *NEC* did not adequately address electrical safety as it related to *people* (the *NEC* primarily covered electrical equipment, installations, and designs), so the administration again asked the National Fire Protection Association (NFPA) for help. As a result, in January 1976, the NFPA chartered and assembled a technical committee to review the content of the 1972 *NEC* and to write a national consensus standard that covered only the personal safety aspects of interacting with electrical energy.

The document that the committee produced was *NFPA 70E*®, an American National Standard. It was the first national consensus standard that covered work practices. OSHA embraced that standard and promulgated requirements based on it in the Federal Register. At that point, the requirements became federal law.

 Evolution of *NFPA 70E*

The first edition of *NFPA 70E* consisted only of installation requirements, which represented only a fraction

of the provisions contained in the *NEC*. Only those requirements from the *NEC* considered directly related to employee safety were included in *NFPA 70E*. In the first edition, these appeared in Chapter 1. (Those installation requirements were relocated to Chapter 4 in the 2004 edition. Chapter 4 and its installation requirements were deleted from *NFPA 70E* in the 2009 edition to make clear that *NFPA 70E* does not address installation requirements.)

Subsequent editions saw the evolution of the document add numerous concepts and requirements. The second edition of *NFPA 70E* was published in 1981 and added safety-related work practice requirements. The third edition, published in 1983, added safety-related maintenance requirements. The fourth edition was released in 1988 with only minor revisions. This edition became the primary source for OSHA requirements defined in the 1990 version of Subpart S.

The Fifth Edition

Because the *NEC* has a review cycle of three years, the basic information for *NFPA 70E* and, consequently, the OSHA law, became outdated. On September 14, 1990, Gerard F. Scannel of OSHA asked the NFPA to request that the *70E* Technical Committee be reconstituted to update, revise, rewrite, and redevelop the content of 70E. OSHA and NFPA officials agreed to do so.

Scannel specifically requested that *NFPA 70E* be revised to parallel the content of OSHA *Electrical Safety-Related Work Practices* more closely. In addition, he asked that *NFPA 70E* include sections on electrical protective clothing, electrical safe approach distances for all workers, and training requirements. Scannel also asked the Technical Committee to update the document to reflect any new information, issues, and concerns that would ensure a safe work environment for employees.

The result of this request was the fifth edition of *NFPA 70E*, *Standard for Electrical Safety-Related Workplaces*.

> "OSHA is issuing a new standard on electrical safety-related work practices for general industry. These performance-oriented regulations complement the existing electrical installation standards. The new standard includes requirement for work performed on or near exposed energized and deenergized parts of electric equipment; use of electrical protective equipment; and the safe use of electrical equipment. Compliance with these safe work practices will reduce the number of electrical accidents resulting from unsafe work practices by employees . . . ."

> *Source*: Preamble to OSHA Subpart S, Final Rule, issued Monday, August 6, 1990.

The fifth edition, issued in 1995, contained significant changes, including a revision of the installation requirements to adopt the language contained in the 1993 *NEC*, as well as the addition of requirements addressing limits of approach and the flash protection boundary.

The Sixth Edition

The 2000 version of *NFPA 70E* (), the sixth edition, again saw an update of the installation standards, this time reflecting the provisions contained in the 1999 *NEC*. In addition, this version included a chapter addressing safety requirements for special equipment. Another major change to this sixth edition, and considered by many to be the most popular and useful, was the inclusion of tables intended to provide a means of selecting personal and other protective equipment, based on a number of common work tasks within the parameters contained in notes to the tables.

The Seventh Edition

The 2004 edition, the seventh, continued the evolution of the standard. It contained many new important and exciting concepts and procedures, as well as a new name, layout, look, and numbering format. One significant addition to the standard for the 2004 edition was the requirement for work to be performed under an energized electrical work permit if live parts were not placed in an electrically safe work condition.

Layout of *NFPA 70E*

In addition to numerous revisions, the year 2004 brought a new name to the document: *NFPA 70E, Standard for Electrical Safety in the Workplace* (Figure 5.3). In addition, *NFPA 70E* was reformatted to comply with the *NEC Style Manual*, a format to which users of the *NEC* have long been accustomed.

Purpose

The 2009 edition also brought change. In this eighth edition, Section 90.1 identifies the purpose of *NFPA 70E*, which is to provide a practical safe working area for employees relative to the hazards arising from the use of electricity.

Scope

The document scope has been relocated to Section 90.2 in the 2009 Edition. In addition to being relocated, the scope has been editorially modified. That section indicates that the *NFPA 70E* scope addresses electrical safety requirements for employee workplaces that are necessary for the practical safeguarding of employees during

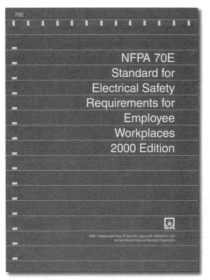

Figure 5.2 The *NFPA 70E*, 2000 Edition was the first to include tables as a means to select personal and other protective equipment.

Figure 5.3 The *NFPA 70E*, 2004 Edition was the first to comply with the *NEC Style Manual*.

activities such as the installation, operation, maintenance, and demolition of electric conductors, electric equipment, signaling and communications conductors and equipment, and raceways. The scope was changed to make it clear that *NFPA 70E* does not address installation requirements due to the deletion of the Chapter 4 installation requirements. A question that is often asked related to scope is whether *NFPA 70E* "applies to utility work or work that linemen do." That question can be answered by looking at Section 90.2:

90.2 Scope.

(A) Covered. This standard addresses electrical safety requirements for employee workplaces that are necessary for the practical safeguarding of employees during activities such as the installation, operation, maintenance, and demolition of electric conductors, electric equipment, signaling and communications conductors and equipment, and raceways for the following:

(1) Public and private premises, including buildings, structures, mobile homes, recreational vehicles, and floating buildings

(2) Yards, lots, parking lots, carnivals, and industrial substations

(3) Installations of conductors and equipment that connect to the supply of electricity

(4) Installations used by the electric utility, such as office buildings, warehouses, garages, machine shops, and recreational buildings, that are not an integral part of a generating plant, substation, or control center

(B) Not Covered. This standard does not cover the following:

(5) Installations under the exclusive control of an electric utility where such installations:

(a) Consist of service drops or service laterals, and associated metering, or

(b) Are located in legally established easements or rights-of-way designated by or recognized by public service commissions, utility commissions, or other regulatory agencies having jurisdiction for such installations, or

(c) Are on property owned or leased by the electric utility for the purpose of communications, metering, generation, control, transformation, transmission, or distribution of electric energy.

It is important to note that *NFPA 70E* "addresses electrical safety requirements for employee workplaces that are necessary for the practical safeguarding of employees during activities such as the installation, operation, maintenance, and demolition of electric conductors, electric equipment, signaling and communications conductors and equipment, and raceways . . .". Users also should note that Section 90.2(B), "Not Covered," indicates that "installations under the exclusive control of an electric utility" are not covered by *NFPA 70E*. Therefore, work performed on installations that are *not* under the exclusive control of an electric utility would be covered by the scope of *NFPA 70E*.

Arrangement

The arrangement of the standard is described and illustrated in Section 90.3 and Figure 90.3, respectively.

Chapter 1 applies generally to electrical safety in the workplace. Chapter 2 covers safety-related maintenance requirements. Chapter 3 provides for safety requirements for special equipment. Its provisions supplement and/or modify those contained in Chapter 1.

Organization

Section 90.4 indicates that *NFPA 70E* comprises three chapters and fifteen annexes. Those three chapters are as follows:
- Chapter 1, Safety-Related Work Practices
- Chapter 2, Safety-Related Maintenance Requirements
- Chapter 3, Safety Requirements for Special Equipment

The fifteen annexes are as follows:
- Annex A, Referenced Publications
- Annex B, Informational References
- Annex C, Limits of Approach
- Annex D, Incident Energy and Flash Protection Boundary Calculation Methods
- Annex E, Electrical Safety Program
- Annex F, Hazard/Risk Evaluation Procedure (**Figure 5.4**)
  - *NFPA 70E* Annex F.2 explains that a **hazard/risk evaluation** is "an analytical tool consisting of a number of discrete steps intended to ensure that hazards are properly identified and evaluated, and that appropriate measures are taken to reduce those hazards to a tolerable level."
- Annex G, Sample Lockout/Tagout Procedure
- Annex H, Simplified, Two-Category, Flame-Resistant (FR) Clothing System
- Annex I, Job Briefing and Planning Checklist
- Annex J, Energized Electrical Work Permit
- Annex K, General Categories of Electrical Hazards
- Annex L, Typical Application of Safeguards in the Cell Line Working Zone
- Annex M, Layering of Protective Clothing and Total System Arc Rating
- Annex N, Example Industrial Procedures and Policies for Working Near Overhead Electrical Lines and Equipment
- Annex O, Safety Related Design Requirements

Numbering Style

In accordance with the *NEC Style Manual*, the three chapters in *NFPA 70E* are further broken down into articles. The articles are broken down into sections. The sections are broken down into subdivisions.

HAZARD/RISK EVALUATION PROCEDURE

Task: Voltage testing _____   Document no.: _____

Equipment: _____   Part of: _____

Issued by: _____   Pre-risk assessment

Date: _____   Intermediate risk assessment

Follow-up risk assessment

Black area = Safety measures required   Gray area = safety measures recommended

Consequences	Severity Se	Class CI					Frequency Fr	Probability Pr	Avoidance Av
		3-4	5-7	8-10	11-13	14-15			
Irreversible trauma, death	4						Daily 5	Common 5	
Permanent, third degree burn	3						Weekly 4	Likely 4	
Reversible, second degree burn	2						Monthly 3	Possible 3	Impossible 5
Reversible, first aid	1						Yearly 2	Rarely 2	Possible 3
							Less 1	Negligible 1	Likely 1

Hzd. No.	Hazard	Se	Fr +	Pr +	Av +	= CI	Severity Mitigators	Safe
1	Human factors	4	5	3	5	13	Use appropriate PPE and follow established safety procedures.	Y
2	Shortened test loads	3	5	2	5	12	Inspect leads before each use.	Y
3	Meter misapplication	4	5	3	5	13	Ensure that the meter is rated for the level of voltage being tested.	Y
4	Meter malfunctions	3	5	2	5	12	Ensure that the meter is CAT rated to the appropriate hazard level.	Y

Comments:

PPE required: Voltage rated gloves and leather protectors, face and head protection, clothing rated for the incident energy exposure.

© 2008 National Fire Protection Association   NFPA 70E (p. 1 of 1)

Figure 5.4 Figure F.2.1 from *NFPA 70E* Annex F.2 illustrates examples of the steps of a hazard/risk analysis procedure that is required by Section 110.7(F).

The following is a look at how Chapter 1 is broken down into **articles**:

Chapter 1, Safety-Related Work Practices

Article 100: Definitions

Article 110: General Requirements for Electrical Safety-Related Work Practices

Article 120: Establishing an Electrically Safe Work Condition

Article 130: Work Involving Electrical Hazards

An example of the breakdown of the Articles into **sections** is in the following example:

Article 110: General Requirements for Electrical Safety-Related Work Practices

110.1 Scope

110.2 Purpose

110.3 Responsibility

110.4 Organization

110.5 Relationships with Contractors

110.6 Training Requirements

110.7 Electrical Safety Program

110.8 Working While Exposed to Electrical Hazards

110.9 Use of Equipment

An example of the breakdown of Sections into **level 1 subdivisions** is in the following example:

110.6 Training Requirements

(A) Safety Training

(B) Type of Training

(C) Emergency Procedures

(D) Employee Training

(E) Training Documentation

An example of the breakdown into **level 2 subdivisions** is in the following example:

110.6(D) Employee Training

(1) Qualified Person

(2) Unqualified Persons

(3) Retraining

An example of the breakdown into **level 3 subdivisions** is in the following example:

110.6(D) Employee Training

(1) Qualified Person

(a) Such persons shall also be familiar with ...

(b) Such persons permitted to work within ...

(c) An employee who is undergoing on-the-job training ...

(d) Tasks that are performed less often than once per year ...

(e) Employees shall be trained to select an appropriate voltage detector ...

An example of the breakdown into list items is as follows:

110.6(D) Employee Training

(1) Qualified Person

(a) Such persons shall also be familiar with ...

(b) Such persons permitted to work within ...

(1) The skills and techniques necessary to distinguish ...

(2) The skills and techniques necessary to determine ...

(3) The approach distances specified ...

(4) The decision-making process necessary ...

The layout of the entire document can be found in the Table of Contents of *NFPA 70E*. The Table of Contents is always the right place to begin a search and can prove to be an efficient tool to help locate a particular requirement. As an example, using the layout of Article 110 in the Table of Contents in *NFPA 70E*, or in the information just covered above, a user can quickly

Figure 5.5 *NFPA 70E* continues to grow in recognition and acceptance as a consensus standard assisting employers in understanding how to comply with OSHA.

and efficiently locate the requirements applicable to an Electrical Safety Program in Section 110.7 or Qualified Person Employee Training in Section 110.6(D)(1).

NFPA 70E Today

NFPA 70E has an interesting history and evolution. The 2009 edition (Figure 5.5) has become an invaluable tool to help employers comply with the requirements of OSHA. OSHA requirements are, in many cases, written in **performance language**, requiring compliance without necessarily stating how to comply. *NFPA 70E* generally is not written in performance language. Accordingly, 70E is growing in recognition and acceptance as a consensus standard that helps employers understand how to comply with OSHA's performance-based requirements. OSHA has referenced the provisions of *NFPA 70E* in citations issued to employers.

Summary

NFPA 70E, Standard for Electrical Safety in the Workplace, has become a vital tool to help employers comply with the performance-based requirements of OSHA in an effort to provide a workplace free from recognized hazards. Understanding the layout and the content of *NFPA 70E* is an important part of achieving those goals.

Chapter 3 of this textbook contains a number of examples of OSHA requirements that address the hazards of working on or near energized electrical parts. This includes Section 5(a)(1) of the Occupational Safety and Health Act. That chapter should be reviewed to gain a greater understanding of a number of OSHA's requirements that are applicable to protecting workers from the hazards associated with electrical work.

Lineman Electrocuted While Working from the Bucket of an Aerial Lift Truck

On September 16, 1991, a lineman was electrocuted when he contacted an energized power line while working from the bucket of an aerial lift truck.

The employer in this incident was a company that had been in operation for 90 years and employed about 13,000 workers, including 1800 linemen. The company had a written safety policy and written safety rules and procedures that were administered by the managers of the safety department and by job site foremen. The company had established a corporate safety department and five divisional safety departments, staffed with safety managers, safety engineers, industrial hygienists, fire safety personnel, and related clerical staff. The company maintained a video training library that covered all aspects of overhead and underground transmission and distribution.

The company held and documented weekly safety meetings and promoted safety through the use of written tests, work performance evaluations, communications, and incentive programs. Cardiopulmonary resuscitation and first aid training were mandatory for all field personnel. Also, prior to each job, the company conducted specific tailgate conferences at each job site. The victim had worked for the company for 15 years, including about 8 years as a lineman.

The employer had received a request from a local communications company to install a number of new utility poles that would be 5 feet taller than the existing ones, and to transfer the 3-phase, 34,500-volt (phase-to-phase) overhead power line onto the new poles. The taller poles would provide additional clearance for the communications company transmission lines.

On the morning of the incident, a pole crew arrived at the job site, erected a taller pole about 4 feet away from the existing pole, and moved to the next site. Later that morning a five-man crew—a job site foreman, two linemen (including the victim), a lead lineman, and a groundsman arrived at the job site to make the transfer of the power lines. The crew positioned insulating line sleeves and blankets over the energized power lines, ground wire, and insulators, and also over the crossarm attached to the shorter pole. Crew members further protected the top section of the taller pole with an insulating blanket.

The victim proceeded to move the power lines, one at a time, to the taller pole. First, he transferred the middle phase to an insulator attached to a ridge pin located at the top of the taller pole. He used an aluminum wire to secure the power line to the insulator. Second, he attached a crossarm to the taller pole and covered it with an insulating blanket. Next, he transferred the outside phases, one at a time, to the insulators located on opposite ends of the crossarm and secured in place. Finally, the victim began removing the insulating sleeves and blankets.

He first removed the line sleeves and insulating blanket from one of the outside phases. Next, the victim positioned the lip of the bucket next to the crossarm, between the other outside phase and middle phase. He then removed the insulating blanket from the crossarm and middle phase insulator. While trying to remove the middle phase line sleeve, located farthest away from the bucket, the victim snagged his rubber glove on the aluminum wire that secured the power line to the insulator. When he pulled his arm back, the rubber glove was pulled partially off, and his exposed right wrist contacted the power line. Current passed through the victim's right arm and exited the body at the lower abdomen on the left side, where he was in contact with the crossarm attached to the utility pole.

Although the incident was unwitnessed, the other crew members heard a buzzing sound, looked up toward the victim, and saw him slump down into the bucket. The groundsman immediately lowered the bucket, while another lineman radioed the dispatcher to call the rescue squad. The crew members removed the victim from the bucket and began cardiopulmonary resuscitation. The rescue squad arrived in about 12 minutes, performed advanced cardiac life support, and transported the victim to the local hospital, which was approximately 20 minutes away. The victim was pronounced dead by the emergency room physician 80 minutes after the incident occurred.

The medical examiner's report listed the cause of death as acute cardiac arrhythmia due to electrocution.

Recommendations/Discussion

Recommendation 1: Employers should stress the importance of adherence to established safe work procedures.

Discussion: In addition to requiring the use of personal protective equipment (PPE), established safe work practices required employees to maintain adequate clearance between themselves and objects with ground potential while working with energized power lines. The pole crew was working from an insulated bucket, wearing appropriate PPE. They had covered the power lines, insulators, crossarms, and pole top with insulating materials, as required by applicable safe work procedures. Formulating safe work procedures is only the first step in injury prevention. For safe work procedures to be effective, they must be clearly communicated and implemented by all employees and supervisors. Supervisors should ensure that established work procedures are followed at all times.

Recommendation 2: Employers should review and revise, where applicable, safe work procedures regarding the removal of insulating materials, the positioning of aerial buckets, and the procedure used in securing power lines to insulators.

Discussion: Safe work procedures regarding the removal of the insulating blanket from the crossarm before the removal of the line sleeves and insulator blankets should be reviewed and revised, where appropriate. A work procedure that specifies leaving the crossarm-insulating blanket in place until after the removal of line sleeves and insulator blankets may help to reduce inadvertent contact with energized conductors. Safe work procedures regarding the positioning and repositioning of the aerial bucket should be periodically reviewed and revised where indicated, to assure the procedure is appropriate in providing access to the middle powerline sleeves, while allowing adequate clearances from ground potential items. The procedure used in securing power lines to utility pole insulators should be reviewed and revised, where applicable, to ensure the procedure adequately addresses the appropriate manner in which the power line is secured to the insulator. The tie down (i.e., a piece of wire used to secure the power line to the insulator), should be fastened in such a manner that the ends of the tie down do not protrude from the power line.

Adapted from: FACE 92-02. Accessed 5/9/08 at http://www.cdc.gov/niosh/face/In-house/full9202.html.

Questions

1. The case suggests that the snagging of the rubber glove was a contributing factor in the incident. What required component of PPE is not discussed in the case?
   A. Insulated dielectric boots or shoes
   B. An adequately rated, Category 4 voltmeter
   C. Leather protectors
   D. A completed energized work permit

2. In which of the following sequences should the victim have removed the insulating blanket?
   A. The blankets installed on the highest point on the new pole first
   B. The blankets installed on the old pole's crossarm first
   C. The blankets installed on the outside insulators first
   D. The blankets installed on the new crossarm first

3. Why was the victim in the bucket by himself?
   A. The crew groundsman had other duties.
   B. The other lineman was on a break.
   C. The foreman was lining up the work on the next pole for the crew.
   D. The bucket had room for only one person.

4. The tailgate conference should have discussed which of the following subjects?
   A. Which voltmeter to use
   B. The location of the nearest first-aid office
   C. The sequence for installing and removing insulating blankets
   D. What time the lunch break should occur

5. The case suggests that the aluminum wire tie down was a contributing factor. What would have prevented the aluminum wire from becoming a snag point?
   A. The tie down should be fastened to the power line in such a manner that the ends do not protrude from the power line.
   B. Only number-11 iron wire should be used for tie downs.
   C. Only rope constructed from hemp should be used for tie downs.
   D. Tie downs are unnecessary, and the aluminum wire should not have been used.

Ready for Review

- The *NEC*, OSHA, and *NFPA 70E* all relate to electrical safety. In some cases, *NFPA 70E* evolved from the *NEC* and OSHA requirements, while in other cases, OSHA requirements evolved from *NFPA 70E*.
- The *NEC* covers the practical safeguarding of persons and property from hazards arising from the use of electricity and is primarily an installation document.
- OSHA outlines electrical requirements for employees and employers in the workplace and is, in many cases, written in performance language.
- *NFPA 70E* addresses electrical safety-related work practices and maintenance of electrical systems and is, in many cases, written in prescriptive language.
- *NFPA 70E* began as coverage of installation requirements and has evolved significantly. It has been retitled *NFPA 70E, Standard for Electrical Safety in the Workplace*. Its primary link to the *NEC*, now that Chapter 4 has been deleted, is the common definitions located in Article 100.
- The scope of *NFPA 70E* includes work performed on installations that are not under the exclusive control of an electric utility.
- *NFPA 70E* contains chapters and annexes. The chapters are organized into articles, sections, subdivisions, and lists.

Vocabulary

American National Standard An accreditation provided by the American National Standard Institute, an organization that approves standards and ensures that they can be used internationally.

article A component of *NFPA 70E* that appears second (after chapter) and bolded in this example: **110**.6.(D)(1)(a).

hazard/risk evaluation An analytical tool consisting of a number of discrete steps intended to ensure that hazards are properly identified and evaluated, and that appropriate measures are taken to reduce those hazards to a tolerable level (adapted from ANSI/ASSE Z244.1).

level 1 subdivision The fourth component of *NFPA 70E*, bolded in this example: 110.6.**(D)**(1)(a).

level 2 subdivision The fifth component of *NFPA 70E*, bolded in this example: 110.6.(D)**(1)**(a).

level 3 subdivision The sixth component of *NFPA 70E*, bolded in this example: 110.6.(D)(1)**(a)**.

National Electrical Code® A standard published by the National Fire Protection Association (*NFPA*) primarily addressing electrical installations.

Occupational Safety and Health Administration (OSHA) The U.S. government agency that sets standards for worker safety and health protection, including outlining requirements for electrical safety for employers and employees.

performance language Wording that indicates that something must be done, but not how to do it. For example, ". . . the authorized employee shall have knowledge of the type and magnitude of the energy, the hazards of the energy to be controlled, and the method or means to control the energy."

section The third component of *NFPA 70E*, bolded in this example: 110.**6**.(D)(1)(a).

OPENING CASE

On March 16, 1992, a 60 year-old male Journeyman electrician was electrocuted when he contacted an energized electrical cable carrying 277 volts. The incident occurred in an office building that was being renovated. The victim was an electrician who had been hired from the union hall and had worked for the employer for less than a week. The employer was an electrical subcontractor who employed about 50 workers. The employer was under contract with the site owner for more than 5 years and was responsible for maintaining the building's electrical systems.

The employer had no electrical training, nor did he have a safety program, explaining that the workers were hired from the trade union halls and were certified by the union as trained Journeymen or apprentices. As part of the contract with the site owner, the employer was required to follow all of the site owner's safety rules, including the enforcement of a lockout/tagout procedure.

A part of the building's second floor was being renovated to enlarge some of the private offices, a job that required repositioning the partition walls and moving some of the overhead fluorescent light fixtures. The employees in the offices were moved out prior to the renovation, and the area was isolated by hanging plastic sheets between the construction and office areas.

Three employees of the electrical contractor were at the site on the morning of the incident: the foreman, a Journeyman electrician (the victim), and an apprentice electrician. At 7:15 a.m., the foreman gave the victim a set of blueprints and explained the job to him. He was told to install new 2 by 2 ft fluorescent lighting fixtures in two private offices that were being expanded. To do this, he was instructed to install a fixture tail to one fixture and connect it to the existing lights. He also was told to reconnect the power on a second bank of lights in an adjacent open area that had been disconnected the week earlier. The foreman did not instruct the victim to deenergize the circuit breakers.

Overview of *NFPA 70E*® Concepts

CHAPTER 6

Introduction

An important provision from Section 5 of the Occupational Safety and Health Act of 1970 (OSH Act) requires that workers be provided with a workplace free from recognized hazards. Many people consider the requirements defined in *NFPA 70E* as a means to comply with the OSHA requirements related to the hazards of working on or near exposed energized parts. This chapter provides a brief overview on the entire *NFPA 70E* standard, with the primary focus on the Article 100 and 110 provisions of Chapter 1.

In many instances, Article 110 serves as an introduction to concepts addressed in greater detail in other portions of *NFPA 70E*, such as those contained in Articles 120 and 130. Although many of the concepts from Article 110 are addressed by more specific requirements later in Chapter 1, it is important to recognize that many concepts are addressed only in Article 110. These include, but are not limited to, host and contract employer responsibilities (110.5), training requirements (110.6), electrical safety program considerations (110.7), and use of equipment (110.9).

OPENING CASE, CONT.

The Journeyman and the apprentice electrician then went to the job site. After setting up an 8-ft wooden stepladder, the victim began installing the new fixtures in the corner office. At some point, the victim asked the apprentice "What's up there?" The apprentice replied, "There's a hot switch leg and another fixture tail to be tied into the new fixtures," to which the victim said "Okay." The switch leg was an energized 277-volt electrical cable that had been previously connected to a wall switch in a partition wall. After the partition was moved, the cable was left hanging from a junction box in the ceiling. The fixture tail was an electrical cable from another light fixture that had not yet been connected and was deenergized. While working on the fixtures, the victim hung his tool belt on the pipes above the suspended ceiling and placed a voltmeter on a ceiling panel near the cables.

No one witnessed the incident. At about 9:30 a.m., after a coffee break, the victim resumed work on wiring the fixtures and was last seen working on an electrical cable while standing on the fourth rung of the ladder. He contacted the 277 volts apparently while stripping the insulation from the switch leg wires with a wire cutter. The electric shock burned the victim's left hand and knocked him from the ladder. The apprentice, who was working at another end of the room, and several office workers heard the victim fall and went to his aid. The apprentice checked the victim for a pulse and found none. An office employee then started cardiopulmonary resuscitation (CPR), which was continued after the company

Chapter Outline

- **Introduction**
- **Organization**
- **Scope**
- **Purpose and Responsibility**
- **Host and Contractor Responsibilities**
- **Training Requirements**
- **Electrical Safety Program**
- **Working While Exposed to Electrical Hazards**
- **Use of Equipment**
- **Summary**
- **Lessons Learned**
- **Current Knowledge**

Learning Objectives

1. Become familiar with a number of the *NFPA 70E* definitions found in Article 100.
2. Understand the layout and topics covered in Article 110.
3. Recognize that Article 110 introduces many of the electrical safety-related work practices in Articles 120 and 130 and defines some requirements addressed in Article 110 only.

References

1. *NFPA 70E*, 2009 Edition
2. OSHA 29 CFR Part 1926
3. *National Electrical Code*®

As discussed in Chapter 3, OSHA requires that workers be provided with a workplace free from recognized hazards. What is not always necessarily clear is how that "hazard-free" workplace is to be provided. Chapter 3 explained that OSHA requirements addressing electrical hazards are often written in performance language. Rules written in performance language define a result without necessarily providing details of how to accomplish it. As discussed in Chapter 5, *NFPA 70E* can help.

To begin understanding how *NFPA 70E* can help accomplish the performance requirements of OSHA, an overview of the concepts contained in the *NFPA 70E* standard is necessary.

As discussed in the overview in Chapter 5, *NFPA 70E* contains three chapters and fifteen annexes. The primary focus of this textbook is the safety-related work practices contained in Chapter 1 and the safety-related maintenance requirements in Chapter 2. However, it is important to also recognize that Chapter 3 illustrates numerous additional electrical safety issues. All of the requirements of *NFPA 70E* must be reviewed and implemented.

Organization

The organization of *NFPA 70E* Chapter 1 is outlined in Section 110.4, which points out that Chapter 1 is divided into four articles. The order of these articles is deliberate. When they reformatted and reorganized *NFPA 70E* for the 2004 edition, the NFPA 70E Technical Committee changed the order of those articles to reflect uses of the standard.

The first of these articles is Article 100, Definitions. The second, Article 110, covers general requirements for electrical safety-related work practices as the title of the article suggests. Article 120 discusses establishing an electrically safe work condition. The fourth and last article of Chapter 1 is Article 130. It covers work on circuits or equipment that is not in an electrically safe work condition, which involves electrical hazards. Article 130 is last of the four articles for a specific reason. Energized work is permitted only after an employer has demonstrated justification for not placing equipment in an electrically safe work condition. The *NFPA 70E* Technical Committee wanted users to recognize that energized work is a last resort, and thus put the article on work involving electrical hazards last.

OPENING CASE, CONT.

nurse arrived. The police and paramedics arrived soon after, and the victim was transported to the local hospital where he was pronounced dead at 10:36 a.m.

It is not known why the victim was working on the energized switch leg. The apprentice thought that the victim grabbed the wrong cable and failed to test both the black and white wires after removing the wire nuts. The foreman later stated that there was no reason for the circuits to be energized and that they always test every circuit beforehand. Although the contractor had access to the breaker box, they did not deenergize the circuits in that area. It was noted by Occupational Safety and Health Administration (OSHA) that the second bank of lights adjacent to the accident site was deenergized by taping off the wall switch.

The county medical examiner attributed the cause of death to electrocution. The medical examiner's report stated that there were third-degree burns on the first three fingers of the victim's left hand. No other electrical burns were found.

Adapted from: New Jersey Case Report 92NJ007. Accessed 7/5/08 at http://www.cdc.gov/niosh/face/stateface/nj/92nj007.html.

1. How can electrical workers ensure that all electrical circuits are deenergized and thoroughly tested before working on them?

2. What actions could be taken to ensure that employees do not work on energized circuits except when absolutely necessary and that when energized work must be performed, adequate safety measures are taken?

Figure 6.1 An electrical hazard is defined in *NFPA 70E* as "a dangerous condition such that contact or equipment failure can result in electric shock, arc flash burn, thermal burn, or blast" and is an example of a defined term that has a specific meaning in *NFPA 70E*.

Chapter 1 begins appropriately with Article 100, Definitions. A review of the scope of Article 100 and the definitions contained therein is essential. While many definitions contained in Article 100 of *NFPA 70E* are identical to those found in the *National Electrical Code* (*NEC®*), a thorough review and understanding of all of the *NFPA 70E* Article 100 definitions is essential to properly understand and apply the standard. This is especially important since a number of definitions are unique to *NFPA 70E* (**Figure 6.1**, **Figure 6.2**, and

Figure 6.2 "Arc rating" is an example of a term defined in *NFPA 70E* and is expressed in calories per square centimeter (cal/cm^2).

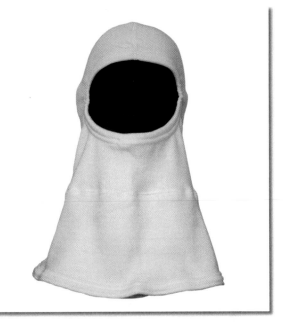

Figure 6.3 A balaclava, also known as a sock hood, is one form of arc-rated flame-resistant equipment that can be used to protect the neck and portions of the head.

Figure 6.3). Terms defined in Article 100 will be covered in subsequent chapters.

Article 110 follows Article 100 and provides an overview of the requirements located throughout the balance of Chapter 1. Article 110 includes the following topics:
- Scope
- Purpose
- Responsibilities
- Organization
- Relationships with contractors
- Training requirements
- Electrical safety program considerations
- Working while exposed to electrical hazards
- Use of equipment

A closer examination of each of these topics is included in this and subsequent chapters.

Scope

Section 90.2 of *NFPA 70E* defines the scope of the standard. Said another way, Section 90.2 describes where the requirements of the standard are intended to apply and where the requirements might not apply. It is important to read the scope to become familiar with what is covered. A careful review of *NFPA 70E* Sections 90.2 and 110.1 is essential to becoming familiar with what is and what is not covered by the provisions contained

Figure 6.4   *NFPA 70E* is not intended to apply to installations under the exclusive control of a utility.

in *NFPA 70E*. These sections must be understood for the requirements to be properly applied.

NFPA 70E Sections 90.2(A) and (B) indicate that the scope is similar to that of the *NEC*. Like the *NEC*, *NFPA 70E* is not intended to apply to installations that are under the exclusive control of a utility (**Figure 6.4**).

Unlike the *NEC*, *NFPA 70E* addresses electrical safety requirements to protect employees during activities including, but not limited to, the installation and maintenance of electric equipment. The following excerpts are extracted from Sections 90.2(A) and (B):

90.2 Scope.

(A) Covered. This standard addresses electrical safety requirements for employee workplaces that are necessary for the practical safeguarding of employees during activities such as the installation, operation, maintenance, and demolition of electric conductors, electric equipment, signaling and communications conductors and equipment, and raceways for . . .

(B) Not Covered. This standard does not cover . . .

(5) Installations under the exclusive control of an electric utility where such installations:

(a) Consist of service drops or service laterals, and associated metering, or

(b) Are located in legally established easements or rights-of-way designated by or recognized by public service commissions, utility commissions, or other regulatory agencies having jurisdiction for such installations, or

(c) Are on property owned or leased by the electric utility for the purpose of communications, metering, generation, control, transformation, transmission, or distribution of electric energy.

NFPA 70E Section 110.1 describes when the electrical safety-related practices and procedures in *NFPA 70E* Chapter 1 are intended to apply. The phrase "exposed to an electrical hazard" has a very specific meaning and contains the defined term electrical hazard. Therefore, Chapter 1 covers electrical safety-related practices and procedures for employees who are exposed to hazards associated with contacting an exposed energized conductor or equipment failure that might result in electric shock, arc flash burn, thermal burn, or blast injuries.

Purpose and Responsibility

Section 110.2 indicates that the purpose of the Chapter 1, Safety-Related Work Practices, is to protect workers from electrical hazards. Since an electrical hazard is a term defined in Article 100, *NFPA 70E*'s safety-related work practices and procedures are intended to protect employees from electric shock, arc flash burn, thermal burn, or blast caused by contact or equipment failure.

Section 110.3 identifies the responsibilities of employers and employees similar to those required by OSHA regulations. *NFPA 70E*, like OSHA, requires employees to put into practice the safety-related work practices that their employers have provided. Employers must train their employees to understand the work practices. Developing, providing, and implementing work practices and training is a considerable and worthwhile investment. Training and protective equipment are not effective if they are not used. Leaving a lock and a tag or flame-resistant personal protective equipment (PPE) in

70E Highlights

NFPA 70E Section 110.1

Scope. Chapter 1 covers electrical safety-related work practices and procedures for employees who are exposed to an electrical hazard in workplaces covered in the scope of this standard. Electric circuits and equipment not included in the scope of this standard might present a hazard to employees not qualified to work near such facilities. Requirements have been included in Chapter 1 to protect unqualified employees from such hazards.

the truck or job box and working without the protection that has been provided is like asking for an injury. Once an incident is underway, workers have no time to put on PPE. Any worker who is not wearing PPE when an incident begins might be injured, sometimes fatally.

Host and Contractor Responsibilities

Host and contractor responsibilities are defined in Section 110.5. The 2009 edition of *NFPA 70E* includes a significant change related to this topic. Section 110.4 in the 2004 edition of *NFPA 70E* was entitled "Multi-employer Relationship" and required a documented meeting between the on-site employer and any outside employers to cover such topics as hazards and work practices. Those requirements have been entirely replaced by Section 110.5 in the 2009 edition. Requirements that must be met by both the host and contract employer are outlined in 110.5(A) and (B), respectively.

Host Employer Responsibilities

The host employer is required to notify the contractor if any of the contractor's employees are witnessed violating the provisions of *NFPA 70E*. The host employer's responsibilities include, but are not limited to, making sure that contractors are made aware of hazards related to the work they are performing. The host employer must provide all information necessary to perform an electrical hazard/risk analysis. Information that a contractor needs to make the assessments required in Chapter 1 include, but are not limited to, the electrical hazard analysis required by Section 110.8, the shock hazard analysis required by Section 130.2, and the arc flash hazard assessment required by Section 130.3. These assessments require knowledge of circuit and system factors, including the voltage, clearing time, available fault current, and condition of maintenance of the overcurrent protective device (OCPD). Section 130.3 also requires that the arc flash hazard analysis take into consideration the condition of the OCPD and if it has been maintained adequately.

Contractor Responsibilities

Contractor responsibilities include, but are not limited to, making sure that their employees are trained to understand and follow the applicable provisions of both *NFPA 70E* and the procedures provided to them. The contractor must ensure that employees are trained to understand and react to all information reported by the host employer.

Training Requirements

Training is covered in Section 110.6. Requirements are broken down into five major categories:
- Safety training
- Type of training
- Emergency procedures
- Qualified and unqualified person training
- Documentation of training

These training requirements reflect a significant number of new and revised requirements as compared with the 2004 edition of *NFPA 70E*. The 2009 edition of the *NFPA 70E* requires annual recertification of CPR training, documentation of training, training about how to select the appropriate voltage detector, and provisions for retraining. Training requirements are explored in detail in Chapter 8, which specifically covers *NFPA 70E* training requirements.

Electrical Safety Program

Section 110.7 covers an electrical safety program. An electrical safety program is an essential part of providing a workplace free from recognized hazards. Section 110.7 begins with a general requirement for an employer to develop, document, and put into practice an electrical safety program. The program must provide work practices and procedures that protect workers from the voltage, energy level, and circuit conditions to which they might be exposed. The term energy level might include, but is not limited to, knowledge of the degree of electrical hazards to which a worker could be exposed. Since the term electrical hazard is defined, the degree of electrical hazard includes electric shock, arc flash, or blast.

Incident energy is the amount of thermal energy generated during an electric arc. Knowledge of the incident energy level is instrumental in providing workers with PPE for protection from a specific arc flash hazard. Incident energy is determined from variables, such as the available short-circuit current at the point of exposure and the clearing time of the OCPD that must clear to open the circuit. One of the many important variables for determining the clearing time of an OCPD is its condition (how well it has been maintained). This following sentence is extracted from Section 130.3 in the 2009 edition of *NFPA 70E*:

> The arc flash hazard analysis shall take into consideration the design of the overcurrent protective device and its opening time including its condition of maintenance.

The remaining requirements for what *NFPA 70E* requires to be part of an electrical safety program includes the following:

- Awareness and self-discipline
- Electrical safety program principles
- Controls and procedures
- A procedure to evaluate hazards and associated risk
- A job briefing
- Auditing

These requirements will be explored in Chapter 7, which specifically covers the *NFPA 70E* electrical safety program requirements.

Working While Exposed to Electrical Hazards

As the title suggests, Section 110.8 of *NFPA 70E* introduces concepts related to performing tasks while exposed to electrical hazards. The definition of an electrical hazard includes electric shock, arc flash burn, thermal burn, or blast caused by contact or equipment failure. Therefore, Section 110.8 requires workers to be protected from the hazards of shock, arc flash burn, thermal burn, and blast.

Chapter 1 discusses safety-related work practices, and Chapter 1, Article 110, defines general requirements for electrical safety-related work practices. Accordingly, the requirements of Article 110 are applicable throughout Chapter 1. Much of the information introduced in Section 110.8 is covered in greater detail elsewhere in Chapter 1. Likewise, much of the information introduced in this section of this textbook will be covered in greater detail. Section 110.8 is broken down into two major categories: The general requirements introduced in Section 110.8(A) and requirements related to working inside the limited approach boundary in Section 110.8(B).

General Requirements

The requirements of Section 110.8(A) recognize that the protective nature of work practices must provide protection that is equal to or greater than the degree of the shock, arc flash burn, thermal burn, or blast hazard that might be encountered by workers. The general requirements of Section 110.8(A) are further broken down into two categories. The first requirement is that electrical **equipment** must be put into an **electrically safe work condition** before a worker is permitted to cross the **limited approach boundary**, unless the employer demonstrates that it is infeasible or a greater hazard to deenergize.

CAUTION

It is important to recognize that a worker might have crossed the AFPB and, therefore, requires arc flash protection before the limited approach boundary is reached.

However, even if adequate justification exists for the equipment to remain energized while a work task is performed, the employer must provide appropriate personal and other protective equipment, and the employee must use it when the work task is performed. Only a **qualified person** who has received the training defined in Section 110.6 is permitted to work on electrical equipment that does not comply with the applicable provisions of Article 120.

Requirements for Working Inside the Limited Approach Boundary

Section 110.8(B) reiterates that establishing an electrically safe work condition is the preferred work practice when the circuit voltage is 50 volts or more. However, if an electrically safe work condition cannot be established, the remaining requirements of Section 110.8(B) apply for work performed inside the limited approach boundary.

Section 110.8(B) is further broken down into four categories. These include, but are not limited to, performing an electrical hazard analysis, implementing an energized electrical work permit, prohibiting **unqualified persons** from entering locations they are not permitted and trained to enter, and allowing only qualified persons under the specific guidelines outlined to bypass an electrical safety interlock.

An electrical hazard analysis consists of both a shock hazard analysis and an arc flash hazard analysis. A shock hazard analysis is performed to determine the voltage to which a worker might be exposed; establish the limited approach boundary, **restricted approach boundary**, and **prohibited approach boundary**; and the PPE necessary for protection from a **shock hazard**. This information is covered again in Section 130.2 of *NFPA 70E* and in the related chapter in this textbook. This information is necessary for the job briefing required in Section 110.7 and in the energized electrical work permit required in Section 130.1(B).

An **arc flash hazard analysis** is performed to determine the **arc flash protection boundary (AFPB)** and the PPE necessary to protect against an **arc flash**

hazard if the AFPB will be crossed. This information is covered again in Section 130.3 of *NFPA 70E* and in the related chapter in this textbook. This information is necessary for the job briefing required by Section 110.7, in the energized electrical work permit required by Section 130.1(B), and on a field-installed label on equipment where required by Sections 130.3 and 130.3(C).

Tasks performed while **working on** energized equipment that is not in an electrically safe work condition is considered **energized** work and requires an energized electrical work permit, as explained in Section 130.1(B). The information on the energized electrical work permit must be discussed in a job briefing as required in Section 110.7. The information listed on the energized electrical work permit includes, but is not limited to, the following:

- Voltage
- Shock and arc flash protection boundaries
- Safe work practices to be used
- Personal and other protective equipment for shock and arc flash
- Signature of the person authorizing the energized work

Use of Equipment

Understanding *NFPA 70E* requirements related to the use of equipment is important since these tasks are performed with some frequency. Section 110.9 addresses four major topics:

- Test instruments and equipment
- Portable electric equipment
- Ground-fault circuit interrupter (GFCI) protection devices
- Overcurrent protection modification

Test instruments, equipment, and their accessories must be rated for circuits and equipment to which they will be connected. These devices must be selected for use in the environment for which they were designed and used in the manner intended by the manufacturer. The devices must be visually inspected for external defects and damage before the equipment is used and removed from service if any defect is found. When testing for absence of voltage, proper operation of the test equipment must be verified by using it on a known source both before and after using it [Section 110.9(A)].

The meaning of the requirement in Section 110.1(A) for test instruments, equipment, and their accessories to be rated for circuits and equipment to which they will be connected might not be obvious. A

note that refers to an industry consensus standard was added to Section 110.9(A)(1) by the *NFPA 70E* Technical Committee, as follows:

> FPN: See ANSI/ISA-61010-1 (82.02.01)/UL 61010-1, *Safety Requirements for Electrical Equipment for Measurement, Control, and Laboratory Use—Part 1: General Requirements*, for rating and design requirements for voltage measurement and test instruments intended for use on electrical systems 1000 volts and below.

The need for (and intent of) this Fine Print Note (FPN) is found in the substantiation for proposal 70E-115, which is published in Volume 1 of the National Fire Protection Association's (NFPA's) *2008 Annual Revision Cycle Report on Proposals*. H. Landis Floyd, the submitter of the proposal, indicates that "the addition of information on how to determine the design and rating requirements for voltage test instruments will help assure that people who are involved in some of the most common and routine activities that place them in proximity of energized components will have instruments designed for the application." Mr. Floyd goes on to say that "…this standard requires voltage test instruments [to] have equivalent surge transient protection (described as Category II, III, and IV) as the installed equipment to which they will be applied. This will serve to reduce arc flash injuries due to test instrument failure from surge transients that occur at the time the test instruments are in contact with the installed equipment under test" (Figure 6.5 and Figure 6.6).

Portable electric equipment, such as portable band saws, drill motors, reciprocating saws, hammer drill, extension cords, and the like, are among the most common equipment used in the workplace. Not surprisingly, *NFPA 70E* contains requirements addressing the use of portable electric equipment. Topics that *NFPA 70E* addresses in Section 110.9(B) include the following:

- The handling of portable electric equipment and prohibited uses for flexible cords
- Considerations for flexible cord, attachment plug, and receptacle use in conjunction with grounding-type equipment
- The visual inspection of portable cord-and-plug connected equipment and flexible cord sets
- Considerations related to the connection of attachment plugs

Recall from Chapter 3 that OSHA requires ground-fault protection for temporary wiring (including extension cords) and that ground-fault protection for 125-volt, single-phase, 15- and 20-ampere receptacle outlets generally be provided by either GFCIs or by an assured equipment grounding conductor program. If a GFCI is not used for ground-fault protection, an assured equipment grounding conductor program is required on construction sites for the following:

- All cord sets
- Receptacles that are not a part of the building or structure
- Equipment connected by cord and plug

The GFCI is the option often used for 120-volt, single-phase 15- and 20-ampere receptacle outlets. *NFPA 70E* 110.9(C) reminds us that GFCI devices are required to be tested in accordance with the manufacturer's instructions. These instruction require both

the circuit breaker and receptacle type GFCIs to be tested monthly.

Article 110 concludes by discussing modification of overcurrent protection. This section never permits overcurrent protection to exceed the requirements in Article 240 of the *NEC* or *Code*, regardless of the reason, even temporarily. This requirement prohibits, for example, replacing fuses or changing adjustable settings on circuit breakers beyond the maximum permitted by *Code* for the circuit being protected. When replacing an overcurrent device (fuse or circuit breaker), the ampacity and interrupting rating of the replacement device must not exceed the capacity of the circuit.

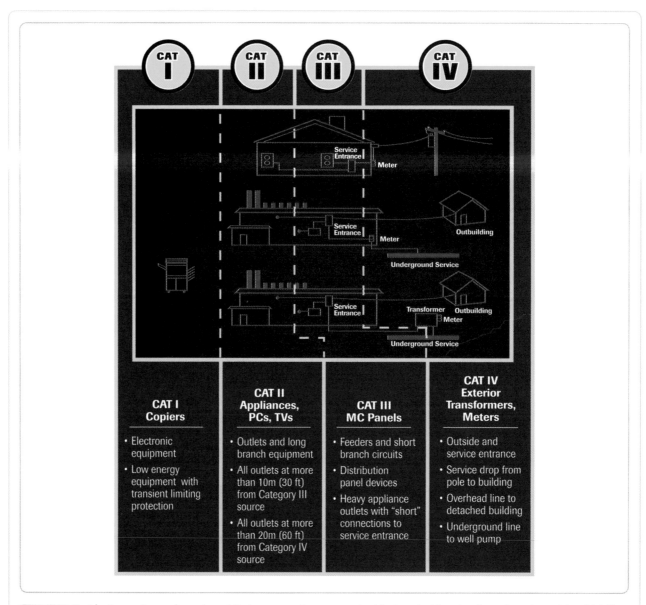

Figure 6.5 Test instruments, equipment, and their accessories are required to be rated for circuits and equipment to which they will be connected.

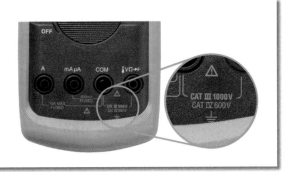

Figure 6.6 Test instruments must have surge transient protection equivalent to the installed equipment to which they will be applied.

Summary

Understanding the scope defined in Sections 90.2 and 110.1, the definitions from Article 100, and the general requirements of Article 110, is essential to comply with the requirements of *NFPA 70E*. An electrical safety program and effective training provide a foundation for workplace free from recognized hazards. Achieving an electrically safe work condition is the primary safety-related work practice and must be used unless the employer demonstrates infeasibility or a greater hazard. Additional safety-related work practices must be determined by a hazard/risk analysis, which includes both a shock hazard analysis and an arc flash hazard analysis before work is performed. Energized work includes tasks such as verifying absence of voltage even to place equipment in an electrically safe work condition. Only qualified persons are permitted to work on electrical conductors or circuit parts that have not been put into an electrically safe work condition and an Energized Electrical Work Permit generally is required if live parts are not placed in an electrically safe work condition.

Much of Article 110 is an introduction to subjects that are discussed later in *NFPA 70E*. Keep in mind that the primary focus of this textbook is the safety-related work practice requirements contained in Chapter 1 of *NFPA 70E*, but Chapters 2 and 3 also contain numerous additional electrical safety requirements. All OSHA requirements and the entire *NFPA 70E* standard are important. An effective electrical safety program adopts and implements appropriate requirements from both sources of information.

Worker Electrocuted After Contacting Energized Conductor While Working From the Bucket of an Aerial Lift Truck

A 46-year-old electrician was electrocuted when he contacted an energized power line while working from the bucket of an aerial lift truck. The employer had been contracted to replace 12 fused electrical cutout switches at a housing project. The switches were located on the crossarms of the utility poles, between the transformers and the 7620-volt energized power lines. Workers had installed five switches the day before and had been at the site 1 hour on the day when the incident occurred.

On the morning of the incident, two workers (the victim and his coworker), arrived at the work site to continue replacing switches. It was determined that the victim would work from the bucket of the aerial lift truck while he replaced the switches, and the coworker would remain on the ground performing other tasks. The victim positioned himself in the bucket and maneuvered the bucket adjacent to the power line in proximity to the switch. He used a "hot stick" to disconnect the jumper wire from the switch to the power line. Without donning PPE or covering the power lines with insulating line sleeves or blankets, he removed one of the bolts, securing the switch to the crossarm, as the coworker watched from the ground.

At that time, the company owner arrived and began a conversation with the coworker. As the victim either began to remove another bolt or tried to reposition the bucket, he contacted the power line. While the owner and coworker were talking, they heard an arcing and popping sound and looked up to see the victim slump down into the bucket. The company owner immediately jumped onto the truck and used the controls mounted on the pedestal to lower the bucket, while the coworker called 911 for help. The paramedics arrived within 6 minutes and determined the victim had died. The medical examiner was notified, and he directed the paramedics to transport the body to a local mortuary.

The employer in this incident was an electrical contracting company that had been in operation for 17 years and employed 10 workers, all of whom were electricians. The company had no written safety or training programs. The victim had worked for the company for 17 years, and this incident was the first fatality the company had experienced. The medical examiner's report listed the cause of death as electrocution.

Recommendations/Discussion

Recommendation 1: Employers should require that workers wear appropriate PPE when they are exposed to hazardous conditions and ensure that energized power lines are insulated or guarded before work is performed on or near them.

Discussion: 29 Code of Federal Regulations (CFR) 1926.28 (a) states "The employer is responsible for requiring the wearing of appropriate personal protective equipment in all operations where there is an exposure to hazardous conditions or where this part indicates the need for using such equipment to reduce the hazards to the employees." In addition, 29 CFR 1926.950 (c)(1)(ii) states "The energized part is insulated or guarded from him and any other conductive object at a different potential." Work was being performed from an aerial lift truck bucket near energized, unguarded, or uninsulated power lines, and appropriate PPE was not being worn. The rubber gloves, sleeves, and line hoses used for this type of work were later found in the aerial bucket and bins of the truck.

Recommendation 2: Employers should develop and implement safety programs designed to enable workers to recognize and avoid hazards, especially electrical hazards.

Discussion: The danger of overhead power lines appears to be obvious; however, contact with overhead power lines and the subsequent occupationally related fatalities continue. OSHA 29 CFR 1926.21(b)(2) states that "The employer shall instruct each employee in the recognition and avoidance of unsafe conditions and the regulations applicable to his work environment to control or eliminate any hazards or other exposure to illness or injury." Employers should develop and implement comprehensive safety programs implemented with particular emphasis on detailed safety procedures specific for all tasks and job categories. Employers also should provide employees with adequate training to ensure that they can recognize potential hazardous exposures and are familiar with the company's safety program and procedures. Evidence suggests that the worker, although an experienced electrician, elected to work near energized overhead power lines without using PPE and without insulating or guarding the power lines.

Adapted from: Fatal Accident Circumstances and Epidemiology (FACE) 92-25. Accessed 7/2/08 at http://www.cdc.gov/niosh/face/In-house/full9225.html.

Questions

1. The victim was working from an aerial bucket truck. Which of the following statements describe a possible cause of the fatality?
   A. The bucket was not insulated from the truck.
   B. The worker contacted an energized conductor and the pole ground.
   C. The helper was not watching the worker in the bucket.
   D. The victim was not guided by a procedure.

2. Which of the following might have prevented the fatality?
   A. Using an adequately rated voltmeter to identify energized conductors
   B. Installing line hose on the overhead conductors
   C. Following the procedure provided by the employer
   D. Wearing adequate voltage-rated gloves

3. Which of the following practices might have prevented the incident?
   A. Continuous improvement of the employer's electrical safety program
   B. Performing a hazard/risk analysis

C. Removing the pole ground before ascending to the elevation of the cutout

D. Installing a ground on the frame of the truck

4. The electrical contractor had no safety program. Which of the following statements about a safety program is accurate?

   A. *NPFA 70E* does not require employers to define and implement an electrical safety program.

   B. 29 CFR 1926 does not require employers to implement an electrical safety program intended to help provide a workplace free from electrical hazards.

   C. 29 CFR 1910 does not require employers to implement a safety program intended to help provide a workplace free from electrical hazards.

   D. Both *NFPA 70E* and OSHA rules require employers to define and implement an electrical safety program.

5. Who is responsible for ensuring that workers have received the necessary training to understand electrical hazards?

   A. The worker's direct employer (the contractor)

   B. The host employer

   C. The local union hall

   D. The worker

Ready for Review

- OSHA requires that workers be provided with a workplace free from recognized hazards. *NFPA 70E* is considered a means to help comply with the OSHA requirements related to the hazards of working on or near exposed energized parts.
- *NFPA 70E* Article 110 covers host and contract employer responsibilities (Section 110.5), training requirements (Section 110.6), electrical safety program considerations (Section 110.7), and use of equipment (Section 110.9), as well as other topics that are additionally covered in Articles 120 and 130.
- OSHA requirements are often written in performance language (require that something be done without necessarily explaining how). *NFPA 70E* helps clarify how.
- *NFPA 70E* Article 100 contains definitions. Note that while many of these definitions are identical to those found in the *NEC*, there are a number of definitions in the *NFPA 70E* that are not in the *NEC*. Therefore, it is important to become familiar with *NFPA 70E* definitions as well as those in the *NEC*.
- *NFPA 70E* Article 110 covers general requirements for electrical safety-related work practices. It covers scope, purpose, responsibilities, organization, relationships with contractors, training requirements, electrical safety program considerations, working while exposed to electrical hazards, and use of equipment
- *NFPA 70E* Article 120 covers the considerations for establishing an electrically safe work condition.
- *NFPA 70E* Article 130 covers work involving electrical hazards. This article is placed last in *NFPA 70E* Chapter 1 to reinforce that energized work is a last resort. It is permitted only after an employer has demonstrated justification for not placing equipment in an electrically safe work condition.
- Becoming familiar with Sections 90.2 and 110.1 helps a worker understand what is and is not covered by the provisions contained in *NFPA 70E*.

- Like the *NEC*, *NFPA 70E* is not intended to apply to installations under the exclusive control of a utility.
- *NFPA 70E* addresses electrical safety requirements to protect employees during activities including, but not limited to, the installation, operation, and maintenance of electric equipment.
- "Exposed to an electrical hazard" has a very specific meaning in *NFPA 70E*; it covers electrical safety-related practices and procedures for employees who are exposed to contact or equipment failure resulting in electric shock, arc flash burn, thermal burn, or blast.
- Employers and employees have a dual responsibility. Employees must put into practice the safety-related work practices that they have been provided with and trained in. Training and protective equipment are not effective if they are not used.
- Host employer responsibilities include notifying the contractor if any of the contractor's employees are witnessed violating the provisions of *NFPA 70E*. In addition, the host employer responsibilities include, but are not limited to, making sure that they make contractors aware of hazards related to the work they are performing and that they are provided with the information necessary to perform an electrical hazard assessment.
- Contractor responsibilities include, but are not limited to, making sure that their employees are trained in and follow the applicable provisions of both *NFPA 70E* and those of the host employer. In addition, contractors must ensure that their employees are aware of and trained to recognize and avoid the hazards reported by the host employer.
- Training requirements in *NFPA 70E* are broken down into five major categories: Safety training; type of training; emergency procedures; qualified and unqualified person training; and documentation of training.
- Requirements related to an electrical safety program in *NFPA 70E* include eight major categories. They discuss the general requirement for

the employer to develop, document, and put into practice an electrical safety program that provides work practices and procedures that can protect workers from the voltage, energy level, and circuit conditions to which they are exposed.

- *NFPA 70E* has requirements relating to working while exposed to electrical hazards. It requires workers to be protected from the hazards of shock, arc flash burn, thermal burn, and blast. These requirements are broken down into two major categories—general requirements and requirements related to working inside the limited approach boundary.

- Work practices must be at least equal to or greater than the degree of the shock, arc flash burn, thermal burn, or blast hazard that could be encountered by workers when they are exposed to electrical equipment that is or can become energized.

- Electrical equipment must be put into an electrically safe work condition before a worker is permitted to cross the limited approach boundary unless the employer demonstrates that it is infeasible or a greater hazard to deenergize.

- If energized work is justified, the employee must be provided with and use the appropriate personal and other protective equipment necessary to perform the task.

- Only a qualified person meeting the applicable requirements of *NFPA 70E* Section 110.6 is permitted to work on electrical equipment that does not comply with the applicable provisions of Article 120 (establishing an electrically safe work condition).

- Requirements for working inside the limited approach boundary require that the lockout/tagout provisions of Article 120 be met before a worker crosses the limited approach boundary to work on energized equipment operating at 50 volts or more.

- If it is infeasible or a greater hazard to deenergize and the lockout/tagout provisions are not implemented, then other safety-related work practices must be put into practice.

- An electrical hazard analysis consists of both a shock and an arc flash hazard analysis. An arc flash hazard analysis is performed to determine the AFPB, and the PPE necessary to protect against a flash hazard if the AFPB will be crossed.

- Tasks performed while working on energized equipment that is not in an electrically safe work condition are considered energized work and would require compliance with the energized electrical work permit requirements of Section 130.1(B).

- *NFPA 70E* also has requirements related to the use of equipment, covered in four categories: (1) test instruments and equipment; (2) portable electric equipment; (3) GFCI protection devices; and (4) overcurrent protection modification.

- Test instruments, equipment, and their accessories are required to be rated for circuits and equipment to which they will be connected, be designed for the environment to which they will be exposed and for the manner in which they will be used, and be visually inspected for external defects and damage before the equipment is used and removed from service if warranted.

- The proper operation of the test equipment must be verified before and after a test whenever testing for the absence of voltage is performed on circuits operating at 50 volts or more.

- Topics relating to portable electric equipment include the handling of portable electric equipment and prohibited uses for flexible cords, considerations for flexible cord, attachment plug, and receptacle use in conjunction with grounding-type equipment, the visual inspection of portable cord-and-plug connected equipment and flexible cord sets, and considerations related to the connection of attachment plugs.

Vocabulary

arc flash hazard* A dangerous condition associated with the possible release of energy caused by an electric arc.

arc flash hazard analysis* A study investigating a worker's potential exposure to arc-flash energy, conducted for the purpose of injury prevention and the determination of safe work practices, arc flash protection boundary, and the appropriate levels of PPE.

arc flash protection boundary (AFPB)* When an arc flash hazard exists, an approach limit at a distance from a prospective arc source within which a person could receive a second degree burn if an electrical arc flash were to occur.

electrically safe work condition* A state in which an electrical conductor or circuit part has been disconnected from energized parts, locked/tagged in accordance with established standards, tested to ensure the absence of voltage, and grounded if determined necessary.

energized* Electrically connected to, or is, a source of voltage.

equipment* A general term, including material, fittings, devices, appliances, luminaires, apparatus, machinery, and the like used as a part of, or in connection with, an electrical installation.

limited approach boundary* An approach limit at a distance from an exposed energized electrical conductor or circuit part within which a shock hazard exists.

prohibited approach boundary* An approach limit at a distance from an exposed energized electrical conductor or circuit part within which work is considered the same as making contact with the electrical conductor or circuit part.

qualified person* One who has skills and knowledge related to the construction and operation of the electrical equipment and installations and has received safety training to recognize and avoid the hazards involved.

restricted approach boundary* An approach limit at a distance from an exposed energized electrical conductor or circuit part within which there is an increased risk of shock, due to electrical arc over combined with inadvertent movement, for personnel working in close proximity to the energized electrical conductor or circuit part.

shock hazard* A dangerous condition associated with the possible release of energy caused by contact or approach to energized electrical conductors or circuit parts.

unqualified person* A person who is not a qualified person.

working on (energized electrical conductors or circuit parts)* Coming in contact with energized electrical conductors or circuit parts with the hands, feet, or other body parts, with tools, probes, or with test equipment, regardless of the personal protective equipment a person is wearing. There are two categories of "working on":

Diagnostic (testing) is taking readings or measurements of electrical equipment with approved test equipment that does not require making any physical change to the equipment, *repair* is any physical alteration of electrical equipment (such as making or tightening connections, removing or replacing components, etc.).

*This is the *NFPA 70E* definition.

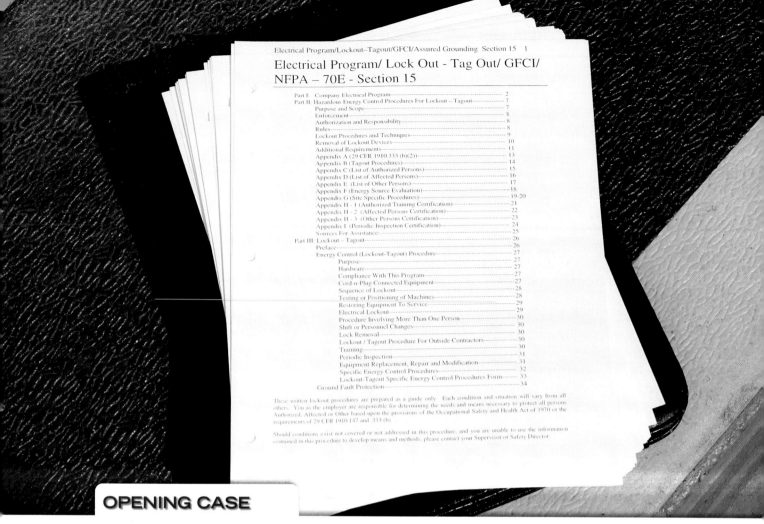

Electrical Program/Lockout–Tagout/GFCI/Assured Grounding Section 15   1

Electrical Program/ Lock Out - Tag Out/ GFCI/
NFPA – 70E - Section 15

Part I: Company Electrical Program ---- 2
Part II: Hazardous Energy Control Procedures For Lockout – Tagout ---- 7
    Purpose and Scope ---- 7
    Enforcement ---- 8
    Authorization and Responsibility ---- 8
    Rules ---- 8
    Lockout Procedures and Techniques ---- 9
    Removal of Lockout Devices ---- 10
    Additional Requirements ---- 11
    Appendix A (29 CFR 1910.333 (b)(2)) ---- 13
    Appendix B (Tagout Procedures) ---- 14
    Appendix C (List of Authorized Persons) ---- 15
    Appendix D (List of Affected Persons) ---- 16
    Appendix E (List of Other Persons) ---- 17
    Appendix F (Energy Source Evaluation) ---- 18
    Appendix G (Site Specific Procedures) ---- 19-20
    Appendix H - 1 (Authorized Training Certification) ---- 21
    Appendix H - 2 (Affected Persons Certification) ---- 22
    Appendix H - 3 (Other Persons Certification) ---- 23
    Appendix I (Periodic Inspection Certification) ---- 24
    Sources For Assistance ---- 25
Part III: Lockout – Tagout ---- 26
    Preface ---- 26
    Energy Control (Lockout-Tagout) Procedure ---- 27
        Purpose ---- 27
        Hardware ---- 27
        Compliance With This Program ---- 27
        Cord-n-Plug Connected Equipment ---- 27
        Sequence of Lockout ---- 28
        Testing or Positioning of Machines ---- 28
        Restoring Equipment To Service ---- 29
        Electrical Lockout ---- 29
        Procedure Involving More Than One Person ---- 30
        Shift or Personnel Changes ---- 30
        Lock Removal ---- 30
        Lockout / Tagout Procedure For Outside Contractors ---- 30
        Training ---- 30
        Periodic Inspection ---- 31
        Equipment Replacement, Repair and Modification ---- 31
        Specific Energy Control Procedures ---- 32
        Lockout-Tagout Specific Energy Control Procedures Form ---- 33
    Ground Fault Protection ---- 34

These written lockout procedures are prepared as a guide only. Each condition and situation will vary from all others. You as the employer are responsible for determining the needs and means necessary to protect all persons Authorized, Affected or Other based upon the provisions of the Occupational Safety and Health Act of 1970 or the requirements of 29 CFR 1910.147 and .333 (b)

Should conditions exist not covered or not addressed in this procedure, and you are unable to use the information contained in this procedure to develop means and methods, please contact your Supervisor or Safety Director.

OPENING CASE

Photo courtesy of National Electrical Contractors Association (NECA).

A 46-year-old male electrical project supervisor died when he contacted an energized conductor inside a control panel. The employer was an industrial electrical contracting company that had been in operation for 10 years. The company employed 20 workers, including three electrical project supervisors. The company's written safety program, administered by the president/CEO and the project supervisors, included disciplinary procedures specifying that three reprimands would result in termination. The safety program mandated preemployment drug screening and on-the-job random drug testing. The president/CEO served as safety officer on a collateral duty basis, and the supervisors held monthly safety meetings with all crew members.

The victim had worked for this employer for 5 years and 3 months as a project supervisor and had approximately 27 years of electrical experience. The company and victim had been working at the packaging plant for 6 months before the incident; this was the company's first fatality.

The company had been contracted to install two control cabinets, conduit, wiring, and two solid-state compressor motor starters for two 400-horsepower air compressors. On the day of the incident, the victim and three coworkers (one electrician and two helpers) arrived at the plant at 7 a.m. They were scheduled to install the last starter and to complete the wiring from the compressor motor to the starter in the control panel, and from the starter control panel to the main distribution panel. Once installation was completed, they were to check the operation of the unit.

At approximately 3:15 p.m., the starter had been installed, and all associated wiring had been completed. The victim directed a helper to turn the switch to the "on" position at the main distribution panel, approximately 6 ft away, to check the starter's operation. The helper turned the switch to the "on" position, energizing the components inside the starter control panel. The victim pushed the starter "start" button, and the starter indicator light activated, but the compressor motor did not start. When the

Electrical Safety Program

CHAPTER 7

Introduction

Both the Occupational Safety and Health Administration (OSHA) and *NFPA 70E*® require the employer to implement an electrical safety program. A **safety program** is an organized effort to reduce injuries. An **electrical safety program** should be a subset of an overall safety program, is an integral part of an overall safety and health program, and is a vital part of establishing an electrically safe workplace. However, it must be understood that an effective electrical safety program cannot be duplicated from any source and implemented directly. While satisfying regulatory requirements might be possible by duplicating a program from an external source, each element of the program must be discussed and analyzed on the basis of the physical environment and experience level of both supervisors and workers for an electrical safety program to be effective. This chapter examines the electrical safety program considerations and requirements set forth in *NFPA 70E*.

Chapter Outline

- **Introduction**
- **Establishing an Effective Electrical Safety Program**
- **NFPA 70E and the Electrical Safety Program**
- **Groundbreaking Development**
- **Summary**
- **Lessons Learned**
- **Current Knowledge**

Learning Objectives

1. Understand the importance of, and challenges implementing, an electrical safety program.
2. Identify the requirements that *NFPA 70E* requires be included in an electrical safety program.
3. Recognize how *NFPA 70E* Annex information may apply in an electrical safety program.

References

1. Appendix B: Partnership Agreement
2. *NFPA 70E*, 2009 Edition

OPENING CASE, CONT.

compressor motor did not engage, the victim concluded that a problem existed inside the starter control panel. The victim directed the electrician to retrieve a voltmeter so that he could check the continuity of the wiring inside the starter control panel. In the interim, the victim opened the starter control panel door without deenergizing the unit and reached inside to trace the wiring and check the integrity of the electrical leads. In doing so, he contacted the 480-volt primary lead for the motor starter with his left hand.

Current passed through the victim's left hand and body and exited through his feet to the ground. The victim yelled, and the helper immediately turned the main distribution switch to the "off" position as the victim collapsed to the floor. A plant maintenance supervisor walking by the area saw the event and called the emergency medical service (EMS). The helper checked the victim and immediately administered cardiopulmonary resuscitation (CPR). The EMS arrived in 10 to 15 minutes, continued CPR, and transported the victim to the local hospital, where he was pronounced dead 1 hour and 20 minutes after the incident occurred. The county coroner reported the cause of death as cardiac arrest due to electrical shock.

Adapted from: Fatality and Assessment and Control Evaluation (FACE) 92-20. Accessed 7/10/08 at http://www.cdc.gov/niosh/face/in-house/full9220.html.

1. What could have been done differently to prevent this fatality?
2. What can be done to ensure that workers follow safety rules?

Establishing an Effective Electrical Safety Program

The text entitled *The Electrical Safety Program Book* by Jones, Jones, and Mastrullo illustrates some excellent points about establishing an effective electrical safety program. The prose and figures in this section are reprinted with permission from that text.

The electrical safety program will always be a work in progress. Many variables and outside influences will interfere with getting it started or with its forward progress. Most highly successful companies and programs have a history of numerous setbacks and, in some cases, situations where the entire company or project appeared to be destined for disaster. Overcoming negative issues with a positive approach and a dedicated focus on the program goals can achieve success. A critical component of planning a program is to justify the need for one in the minds of company management.

For most organizations, the initial program set up is difficult. Thinking about the entire program is somewhat overwhelming. However, any single piece of the program helps to get it started. Components of an electrical safety program may include the following (Figure 7.1):

- Policies and procedures
- Site assessment
- Task assessment
- Personal protective equipment (PPE) requirements
- Hazardous boundaries and hazard/risk analysis
- Administration
- Lockout/tagout
- Training
- Auditing and recordkeeping
- Budgeting

It is possible for a single segment to exist, but all of these segments must be present and adequately addressed for an electrical safety program to be complete. This and other chapters address a number of these considerations in the manner in which they are addressed in *NFPA 70E*.

Electrical Safety Program Objectives

H. W. Heinrich was a psychologist who changed the way the world considered safety fundamentals in the 1930s. He developed a theory that states for every 300 recordable injuries, approximately 30 lost-time injuries and one fatality will occur. Over the years, these relationships have proven to be relatively accurate. Some people feel that if the energy source is electrical, the numbers would be more accurate if a zero were removed from each—in other words, 3 lost-time injuries and one fatality per every 30 recordable injuries. This relationship might be used to help justify funding for an electrical safety program.

The safety triangle demonstrated by Heinrich's relationships in Figure 7.2 shows the relationship between behaviors and incidents. Developing an understanding of how these segments of the triangle relate to one another brings clarity to the strategies required for a comprehensive program. The triangle shows that, by not having a comprehensive safety program with a strong safety culture, the company will feel the negative effects of these behaviors in a matter of time.

Heinrich's relationship suggests that a program that is focused on eliminating fatalities has a chance to be successful. However, if the program is focused on elimi-

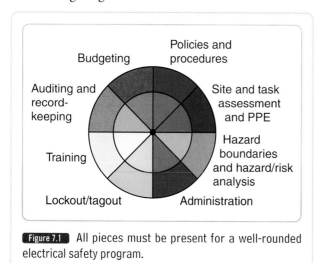

Figure 7.1   All pieces must be present for a well-rounded electrical safety program.

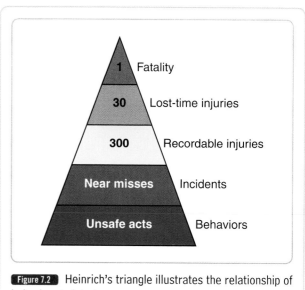

Figure 7.2   Heinrich's triangle illustrates the relationship of behaviors to injuries.

nating lost-time injuries, fatalities also will be eliminated. Likewise, if recordable injuries are eliminated, both lost-time injuries and fatalities will be eliminated. Efforts that are focused lower in the triangle stand a greater chance of success. Note that underpinning Heinrich's triangle are both near misses and unsafe acts. An effective electrical safety program will tend to place emphasis on unsafe acts.

The first priority of the planning process should be to generate an outline that defines the electrical safety goals for the organization. This outline should be consistent and align with the company philosophy on safety to enhance wide acceptance of the program. Electrical safety has obstacles that do not exist in other disciplines. These obstacles include hazards that are not readily visible, along with the stereotypes, misinformation, and lack of knowledge about electrical hazards that exist in the workplace. Many organizations rely on the electrical workers' training and experience as a substitute for safe work practices and proper protective equipment. Typically, a company would not let its employees handle harmful chemicals without proper procedures and protective equipment, but the same company might consistently expect workers to be exposed to energized electrical circuits without safe work practices and protective equipment.

Many organizations believe that all incidents and injuries can be prevented and have the goal of "zero incidents in the workplace." This is a realistic goal that can drive the processes of accident/incident reporting and accident investigation and analysis. Why should an incident in the workplace be an acceptable occurrence as part of a job task? Developing an electrical safety program with a proactive vision can result in lower operating costs and strong employee support. In addition to federal regulations, the company should incorporate into its program best work practices and techniques, both internally and within peer groups and organizations. Improving work practices is a critical process to ensure that the program is effective and up to date with new techniques and procedures.

The goals must be attainable and embraced by senior management for the program to have credibility. In addition, upper management must take an active support role for the program to be successful. The program plan must be attainable within the framework of the organization, and the mission statement should set the tone for the scope of the program and its components. The scope of the program is directly related to the budget and should spell out what steps, procedures, and policies will be required to attain the goal of zero incidents. The philosophy of management toward the electrical safety program should be spelled out in a procedure.

Program Cost Considerations

When planning the electrical safety program, two types of cost benefits must be considered: Direct and indirect. Direct benefits are quantifiable, such as lower insurance cost through safe operations. Measurable benefits are essential to justify expenditure. The indirect benefits can be substantial for a company. Companies want to be perceived as a great place to work. A safe workplace has a far-reaching effect for a company. The concept of a good place to work helps with community relations, hiring and retaining a talented workforce, and municipality issues. Many companies are proactive in promoting good health for employees. Equating safety with good health under one umbrella delivers the message that the company cares about its employees. A well-trained employee performs his or her work with a higher level of confidence and increased productivity. The electrical safety program affects all of the employees, not just the people working on the systems.

NFPA 70E and the Electrical Safety Program

Many who open the pages of *NFPA 70E* expect to find a complete and ready to go electrical safety program laid out for them. What *NFPA 70E* provides is not a complete electrical safety program at all. However, it is not very likely that anyone who performs electrical work is employed by an organization that does not have an electrical safety program in place, even if it is not complete and up to date or only on paper. A safety program cannot be effective unless it is complete and implemented in the workplace as illustrated by the opening case study for this chapter. As mentioned earlier, an electrical safety program will always be a work in progress. What *NFPA 70E* offers is a number of minimum considerations that must be evaluated and integrated into the framework of an existing electrical safety program to make it more comprehensive and complete. Specific requirements are provided as to what must be included in an electrical safety program built around *NFPA 70E* and are located primarily in Section 110.7 and *NFPA 70E* Annex E.

OSHA Tip

OSHA defines *what* is required to prevent injuries, and *NFPA 70E* describes examples of *how* to prevent injuries.

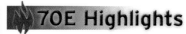

70E Highlights

NFPA 70E Section 110.7 requires an Electrical Safety Program and describes the necessary content.

General Program Requirements

The requirements in Section 110.7 are broken down into eight major categories. Addressed first is the general requirement for the employer to develop, document, and put into practice an electrical safety program that provides work practices and procedures that will protect workers for the voltage, energy level, and circuit conditions to which they will be exposed. Energy level considerations could include, but are not limited to, knowledge of the degree of electrical hazards that workers could be exposed to. Recall that an electrical hazard is a defined term. Therefore, the degree of an electrical hazard includes electric shock, arc flash, or blast. Since incident energy is the amount of energy released during an electric arc, knowledge of the incident energy level would be instrumental in providing workers with PPE for an arc flash hazard. Incident energy would be determined from variables such as the available short-circuit (fault) current at the point of exposure and the clearing time of the overcurrent protective device (OCPD) that would open the circuit supplying the energized electrical conductors or circuit parts where employees are present. One of the many important variables in determining the clearing time of an OCPD is its condition of maintenance. Circuit breakers must be operated and maintained at intervals determined by the manufacturer or consensus standards. This is pointed out in the following sentence from the Section 130.3 arc flash hazard analysis requirements:

> *NFPA 70E*, Section 130.3, in part
>
> The arc flash hazard analysis shall take into consideration the design of the overcurrent protective device and its opening time, including its condition of maintenance.

There are also two Fine Print Notes (FPNs) to Section 110.7(A). The first is a reminder that safety-related work practices are only a piece of a complete electrical safety program. While safety-related work practices such as lockout/tagout, achieving an electrically safe work condition, and the use of PPE are all important pieces of an electrical safety program, this FPN is a reminder that there is much more that needs to be considered. The second FPN informs us that American National Standards Institute/American Industrial Hygiene Association (ANSI/AIHA) Z10-2005,

American National Standard for Occupational Safety and Health Management Systems, is available as a resource in the endeavor to create a complete electrical safety program.

Awareness and Self-Discipline

The second of eight factors that *NFPA 70E* requires to be included in the program is awareness and self-discipline. Section 110.7(B) requires that:

- An electrical safety program must make workers aware of the potential for electric shock, arc flash, or blast and the risk associated with those hazards.
- An electrical safety program must instill self-discipline into the employees who must perform work where the hazards of electric shock, arc flash, and blast are or might be present.
- An electrical safety program must instill safety principles and controls. Among the safety principles and controls to be considered are those outlined in *NFPA 70E* Annex E.

Program Principles, Controls, and Procedures

The third, fourth, and fifth of the eight factors that *NFPA 70E* requires to be included in the electrical safety program are program principles, controls, and procedures. Section 110.7(C), (D), and (E) require that the electrical safety program identify the principles upon which it is based, the controls by which it is measured and monitored, and the procedures that are to be used for work inside the limited approach boundary. No specific requirements indicate what those program principles, controls, and procedures must be. While reference is made to the information in Annex E.1, E.2, and E.3 through the FPNs, FPNs and annex information are nonmandatory. **Figure 7.3** includes examples of what should be included as the principles, controls, and procedures required by Section 110.7(C), (D), and (E), respectively.

Hazard/Risk Evaluation

NFPA 70E Section 110.7(F) requires a means to analyze the risks and hazards associated with each job. Accordingly, the sixth element required by *NFPA 70E* to be included in an electrical safety program is the hazard/risk evaluation procedure.

The hazard/risk evaluation procedure must be included in the electrical safety program and must be used before workers are permitted to cross the limited approach boundary. *NFPA 70E* Annex F offers nonmandatory insight and tools to assist in the understanding of mandatory requirements of Section 110.7(F).

Annex E Electrical Safety Program

This annex is not a part of the requirements of this NFPA document but is included for informational purposes only.

Sec 110.7, Electrical Safety Program.

E.1 Typical Electrical Safety Program Principles.

Electrical safety program principles include, but are not limited to, the following:

(1) Inspect/evaluate the electrical equipment
(2) Maintain the electrical equipment's insulation and enclosure integrity
(3) Plan every job and document first-time procedures
(4) Deenergize, if possible (*see* 120.1)
(5) Anticipate unexpected events
(6) Identify and minimize the hazard
(7) Protect the employee from shock, burn, blast, and other hazards due to the working environment
(8) Use the right tools for the job
(9) Assess people's abilities
(10) Audit these principles

E.2 Typical Electrical Safety Program Controls.

Electrical safety program controls can include, but are not limited to, the following:

(1) Every electrical conductor or circuit part is considered energized until proven otherwise.
(2) No bare-hand contact is to be made with exposed energized electrical conductors or circuit parts above 50 volts to ground, unless the "bare-hand method" is properly used.
(3) Deenergizing an electrical conductor or circuit part and making it safe to work on is in itself a potentially hazardous task.
(4) Employer develops programs, including training, and employees apply them.
(5) Use procedures as "tools" to identify the hazards and develop plans to eliminate/control the hazards.
(6) Train employees to qualify them for working in an environment influenced by the presence of electrical energy.
(7) Identify/categorize tasks to be performed on or near exposed energized electrical conductors and circuit parts.
(8) Use a logical approach to determine potential hazard of task.
(9) Identify and use precautions appropriate to the working environment.

E.3 Typical Electrical Safety Program Procedures.

Electrical safety program procedures can include, but are not limited to, the following:

(1) Purpose of task
(2) Qualifications and number of employees to be involved
(3) Hazardous nature and extent of task
(4) Limits of approach
(5) Safe work practices to be utilized
(6) Personal protective equipment involved
(7) Insulating materials and tools involved
(8) Special precautionary techniques
(9) Electrical diagrams
(10) Equipment details
(11) Sketches/pictures of unique features
(12) Reference data

Figure 7.3 *NFPA 70E* Annex E.1, E.2, and E.3 offer nonmandatory examples of the electrical safety program principles, controls, and procedures required by Section 110.7(C), (D), and (E).

Annex F is broken up into F.1 and F.2, with F.1 being an example flow chart illustrating steps of a hazard/risk evaluation procedure. But what are hazard/risk evaluations and hazard/risk evaluation procedures?

The hazard/risk evaluation procedure illustrated in Figure F.1 is a specific process intended for use by a worker immediately before he or she begins to perform a work task within the limited approach boundary, such as inserting a specific starter unit into a specific motor control center unit and section. This evaluation procedure applies to single work tasks and is not intended for use in assessing general hazards associated with electrical work. Figure F.2 is a general procedure intended for use in assessing hazards associated with a class of work tasks, such as inserting starter units into motor control centers. This is a written assessment that documents acceptable risk associated with a class of work.

According to Annex F.2, a **hazard/risk evaluation** is "an analytical tool consisting of a number of discrete steps intended to ensure that hazards are properly identified and evaluated, and that appropriate measures are taken to reduce those hazards to an acceptable level [adapted from American Society of Safety Engineers (ASSE) Z244.1-2003]," and a **hazard/risk procedure** is "a comprehensive review of the task and associated foreseeable hazards that use event severity, frequency, probability, and avoidance to determine the level of safe practices employed." Annex F.2 also outlines the following nonmandatory steps of a hazard/risk assessment procedure and nonmandatory probability estimators:

F.2 Hazard/Risk Assessment Procedure.

This procedure includes:

(1) Gathering task information and determining task limits.

(2) Documenting hazards associated with each task.

(3) Estimating the risk factors for each hazard/task pair.

(4) Assigning a safety measure for each hazard to attain an acceptable or tolerable level of risk.

While this procedure may not result in a reduction of PPE for a task, it can help in understanding of the specific hazards associated with a task to a greater degree and thus allow for a more comprehensive assessment to occur.

While severity, frequency, and avoidance factors are straightforward, consideration of probability includes the following estimators:

(1) Hazard exposures

(2) Human factors

(3) Task history

(4) Workplace culture

(5) Safeguard reliability

(6) Ability to maintain or defeat protective measures

(7) Preventive maintenance history

Reduction strategies to be employed if an unacceptable risk cannot be achieved include the following hierarchy of controls:

(1) Eliminate the hazard

(2) Reduce the risk by design

(3) Apply safeguards

(4) Implement administrative controls

(5) Use personal protective equipment

Annex F.2 also includes figures to help illustrate the information. **Figure 7.4** illustrates the steps of a hazard/risk evaluation assessment procedure. **Figure 7.5** is a sample form incorporating concepts of Annex F.2 with information completed.

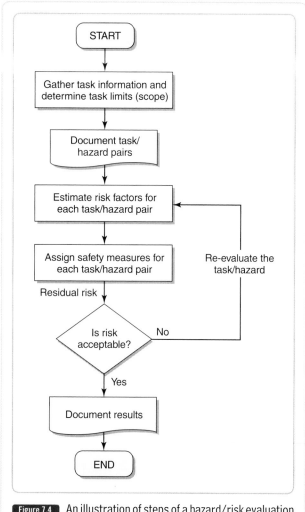

Figure 7.4   An illustration of steps of a hazard/risk evaluation assessment procedure.

HAZARD/RISK EVALUATION PROCEDURE

Task: Voltage testing _____   Document no.: _____

Equipment: _____   Part of: _____

Issued by: _____   Pre-risk assessment

Date: _____   Intermediate risk assessment

Follow-up risk assessment

Black area = Safety measures required   Gray area = safety measures recommended

Consequences	Severity Se	Class CI					Frequency Fr	Probability Pr	Avoidance Av
		3-4	5-7	8-10	11-13	14-15			
Irreversible trauma, death	4						Daily 5	Common 5	
Permanent, third degree burn	3						Weekly 4	Likely 4	
Reversible, second degree burn	2						Monthly 3	Possible 3	Impossible 5
Reversible, first aid	1						Yearly 2	Rarely 2	Possible 3
							Less 1	Negligible 1	Likely 1

Hzd. No.	Hazard	Se	Fr +	Pr +	Av +	= CI	Severity Mitigators	Safe
1	Human factors	4	5	3	5	13	Use appropriate PPE and follow established safety procedures.	Y
2	Shortened test loads	3	5	2	5	12	Inspect leads before each use.	Y
3	Meter misapplication	4	5	3	5	13	Ensure that the meter is rated for the level of voltage being tested.	Y
4	Meter malfunctions	3	5	2	5	12	Ensure that the meter is CAT rated to the appropriate hazard level.	Y

Comments:

PPE required: Voltage rated gloves and leather protectors, face and head protection, clothing rated for the incident energy exposure.

© 2008 National Fire Protection Association   NFPA 70E (p. 1 of 1)

Figure 7.5 Sample form that illustrates a completed hazard/risk evaluation assessment procedure for voltage testing.

Additional important guidance is offered in a FPN to Section 110.7(F). This FPN recommends that consideration be given to when a second person might be required, and what training and equipment the second person would need, when energized work is being performed. Consideration of when a second person is required for energized work to be performed is another example of something that should be evaluated as part of the hazard/risk evaluation procedure for a particular task. This FPN reminder also ties in to a portion of the training requirements of Section 110.6(C), where employees who are exposed to shock hazards are required to be trained in methods of release of victims from contact with exposed energized equipment. Among other things, a second person should be trained in how to safely release a worker who is "hung up." The insulated rescue hook is an example of equipment that could be used to release a victim from contact with energized parts (Figure 7.6).

Job Briefing

The seventh of the eight topics that *NFPA 70E* requires to be included in an electrical safety program is the job briefing. These provisions are outlined in Section 110.7(G) and are broken down into three categories:

1. General job briefing requirements
2. Requirements for repetitive and similar tasks
3. Requirements for routine tasks

The general job briefing requirements mandate that job briefings be conducted before commencement of work and include everyone that will be involved. Note that Annex I offers an example of what job briefing documentation could contain and look like, using a checklist format (Figure 7.7).

Figure 7.6 The insulated rescue hook is an example of equipment that could be used to release a victim from contact with energized parts.

Identify

❏ The hazards
❏ The voltage levels involved
❏ Skills required
❏ Any "foreign" (secondary source) voltage source
❏ Any unusual work conditions
❏ Number of people needed to do the job

❏ The shock protection boundaries
❏ The available incident energy
❏ Potential for arc flash (Conduct a flash-hazard analysis.)
❏ Flash protection boundary

Ask

❏ Can the equipment be deenergized?
❏ Are backfeeds of the circuits to be worked on possible?

❏ Is a "standby person" required?

Check

❏ Job plans
❏ Single-line diagrams and vendor prints
❏ Status board
❏ Information on plant and vendor resources is up to date

❏ Safety procedures
❏ Vendor information
❏ Individuals are familiar with the facility

Know

❏ What the job is
❏ Who else needs to know— Communicate!

❏ Who is in charge

Think

❏ About the unexpected event . . . What if?
❏ Lock — Tag — Test — Try
❏ Test for voltage—FIRST
❏ Use the right tools and equipment, including PPE

❏ Install and remove grounds
❏ Install barriers and barricades
❏ What else . . . ?

Prepare for an emergency

❏ Is the standby person CPR trained?
❏ Is the required emergency equipment available? Where is it?
❏ Where is the nearest telephone?
❏ Where is the fire alarm?
❏ Is confined space rescue available?

❏ What is the exact work location?
❏ How is the equipment shut off in an emergency?
❏ Are the emergency telephone numbers known?
❏ Where is the fire extinguisher?
❏ Are radio communications available?

Figure 7.7 *NFPA 70E* Annex I is a nonmandatory example of a form used to document the topics covered in a job briefing, using a checklist format.

Keep in mind that Annex I, like all *NFPA 70E* annexes, is nonmandatory in both format and content. Therefore, at a minimum, the following must be covered as part of a job briefing by the person in charge in accordance with the general requirements of Section 110.7(G)(1):

- The hazards associated with the task to be performed
- The work procedures that will be used
- Any special precautions that are necessary
- How sources of energy will be controlled
- What PPE will be necessary

Much of the information in a job briefing is also required by the energized electrical work permit in Section 130.1(B)(2). A comparison of the following energized electrical work permit requirements with the subjects required as part of a job briefing make the case that the information on the energized electrical permit should be among the information covered in the job briefing:

130.1(B) Energized Electrical Work Permit

(2) Elements of Work Permit. The energized electrical work permit shall include, but not be limited to, the following items:

(1) A description of the circuit and equipment to be worked on and their location

(2) Justification for why the work must be performed in an energized condition

(3) A description of the safe work practices to be employed

(4) Results of the shock hazard analysis

(5) Determination of shock protection boundaries

(6) Results of the arc flash hazard analysis

(7) The arc flash protection boundary

(8) The necessary personal protective equipment to safely perform the assigned task

(9) Means employed to restrict the access of unqualified persons from the work area

(10) Evidence of completion of a job briefing, including a discussion of any job-specific hazards

(11) Energized work approval (authorizing or responsible management, safety officer, or owner, etc.) signature(s)

NFPA 70E modifies its general job briefing requirements with the provisions in Section 110.7(G)(2) and (3) for "repetitive or similar tasks" and "routine work." A single job briefing that covers the hazards, work procedures, special precautions, energy controls, and PPE required by Section 110.7(G)(1) is the minimum required before work commences for tasks identified as repetitive or similar unless something unforeseen occurs that would jeopardize employee safety and warrant additional job briefings. Section 110.7(G)(3) also modifies the general job briefing requirements by recognizing that a "brief discussion" covering the hazards, work procedures, special precautions, energy controls, and PPE can serve as the job briefing for routine tasks where training and experience make it likely that the worker will recognize and avoid the hazards associated with the task to be performed. However, "a more extensive discussion" is required for "complicated or particularly hazardous" work, or where the employee is not trained to "recognize and avoid" the hazards that they will be exposed to. Being trained to recognize and avoid hazards is a fundamental part of the definition of a qualified person in *NFPA 70E*.

> **Qualified Person.** One who has skills and knowledge related to the construction and operation of the electrical equipment and installations and has received safety training to recognize and avoid the hazards involved.

Classifying tasks as "repetitive or similar" and "routine" and conducting an abbreviated job briefing should be done judiciously. Experience has shown far too many injuries and fatalities related to the performance of routine tasks. This point was made by Danny Liggett from DuPont in substantiation for a proposal he submitted to delete Section 110.7(G)(3) where he indicates that "routine work is where employees are most likely to become complacent and lose awareness."

Auditing

The last of the eight topics that *NFPA 70E* requires to be included in an electrical safety program is program auditing. The need to audit safety programs to have an effective program is well established. As pointed out by Danny Liggett in his substantiation for the proposal adding auditing requirements into *NFPA 70E*, "An employer can establish the very best electrical safety program, but if the employer does not audit the performance of the employees and the procedures, then weaknesses cannot be identified. An effective electrical safety program learns from itself and continues to improve." This requirement was new for the *NFPA 70E*, 2009 Edition, and is addressed in Section 110.7(H).

NFPA 70E requires program audits to ensure that its principles and procedures are being followed. Although 3 years was proposed as the maximum time between audits by Mr. Liggett in his proposal, the *NFPA 70E* Technical Committee reached consensus on requiring audits on a frequency to be determined by the employers, "based on the complexity . . . and type of work . . ." with "appropriate revisions" required when auditing finds deficiencies with adherence to the program's principles and procedures. Again, examples of electrical program principles and procedures can be found in *NFPA 70E* Annex E.

Groundbreaking Development

By now it should be obvious that *NFPA 70E* Section 110.7 and *NFPA 70E* Annex E offer a number of considerations that should be evaluated and integrated into the framework of an existing electrical safety program to make it more comprehensive and complete. Is anyone buying into it? The answer is: Absolutely. The following is one example, provided by Stephen M. Lipster, Training Director of the Columbus Joint Apprenticeship and Training Committee for the Electrical Industry.

In late 2001, representatives from the Columbus offices of Occupational Safety and Health Administration, the National Electrical Contractors Association (NECA), and the International Brotherhood of Electrical Workers (IBEW) met in the boardroom of the IBEW/NECA training facility in Columbus, Ohio. The purpose of this meeting was to explore common ground in an effort to create a safer workplace for electrical workers. Using compiled data, the representatives identified several particular areas to target: Fall protection, worker education, and electrical safe work practices. This was the beginning of what was to become the Central Ohio IBEW/NECA/OSHA Charter.

Once target areas were identified, combining the resources of all the organizations at the table to sharpen the focus made sense. While this made sense philosophically, some substantial barriers had to be overcome before the Charter could come into being. The historical relationship between OSHA and NECA was, not without reason, somewhat confrontational. The real and perceived wounds and wrongs of the past had to be set aside, and a new environment of cooperation fostered. Much to the parties' credit, by small steps, this new environment developed and flourished. The barrier faced by the IBEW was entirely singular. While supporting safety and safety education from its inception, the Columbus area local union had real reservations about becoming a party to the Charter. By law, the safety and well-being of a worker is solely the employer's responsibility. Does it make sense for the local union to become legal partner to a safety program? Happily, the IBEW chose the high road. The local union took the position that no one had a more vested right in the safety and health of their members then the local union itself and signed on as a partner.

Once the formidable barriers had been surmounted, a steering committee was formed to define the partnership and craft a document defining the goals, measures, and outcomes of this yet-to-be named enterprise. Using similar OSHA partnership documents as a template, the steering committee produced a total of 16 drafts before crafting a final document acceptable to all parties. The document and enterprise are called the IBEW/NECA/OSHA Charter. The IBEW participation in the project was groundbreaking and demanded appropriate recognition. The Charter adopted strong language concerning fall protection and required, at a minimum, an OSHA 10-hour course for all workers and an OSHA 30-hour course for all persons trained at the level of foreman or higher (according to the collective bargaining agreement). Perhaps the most remarkable section of the Charter was the adoption of the *NFPA 70E, Standard for Electrical Safety in the Workplace,* as the de facto standard for working on energized circuits. This was the first time this particular standard was voluntarily adopted in a collaborative effort. Finally, the Charter was designed to be a living document that would grow and change with the evolving workplace.

Since the birth of the Charter in 2002, over 700 electrical workers have completed an OSHA 10- or 30-hour course, the *NFPA 70E* standard has undergone two revisions, and the Charter has also undergone a substantial revision. After 6 years, the Charter continues to be a viable enterprise that provides a safer, smarter workplace for central Ohio electrical workers.

The Central Ohio IBEW/NECA/OSHA Charter has served as a model for other organizations in a number of different areas of the country to develop and implement similar agreements. A copy of the second generation of that unique and groundbreaking Charter is included as Appendix B in this textbook for your review.

Summary

The importance of an electrical safety program cannot be overstated. An electrical safety program is essential in an effort to train and protect workers from electrical hazards and a fundamental part of providing an effective overall safety and health program for worker safety. Be sure to review and understand the eight elements that *NFPA 70E* requires to be included in an electrical safety program as minimum considerations for making improvements to a new or existing electrical safety program.

Journeyman Electrician Electrocuted

An electrician was electrocuted while making a connection in a 4 in. junction box at a construction site. The foreman at the job site is responsible for job site safety, and he conducts weekly safety meetings where various safety issues are discussed. The job foreman and job manager were not aware of any safety problems in the past. The company apparently did not have a written safety policy.

The victim worked for an electrical contractor that employs 110 employees, 30 of whom work at the job site where the accident occurred. These employees, who are classified as either Journeymen electricians or electrician apprentices, normally work in pairs. The victim was considered a Journeyman electrician. All Journeyman electricians are expected to have had extensive training before they are hired; therefore, no safety training is provided by the employer.

The construction site is a large office building in a corporate office park. Interior wiring was being completed by the electrical contractor. The victim was pulling electrical wire for the 277-volt emergency back-up lighting system. This task required the installation of a 4 in. junction box in the ceiling. At approximately 8:15 a.m., the victim was standing on a wooden ladder preparing to strip a "hot wire." The victim's partner told him the wire was hot; the partner is certain that the victim understood his warning. (Interviews with the foreman and other electricians revealed that making connections while the wires are "hot" is not an unusual practice and might be done 50 percent of the time.) The coworker walked away from the victim to complete other work. A short time later, the partner heard a groan. He ran back, stepped up on the ladder that the victim was using, and attempted to pull him down. The victim appeared to be caught on the metal support bars for the drop ceiling. Unable to pull the victim down, the partner kicked the ladder out from under him. The victim then fell to the floor.

CPR was started at the accident site and was continued by emergency medical personnel. Burns were noted on the victim's right thumb, where he apparently contacted the metal of the stripping tool. Another burn was noted on the left lateral chest wall. The victim was pronounced dead at a local hospital.

Recommendations/Discussion

Recommendation 1: Wiring should not be done while the lines are energized.

Discussion: The victim was working with an energized wire. It is apparent that this is not an unusual work practice among electricians. Had the system been deenergized, this fatality would not have occurred.

Recommendation 2: The company should develop and implement a comprehensive occupational safety program.

Discussion: Worker safety is a primary responsibility of employers. To optimally carry out this responsibility, an employer should do the following: Develop a company policy that expresses management's commitment to providing a safe workplace, and develop, document, and enforce the adoption of safe work procedures and practices for all employees.

Recommendation 3: Upon initial hiring and at regular intervals thereafter, all workers and supervisors should receive training in hazard recognition and safe work practices.

Discussion: Workers who perform hazardous tasks, who work at precarious workstations, and/or who work in close proximity to sources of hazardous energy can develop a cavalier attitude over time. Therefore, it is particularly important that not only apprentice workers, but also experienced and highly skilled workers, be trained in hazard identification, safe work practices, and emergency response; this training should be periodically repeated.

Recommendation 4: When hiring personnel who are expected to perform jobs or tasks that present high risk, the employer should verify experience and determine skill level.

Discussion: The field evaluation of this incident did not identify this problem area as contributing to this accident; however, several similar accidents have identified employees who were hired to perform hazardous activities but who had not been adequately trained, and the employers were unaware of this lack of training. It is in the best interest of the employer to determine (by certification, training, or demonstration) that newly hired employees can safely complete the duties assigned.

Adapted from: FACE 86-02. Accessed 7/10/08 at http://www.cdc.gov/niosh/face/in-house/full8602.html.

Questions

1. OSHA could have found this company in violation for which of the following reasons?
   A. For allowing "hot work" to be done
   B. Not having a written safety program
   C. Inadequate insulated gloves
   D. Standing on a ladder without fall protection

2. What is an appropriate percentage of jobs for workers to work on energized circuits?
   A. 0 percent
   B. 25 percent
   C. 50 percent
   D. 75 percent

3. How do you explain the fact that the victim had two burns on his body?
   A. The victim must have been smoking a cigarette to exhibit the burn on his thumb.
   B. The victim must have hit the energized conductor with two body parts at the same time.
   C. The wooden ladder is nonconductive.
   D. For the victim to be electrocuted, current had to enter his body at one point and exit at some other point.

4. What was the insulation rating of the material on the handles of the stripping tool?
   A. 150 volts
   B. 277 volts
   C. It has no insulation rating.
   D. 300 volts DC

Ready for Review

- A safety program is an organized effort to reduce injuries.
- An electrical safety program should be a subset of an overall safety program.
- An electrical safety program is an integral part of an overall safety and health program and a vital part of establishing an electrically safe workplace.
- OSHA and *NFPA 70E* require the employer to implement an electrical safety program.
- In the 1930s, psychologist H.W. Heinrich showed that there is a relationship between behaviors and incidents. His theory states that for every 300 recordable injuries, approximately 30 lost-time injuries and one fatality will occur.
- Obstacles to electrical safety include hazards that are not readily visible, along with the stereotypes, misinformation, and lack of knowledge about electrical hazards that exist in the workplace.
- Many organizations rely on the electrical workers' training and experience as a substitute for safe work practices and proper protective equipment; a company might consistently expect workers to be exposed to energized electrical circuits without safe work practices and protective equipment. The electrical safety program should address this.
- Developing an electrical safety program with a proactive vision can result in lower operating costs and strong employee support. In addition to federal regulations, the company should incorporate into its program best work practices and techniques, both internally and within peer groups and organizations.
- Goals must be attainable and embraced by senior management for the program to have credibility. In addition, upper management must take an active support role for the program to be successful. The scope of the program is directly related to the budget and should spell out what steps, procedures, and policies will be required to attain the goal of zero incidents.

- An electrical safety program has direct and indirect cost benefits. Direct benefits are quantifiable, such as lower insurance cost through safe operations. Indirect benefits include items such as company image.
- *NFPA 70E* does not provide a complete electrical safety program, but it does provide a number of minimum considerations that must be evaluated and integrated into the framework of an existing electrical safety program. *NFPA 70E* Section 110.7 and Annex E cover these.
- The requirements in Section 110.7 are broken down into eight major categories. Addressed first is the general requirement for an employer to develop, document, and put into practice an electrical safety program that provides work practices and procedures that protect workers for the voltage, energy level, and circuit conditions to which they are exposed.
- The second of eight factors that *NFPA 70E* requires to be included in the program is awareness and self-discipline.
- The third, fourth, and fifth of the eight factors are program principles, controls, and procedures, respectively. Section 110.7(C), (D), and (E) require that the electrical safety program identify the principles upon which it is based, the controls by which it is measured and monitored, and the procedures that are to be used for work inside the limited approach boundary.
- The sixth element required by *NFPA 70E* to be included in an electrical safety program is the hazard/risk evaluation procedure. This procedure must be used before workers are permitted to cross the limited approach boundary. Annex F offers nonmandatory insight and tools to assist in the understanding of mandatory requirements of Section 110.7(F).
- A FPN to Section 110.7(F) recommends that consideration be given to when a second person might be required, and to what training and equipment the second person would need when energized work is being performed.

- The seventh of the eight topics is the job briefing. These provisions are outlined in Section 110.7(G) and are broken down into three categories:
  - General job briefing requirements
  - Requirements for repetitive and similar tasks
  - Requirements for routine tasks.
  - Annex I offers an example of what job briefing documentation could contain.
- General job briefing requirements are further categorized as "repetitive or similar tasks" and "routine work." Classifying tasks as "repetitive or similar" and "routine" and conducting an abbreviated job briefing should be done judiciously; many injuries and fatalities are related to the performance of routine tasks.
- The last of the eight topics is program auditing. Auditing helps ensure that an established electrical safety program is being used and is assessed to identify areas of weakness that can be improved.

Vocabulary

electrical safety program A subset of an overall safety program, ideally tailored to address the specific issues in a particular workplace, and components of which can include policies and procedures, site assessment, task assessment, PPE requirements, hazardous boundaries and hazard/risk analysis, administration, lockout/tagout, training, auditing and recordkeeping, and budgeting.

hazard/risk evaluation* An analytical tool consisting of a number of discrete steps intended to ensure that hazards are properly identified and evaluated, and that appropriate measures are taken to reduce those hazards to an acceptable level (adapted from ANSI/ASSE Z244.1).

hazard/risk procedure* A comprehensive review of the task and associated foreseeable hazards that use event severity, frequency, probability, and avoidance to determine the level of safe practices employed.

safety program An organized effort to reduce injuries.

*This is the *NFPA 70E* definition.

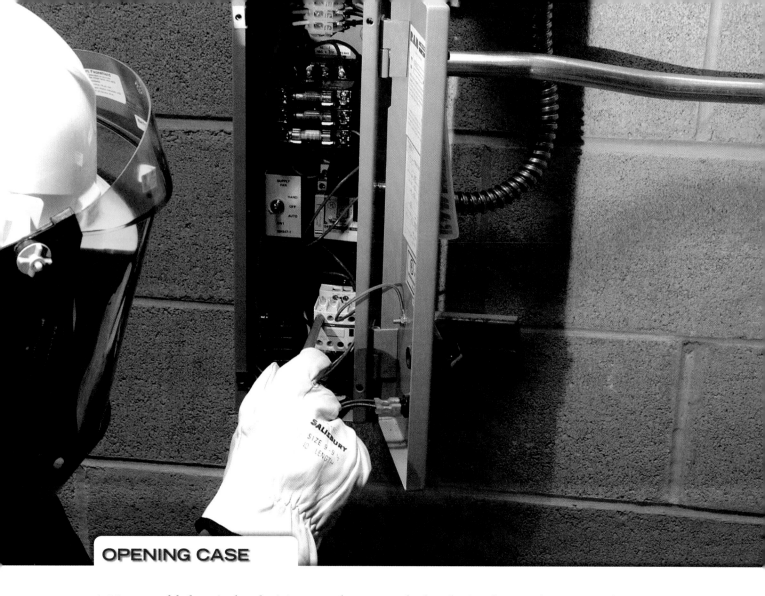

OPENING CASE

A 30-year-old electrical technician was electrocuted when he inadvertently contacted an energized conductor inside a control panel cabinet. The company was in the process of starting up a new plant that employed 35 personnel, including five electrical technicians who installed and maintained the plant's electrical equipment. The company had a written safety policy, a comprehensive safety program with written safety procedures and a full-time director of safety. The victim had worked for this employer for 16 months prior to the incident.

The plant's buildings had been erected, rolling machines and all related equipment had been installed, and the employees were in the process of testing the equipment prior to the actual start-up of the rolling mill. On the day of the incident, the victim met the service representative from the company that manufactured a piece of voltage-regulating equipment. The two men discussed what had to be done, then began checking voltages on the low-voltage (15 volts) wiring from the system. They determined that a problem existed within the equipment, as they could not detect the appropriate amount of voltage.

The victim left the area to get the service manual for the equipment. Without deenergizing the system, the service representative opened the panel cover on the system's control cabinet in preparation for tracing the low-voltage wires in question. The wires were not color-coded. He then climbed onto the compartment adjacent to the capacitor rack to view the wiring.

The victim returned, knelt in front of the opened control cabinet, and positioned his head and shoulders inside the 800-volt resistor compartment. As he began tugging on the wires inside the compartment, the service representative told him to be careful around the conductors. The service representative was

Training

CHAPTER 8

Introduction

One of the cornerstones of an effective safety program is training. Electrical safety training is certainly no exception. As stated in *The Electrical Safety Program Book* by Jones, Jones, and Mastrullo, ensuring that employees understand the intricacies of the program is the primary objective of any training session. Input from the electrical workers themselves is absolutely necessary for effective procedures, as is input from the workers in the actual training program.

Every electrical training program should contain at least one segment that discusses electrical hazards. Regardless of a student's experience and expertise, each training opportunity should review or provide new information about electrical hazards to which the student will or might be exposed. Greater understanding of electrical hazards and the possible consequences have a bearing on the person's behavior.

Although the purpose of electrical hazard training is to discuss the details of each hazard, the objective of the training must be to build understanding about the hazard. If a worker understands electrical arc flash, then he or she is likely to respect the hazard

OPENING CASE, CONT.

watching to see which wire moved when he heard the victim making a gurgling noise. He looked down and saw the victim shaking as though he were being shocked. The service representative jumped down from the cabinet and knocked the victim away from contact with the energized conductor, receiving a shock as he did so.

The service representative ran to his truck and called the plant's main office for assistance. Personnel from the main office called the emergency medical service (EMS) and proceeded to the incident site, where cardiopulmonary resuscitation (CPR) was being administered by one of the plant's managers. EMS arrived approximately 15 minutes after being notified, continued CPR, and transported the victim to the local hospital's emergency room, where he was pronounced dead 1 hour and 45 minutes after the incident occurred. The medical examiner's office reported the cause of death as electrocution.

Adapted from: Fatality and Assessment and Control Evaluation (FACE) 92-07. Accessed 7/15/08 at http://www.cdc.gov/niosh/face/In-house/full9207.html.

1. At what point should the workers have deenergized the equipment?
2. What is the correct procedure for removing a person who is in contact with an energized conductor?

Chapter Outline

- **Introduction**
- **Is Having a Program in Place Enough?**
- **Training Requirements**
- **Summary**
- **Lessons Learned**
- **Current Knowledge**

Learning Objectives

1. Identify the safety training required by *NFPA 70E®*.
2. Recognize the *NFPA 70E* training required for qualified persons and unqualified persons.
3. Understand the additional training, retraining, and documentation of training requirements of *NFPA 70E*.

References

1. *NFPA 70E*, 2009 Edition
2. *The Electrical Safety Program Book* by Jones, Jones, and Mastrullo

and wear protection warranted by the circumstances. As an employee begins to develop a new or renewed understanding of electrical hazards, skepticism will give way to an appreciation of the catastrophic result of an unexpected release of energy.

Recall that electrical hazards were examined in Chapter 2, and electrical safety program requirements were covered in Chapter 7. Keep the information learned in those chapters in mind as the training requirements in *NFPA 70E* Section 110.6 are explored here.

 ## Is Having a Program in Place Enough?

Is simply having an electrical safety program in place enough, or should other elements be in place, such as training and implementation of the electrical safety program? These are good points to consider as we review the findings in the Background box entitled Prevention: Elements of an Electrical Safety Program, from an evaluation of 224 incidents reported by FACE.

BACKGROUND

Prevention: Elements of an Electrical Safety Program

Virgil Casini, B.S.

At least one of the following five factors was present in all 224 incidents evaluated by the FACE program:

1. Established safe work procedures were either not implemented or not followed.

2. Adequate or required personal protective equipment (PPE) was not provided or worn.

3. Lockout/tagout procedures were either not implemented or not followed.

4. Compliance with existing Occupational Safety and Health Administration (OSHA), *National Electrical Code®* (*NEC®*), and National Electrical Safety Code (NESC) regulations were not implemented.

5. Worker and supervisor training in electrical safety were not adequate. These subjects are addressed in various National Institute for Occupational Safety and Health (NIOSH) Alerts[26–36] and related publications.[37]

Most of the 224 occupational electrocution incidents investigated as part of the FACE program could have been prevented through compliance with existing OSHA, *NEC*, and NESC regulations; and/or the use of adequate PPE. All workers should receive hazard awareness training, so that they are able to identify existing and potential hazards present in their workplaces and relate the potential seriousness of the injuries associated with each hazard. Once these hazards are identified, employers should develop measures that allow for their immediate control.

Based on an analysis of these data, to reduce occupational electrocutions, employers should:

- Develop and implement a comprehensive safety program and, when necessary, revise existing programs to thoroughly address the area of electrical safety in the workplace.

- Ensure compliance with existing OSHA regulations Subpart S of 29 Code of Federal Regulations (CFR) 1910.302 through 1910.399 of the General Industry Safety and Health Standards[3] and Subpart K of 29 CFR 1926.402 through 1926.408 of the OSHA Construction Safety and Health Standards.[4]

- Provide all workers with adequate training in the identification and control of the hazards associated with electrical energy in their workplace.

- Provide additional specialized electrical safety training to those workers working with or around exposed components of electrical circuits. This training should include, but not be limited to, training in basic electrical theory; safe work procedures; hazard awareness and identification; proper use of PPE; proper lockout/tagout procedures; first aid training, including CPR; and proper rescue procedures. Provisions should be made for periodic retraining as necessary.

- Develop and implement procedures to control hazardous electrical energy, including lockout and tagout procedures, and ensure that workers follow these procedures.

- Provide those workers who work directly with electrical energy with testing or detection equipment that can ensure their safety during performance of their assigned tasks.

- Ensure compliance with the *NEC* and the NESC.[6]
- Conduct safety meetings at regular intervals.
- Conduct scheduled and unscheduled safety inspections at work sites.
- Actively encourage all workers to participate in workplace safety.
- In a construction setting, conduct a job site survey before starting work to identify any electrical hazards, implement appropriate control measures, and provide training to employees specific to all identified hazards.
- Ensure that proper PPE is available and worn by workers where required (including fall protection equipment).
- Conduct job hazard analyses of all tasks that might expose workers to the hazards associated with electrical energy, and implement control measures that can adequately insulate and isolate workers from electrical energy.
- Identify potential electrical hazards and appropriate safety interventions during the planning phase of construction or maintenance projects. This planning should address the project from start to finish, to ensure that workers have the safest possible work environment.

The FACE data indicate that although many companies had comprehensive safety programs, in many cases, they were not completely implemented. This fact underscores the need for increased management and worker understanding, awareness, and ability to identify the hazards associated with working on or in proximity to electrical energy. It is the responsibility of management to provide a safe workplace for their workers and to develop and implement a comprehensive safety program. In some cases, this might entail the development of additional worker training and/or the evaluation and restructuring of existing safety programs. Management should also provide adequate training in electrical safety to all workers and strictly enforce adherence to established safe work procedures and policies. Additionally, adequate PPE should be available where appropriate. Information or assistance in accomplishing these measures can be provided by OSHA, electrical safety consultants, or other agencies or associations that deal with electrical safety. A strong commitment to safety by both management and workers is essential in the prevention of severe occupational injuries and death due to contact with electrical energy.

Source: Worker Deaths by Electrocution: A Summary of NIOSH Surveillance and Investigative Findings. U.S. Department of Health and Human Services, Public Health Service, Centers for Disease Control and Prevention, National Institute for Occupational Safety and Health, May 1998. Accessed 7/13/08 at http://www.cdc.gov/niosh/docs/98-131/pdfs/98-131.pdf.

Training Requirements

Once again *The Electrical Safety Program Book* by Jones, Jones, and Mastrullo makes important points regarding training requirements, as stated in the next two paragraphs.

Skill training requires a combination of classroom and on-the-job training in most instances. It is critical that the worker understand the limit or boundary of his or her qualification. Skill training should be based upon the role that an employee is expected to play, such as the specialized equipment on which he or she will work. In other words, training should be designed and offered to an employee that is appropriate for his or her job assignment.

Recognize that technology associated with power and control equipment changes dramatically over time. Equipment is modified to accommodate the newer technology. Each equipment construction change affects how an installer or maintenance person will interact with the equipment.

The following considerations are addressed by the training requirements in *NFPA 70E* Section 110.6. Those requirements are broken down into five major categories:

- Safety training
- Type of training
- Emergency procedures training
- Qualified and unqualified person training
- Documentation of training

Safety Training

The safety training requirements of *NFPA 70E* Section 110.6 apply to people subjected to the hazards of electric shock, arc flash burn, or blast from contact or equipment failure. It is unlikely that workers would be exposed to an electric shock hazard where electrical equipment is installed in accordance with the *NEC* and all equipment covers are in place and doors are closed and latched (**Figure 8.1**). However, shock protection would be required if workers are working on exposed energized parts, since they are exposed to a shock hazard (**Figure 8.2**).

This would address shock hazard, but arc flash hazard should be addressed as well. It seems rather obvious that arc flash protection is necessary if an employee is working on live parts, since the person could also be exposed to an arc flash hazard. But is an arc flash hazard present if no energized parts are exposed? Although only offered as nonmandatory information in a Fine Print Note (FPN) to the definition of an arc flash hazard, the *NFPA 70E* Technical Committee offers some guidance related to enclosed energized electrical equipment and potential for an arc flash hazard.

> **Arc Flash Hazard.** A dangerous condition associated with the possible release of energy caused by an electric arc.
>
> FPN No. 1: An arc flash hazard may exist when energized electrical conductors or circuit parts are exposed or when they are within equipment in a guarded or enclosed condition, provided a person is interacting with the equipment in such a manner that could cause an electric arc. Under normal operating conditions, enclosed energized equipment that has been properly installed and maintained is not likely to pose an arc flash hazard.

Figure 8.1 Electrical equipment with all covers in place and doors locked and latched pose a lesser electrical shock hazard.

Figure 8.2 Shock protection such as rubber insulating gloves rated for the exposure with leather protectors is required if working on live parts.

The FPN indicates that an arc flash hazard may exist when energized equipment is exposed. This concept is well understood. In addition, the FPN warns that an arc flash hazard can exist for enclosed equipment when "a person is interacting with the equipment in such a manner that could cause an electric arc" and that "enclosed energized equipment that has been properly installed and maintained is not likely to pose an arc flash hazard." Therefore, consideration of the possibility of an arc flash hazard must be given, even to enclosed energized equipment, with respect to the task at hand. Has the equipment been maintained properly? Was the equipment installed in accordance with the *NEC*? Is the equipment being used in accordance with its listing? Will a worker be interacting with the enclosed energized equipment while performing tasks including, but not limited to the following tasks:

- Operating a circuit breaker or the handle of a safety switch
- Racking a circuit breaker in or out (**Figure 8.3**)
- Inserting or removing a switch from busway
- Inserting or removing a motor control center bucket

The training required by *NFPA 70E* Section 110.6 also requires that workers understand the hazards of energized electrical work. By definition, an electrical hazard consists of potential electric shock, arc flash burn, thermal burn, or blast. Also recall from Chapter 1 that electrical incidents are multihazard, as reported by Mary Capelli-Schellpfeffer, MD, MPA, in the Background box entitled Trauma Following Electrical Events. She reminds us that the trauma associated with

Figure 8.3 Certain tasks may require the worker to interact with enclosed energized equipment, as shown in this photo, increasing the hazard.

Figure 8.4 Electrical safety training may occur in the classroom or on the job, or both.

energized work could include many different kinds of energy. She further reports that the results of trauma following electrical incidents include many different types of injuries.

In addition to being trained to understand the hazards of energized electrical work, training must include the practices and procedures required to guard against those hazards and that injury or death could be the result of being exposed to electrical hazards. Those practices and procedures could include, but are not limited to, lockout/tagout, establishing an electrically safe work condition, and performing an electrical hazards analysis (including a shock and arc flash hazard analysis including labeling, determination of boundaries, and PPE).

> ## OSHA Tip
>
> 1910.332(c) Type of training.
>   The training required by this section shall be of the classroom or on-the-job type. The degree of training provided shall be determined by the risk to the employee.

Type of Training

Like OSHA, *NFPA 70E* Section 110.6(B) recognizes that the required training could be accomplished in a variety of settings and that not all training is adequate for all tasks. Classroom training or on-the-job training only might be appropriate in some cases (**Figure 8.4**). However, for many tasks, both classroom and on-the-job training are necessary for the training to be effective. Regardless, the training must be equal to the task.

It clearly takes a different amount and complexity of training to work safely on a doorbell circuit than it does to work on a 4160-volt branch circuit. The more complex the task becomes, the more thorough and detailed the training must be.

Emergency Procedure Training

Section 110.6(C) addresses emergency procedure requirements. The first of the Section 110.6(C) emergency procedure requirements addresses training on how to release victims from contact. A shock hazard is defined as the possible release of energy caused by contact or approach to energized electrical conductors or circuit parts. Therefore, workers who are exposed to the possible release of energy, either by contact with or approach to energized electrical equipment, must be trained in how to safely remove workers from that exposure. Consider the relationship between this requirement to the FPN to Section 110.7(F), where it is recommended to identify when a second person could be required and the training and equipment that person should have.

> **(F) Hazard/Risk Evaluation Procedure.** An electrical safety program shall identify a hazard/risk evaluation procedure to be used before work is started within the limited approach boundary of energized electrical conductors and circuit parts operating at 50 volts or more or where an electrical hazard exists. The procedure shall identify the hazard/risk process that shall be used by employees to evaluate tasks before work is started.
>
>   FPN No. 1: The hazard/risk evaluation procedure may include identifying when a second person could be required and the training and equipment that person should have.

The second of the emergency procedure requirements addresses first aid and resuscitation training.

While many strive to train all electrical workers in first aid and methods of resuscitation, Section 110.6(C) requires employees to provide first aid and approved resuscitation training "if their duties warrant such training." *NFPA 70E* further requires these workers to be "regularly instructed" in those disciplines, with certification of resuscitation training by the employer required on an annual basis.

Qualified and Unqualified Employee Training

It is important to understand the definitions used in *NFPA 70E*. Among the definitions that require close examination here are those for **qualified person** and **unqualified person**, as the specific training requirements for these two types of employees are examined.

The *NFPA 70E* definition of a qualified person is different from that presently contained in the OSHA 29 CFR Parts 1926 and 1910 requirements, and reads as follows:

> **Qualified Person.** One who has skills and knowledge related to the construction and operation of the electrical equipment and installations and has received safety training to recognize and avoid the hazards involved.

The *NFPA 70E* definition of an unqualified person is fairly straightforward and reads as follows:

> **Unqualified Person.** A person who is not a qualified person.

Clearly, by definition, a person is only one of two things in regards to the construction and operation of the electrical equipment and installations; he or she is either qualified or unqualified (**Figure 8.5**). Note that by

Figure 8.5 An unqualified person is defined by *NFPA 70E* as a person who is not a qualified person.

definition, a qualified person must have received safety training to recognize and avoid the hazards involved. Recall that Chapter 2 examined a number of the hazards for which a qualified person would require safety training. These could include identification and understanding of shock, arc flash, and arc blast, for example.

But what further training do workers need according to *NFPA 70E* Section 110.6? Section 110.6(D) covers employee training, with (D)(1) containing the requirements for qualified persons, (D)(2) containing the requirements for unqualified persons, and (D)(3) containing retraining requirements.

Unqualified Person Training

Recall that the scope of *NFPA 70E* Chapter 1 is outlined in Section 110.1 and that a portion of the Chapter 1 scope specifically addresses unqualified persons.

Section 110.1, in part

Electric circuits and equipment not included in the scope of this standard might present a hazard to employees not qualified to work near such facilities. Requirements have been included in Chapter 1 to protect unqualified employees from such hazards.

Requirements have been included in Chapter 1 to protect unqualified employees from such hazards.

Again, an unqualified person is defined as a person who is not a qualified person. Accordingly, *NFPA 70E* requires that anyone who is not trained as a qualified person must be trained in and familiar with any of the electrical safety-related practices that might not be addressed specifically by *NFPA 70E* Chapter 1, but are necessary for his or her safety.

Qualified Person Training

A number of OSHA and *NFPA 70E* provisions state that only qualified persons are permitted to perform certain tasks. For example, only qualified persons are allowed to work on electrical equipment that is not in an electrically safe work condition, perform testing work inside the limited approach boundary of electrical equipment operating at 50 volts or more, and perform maintenance on electrical equipment and installations. Note the many similarities between what *NFPA 70E* requires for employee training and the related OSHA requirements. These requirements include, but are not limited to, the following:

NFPA 70E Section 110.8 Working While Exposed to Electrical Hazards.

(A) General. Safety-related work practices shall be used to safeguard employees from injury while they are exposed to electrical hazards from electrical conductors or circuit parts that are or can become energized. The specific safety-related work practices shall be consistent with the nature and extent of the associated electrical hazards.

(2) Energized Electrical Conductors and Circuit Parts—Unsafe Work Condition. Only qualified persons shall be permitted to work on electrical conductors or circuit parts that have not been put into an electrically safe work condition.

NFPA 70E **Section 130.4 Test Instruments and Equipment Use.**

Only qualified persons shall perform testing work within the Limited Approach Boundary of energized electrical conductors or circuit parts operating at 50 volts or more.

NFPA 70E **Article 205 General Maintenance Requirements**

205.1 Qualified Persons.

Employees who perform maintenance on electrical equipment and installations shall be qualified persons as required in Chapter 1 and shall be trained in, and familiar with, the specific maintenance procedures and tests required.

OSHA 29 CFR 1910.333(c)(2)

Work on energized equipment. "Only qualified persons may work on electric circuit parts or equipment that have not been deenergized under the procedures of paragraph (b) of this section. Such persons shall be capable of working safely on energized circuits and shall be familiar with the proper use of special precautionary techniques, personal protective equipment, insulating and shielding materials, and insulated tools."

According to Section 110.6(D)(1), a person needs training in the following to be considered a qualified person:

- Understand the construction and operation of equipment.
- Understand specific work methods.
- Be aware of and avoid potential contact or equipment failure that can result in electric shock, arc flash burn, thermal burn, or blast hazards associated with work on equipment.
- Be aware of and avoid potential contact or equipment failure that can result in electric shock, arc flash burn, thermal burn, or blast hazards associated with work practices.
- Select a voltage detector appropriate for the task including, but not limited to, its rating and design.
- Demonstrate proficiency on how to use a device to verify the absence of voltage.
- Interpret indications provided by the device used to verify the absence of voltage.
- Understand the limitations of any voltage detector that may be used.

Figure 8.6 A qualified person must be trained to be familiar with PPE.

Section 110.6(D)(1) also requires qualified persons trained to be familiar with:

- The proper use of the special precautionary techniques
- PPE (**Figure 8.6**)
- Insulating and shielding materials (**Figure 8.7**)
- Insulated tools and test equipment (**Figure 8.8**)

Note that these training requirements are very much related to what must be addressed in the Section 110.7 requirements for an *NFPA 70E*-compliant electrical safety program. Like OSHA, *NFPA 70E* recognizes that a worker can be considered qualified for certain equipment and work methods and still be unqualified with respect to other equipment and work methods.

There are also additional requirements that a qualified person must be trained in if he or she will cross a limited approach boundary to approach energized elec-

OSHA Tip

Note 1 to the OSHA Part 1910.399 definition of a qualified person:

Whether an employee is considered to be a "qualified person" will depend upon various circumstances in the workplace. For example, it is possible and, in fact, likely for an individual to be considered "qualified" with regard to certain equipment in the workplace, but "unqualified" as to other equipment. [See 1910.332(b)(3) for training requirements that specifically apply to qualified persons.]

Figure 8.7 A qualified person must be familiar with insulating materials.

Figure 8.8 A qualified person must be trained in the use of insulated tools.

trical equipment operating at 50 volts or more. These workers must be additionally trained to know how to:

- Determine which parts are exposed energized electrical conductors and circuit parts and which are not
- Determine the operating voltage of exposed energized electrical equipment

These workers must be additionally trained to know:

- The approach distances provided in Table 130.2(C). A qualified person must be trained to determine the approach distances to live parts for shock protection if they will cross a limited approach boundary (**Table 8.1**).
- The voltage to which he or she might be exposed

- The decision-making process necessary to determine the degree and extent of the hazard
- The PPE necessary to protect the worker from the identified hazards
- The planning necessary to safely complete the task

Again, training requirements go hand-in-hand with what must be documented on the energized electrical permit, per Section 130.1(B) and addressed by electrical safety program requirements. For example, recall that the job briefing requires going over task-related hazards,

Table 8.1	Portion of *NFPA 70E* Table 130.2(C): Approach Boundaries to Energized Electrical Conductors or Circuit Parts for Shock Protection (All dimensions are distance from energized electrical conductor or circuit part to employee.)

(1)	(2)	(3)	(4)	(5)
	Limited Approach Boundary[1]		**Restricted Approach Boundary[1]; Includes Inadvertent Movement Adder**	**Prohibited Approach Boundary[1]**
Nominal System Voltage Range, Phase to Phase[2]	**Exposed Movable Conductor[3]**	**Exposed Fixed Circuit Part**		
Less than 50	Not specified	Not specified	Not specified	Not specified
50 to 300	3.05 m (10 ft 0 in.)	1.07 m (3 ft 6 in.)	Avoid contact	Avoid contact
301 to 750	3.05 m (10 ft 0 in.)	1.07 m (3 ft 6 in.)	304.8 mm (1 ft 0 in.)	25.4 mm (0 ft 1 in.)

Note: For Arc Flash Protection Boundary, see 130.3(A).
[1] See definition in Article 100 and text in 130.2(D)(2) and Annex C for elaboration.
[2] For single-phase systems, select the range that is equal to the system's maximum phase-to-ground voltage multiplied by 1.732.
[3] A condition in which the distance between the conductor and a person is not under the control of the person. The term is normally applied to overhead line conductors supported by poles.
Reproduced with permission from *NFPA 70E*®-2009, *Electrical Safety in the Workplace*, Copyright ©2008, National Fire Protection Association. This reprinted material constitutes only a portion of the referenced table, is presented for educational purposes only, and is not the complete and official permission of the NFPA on the referenced subject, which is represented only by the standard in its entirety.

OSHA Tip

Note 2 to the OSHA Part 1910.399 definition of qualified person:

An employee who is undergoing on-the-job training and who, in the course of such training, has demonstrated an ability to perform duties safely at his or her level of training and who is under the direct supervision of a qualified person, is considered to be a qualified person for the performance of those duties.

Figure 8.9 The job briefing requires discussion of task-related hazards and safety-related work practices to be used.

safety-related work practices to be used, the process to achieve an electrically safe work condition and control hazardous energy, PPE use and limitations, and any necessary special precautions (**Figure 8.9**).

Again, like OSHA, *NFPA 70E* recognizes that a worker undergoing on-the-job training is considered a qualified person for the performance of certain duties if he or she is under the direct supervision of a qualified person and has demonstrated competency in the performance of those duties in a safe manner at his or her level of training.

Additional and Retraining Requirements

NFPA 70E also stipulates that an employee must be retrained in the safe performance of a task before he or she is permitted to perform that task again if he or she performs that task less than once per year. In addition to requiring retraining if employees perform tasks less than once a year, Section 110.6(D)(3) requires employees to receive additional training or be retrained in the following instances:

- Employees are not in compliance with the safety-related work practices
- New technology or types of equipment necessitate different safety-related work practices
- Changes in procedures necessitate different safety-related work practices
- Safety-related work practices are necessary that are not customarily used

A review of the proposals and comments published in the National Fire Protection Association (NFPA) *2008 Annual Revision Cycle Report on Proposals* and *Report on Comments* submitted to create this training/retraining requirement reveals information worth reviewing. The *NFPA 70E* committee initially agreed with the submitter of Proposal 70E-80 that refresher training be at least every 3 years. However, Comment 70E-299 modified Proposal 70E-80 and did not include any specific time frame for additional training or retraining except for where a worker performs that task less than once per year.

Documentation of Training

The need to document safety training is not new. This is reinforced in Dennis K. Neitzel's substantiation for Proposal 70E-73, as published in the NFPA *2008 Annual Revision Cycle Report on Proposals*. Here he submits that "... every regulation where training is required, OSHA requires documentation or certification that the training has taken place. Without documentation, there is no evidence that training has taken place." However, there was no requirement for training to be documented prior to the 2009 edition. Neitzel's recommendation from Proposal 70E-73 was accepted by the *NFPA 70E* committee and appears in *NFPA 70E* as Section 110.6(E). The employer must:

- Document that each employee has received the training required by Section 110.6(D). This requirement includes documentation of the qualified person training required by (D)(1), the unqualified person training required by (D)(2), and the additional and retraining training required by (D)(3).
- Document the name of the employee trained.
- Document the date that the training took place.
- Establish the documentation at the time when the employee demonstrates that he or she is proficient in the work practice.
- Maintain the documentation of training for the length of the employment of the worker with the employer.

 ## Summary

Even though many companies have safety programs, in many instances they are not fully implemented, as pointed out in Virgil Casini's evaluation of 224 FACE evaluation incidents. Casini further suggested that "management should also provide adequate training in electrical safety to all workers and strictly enforce adherence to established safe work procedures and policies" and that "a strong commitment to safety by both management and workers is essential in the prevention of severe occupational injuries and death due to contact with electrical energy."

This chapter provided an overview of a number of requirements contained in *NFPA 70E* related to training including safety training, type of training, emergency procedures, qualified and unqualified person training, additional training/retraining, and documentation of training.

It is important to remember that different levels of training are required for different tasks, recognize that a worker could be qualified for one task but not necessarily for another, and that, by definition, a person is either qualified or unqualified.

Remember that the standard recognizes dual responsibility for safety-related work practices, in that the employer is required to provide safety-related work practices and train the employee, who is then required to implement them. Training should never be considered complete. Effective training requires continuing efforts. To maintain worker knowledge at an elevated level, supervisors must review safety concepts and ideas with their workers at frequent intervals.

Lessons *learned*

Electrician Electrocuted in Tennessee

An electrician was electrocuted while troubleshooting a 480-volt direct current (dc) generator that supplied power to a glue machine. The victim was an electrician working for a small company (five employees) that performs industrial electrical wiring and maintenance. The company employs only union electricians, who undergo extensive training prior to reaching the Journeyman level. However, the company has no ongoing safety program.

One of the electrical company's clients asked that a 480-volt dc generator be relocated. The facility engineer agreed to assist the electrician (victim) who was sent to perform the work. On the day this incident occurred, the temperature in the facility was approximately 100°F, and the humidity was high.

After the two men finished moving the dc generator and reconnecting the electrical wiring, the victim turned on the power to the generator and asked the plant engineer to push the "start" button. When the generator failed to start, the victim got his voltmeter and verified that control voltage was reaching the starter. The victim then moved to the rear of the generator with the voltmeter to verify that he had made the electrical connections correctly and to check the two fuses located in the rear of the generator. The victim actually had to reach from the side of the generator, since there was only about 2 ft of clearance between the back of the generator and the glue machine it supplied.

The plant engineer smelled something burning and called to the victim, who did not respond. He then looked behind the generator and saw the victim slumped over, one arm on a transformer, and his head on the floor. He shouted to other employees (working nearby) to call for an ambulance and to cut the power off at the power disconnect switch.

The victim was then pulled from behind the generator, and CPR was initiated. EMS arrived approximately 15 minutes after the incident and began advanced cardiac life support (ACLS). Resuscitation efforts, which were continued en route and after arrival at a local hospital, were unsuccessful. The victim was pronounced dead in the hospital emergency room. The medical examiner ruled that death was due to accidental electrocution. No autopsy was performed, but burns were noted on the victim's left hand.

The victim apparently contacted one phase of the 480-volt alternating current (ac) power supply to the generator, which was energized. When the generator was moved, some nonconductive, particulate matter was apparently dislodged, which prevented the switch from completing the control circuit when the "start" button was initially pressed. After several subsequent attempts, the starter worked and has continued to function properly.

Recommendations/Discussion

Recommendation 1: Decisions about equipment location should include consideration of maintenance requirements.

Discussion: The location chosen for the generator ensured that maintenance to the generator would be from an awkward position at the side of the generator. This increased the possibility of contacting an energized conductor. It should be noted that the facility is voluntarily having the generators relocated away from the work area and that sufficient room for maintenance is being provided.

Recommendation 2: Employers should develop and implement a comprehensive safety program that addresses both the recognition of workplace hazards and procedures to minimize those hazards.

Discussion: The fact that safety training is an integral part of the training required to become a Journeyman electrician is useful and necessary, but not sufficient. This incident occurred at the end of the workday when the victim, who was hot and covered with perspiration (which lowered his electrical resistance), attempted to work in the vicinity of energized equipment in a cramped space. An active company safety program should be developed that stresses hazard recognition and safe work procedures.

Adapted from: FACE 87-69. Accessed 7/15/08 at http://www.cdc.gov/niosh/face/in-house/full8769.html.

Questions

1. A union Journeyman might be assumed to be a qualified person for which of the following reasons?
   A. The company's electrical safety program requires only qualified workers to work in the facility.
   B. In the incident report, the victim had recently purchased computer-based training program.
   C. The worker was provided with extensive skill and knowledge training.
   D. The Journeyman electrician was directly supervised by the facility engineer.

2. What impact did the high temperature and humidity have on the root cause of the incident?
   A. The elevated temperature caused the electrician to perspire heavily, reducing contact impedance, which increased the current through the victim's body.
   B. The high relative humidity reduced the arc-over distance.
   C. No impact at all.
   D. The combination of high temperature and high humidity caused the transformer to jettison electrons onto the electrician.

3. What was the hazard exposure when the worker checked the connections and fuses located in the back of the panel?

   A. The worker was only inspecting the equipment, so no direct exposure to a hazard existed.

   B. The electrician had a voltmeter with him, so the chance of electrocution was reduced, since he could determine which conductors were energized.

   C. The worker was a Journeyman electrician and qualified to work on the equipment while it was energized.

   D. To check connections, the worker probably would need to pull on the conductor or attempt to tighten the bolts holding the termination. If the conductors are energized, the worker could easily be electrocuted.

4. How should the electrician have located the "start" problem?

   A. After verifying that control voltage was being applied to the start circuit, the worker should have deenergized the equipment and verified continuity by measuring resistance with his meter.

   B. The worker should have retrieved the drawings from the office and determined the likely cause from the drawings.

   C. The worker should not have attempted to troubleshoot the problem. The plant engineer was with him and should have located the problem.

   D. The generator should not have been moved in the first place.

Ready for Review

- For an electrical safety program to be effective, employees must be trained in policies and procedures.
- Workers must have a solid understanding of the hazards. Training should review or provide new information about electrical hazards to which the student will or might be exposed.
- An evaluation of 224 incidents reported by FACE found that there were at least one of five factors was present in each incident:
  1. Established safe work procedures were either not implemented or were not followed.
  2. Adequate or required PPE was not provided or worn.
  3. Lockout/tagout procedures were either not implemented or were not followed.
  4. Existing OSHA, *NEC*, and NESC regulations were not implemented.
  5. Neither worker training nor supervisor training were adequate.
- The report found that many companies' electrical safety programs were not completely implemented, rendering them ineffective.
- The report identified 14 measures employers should take to reduce occupational electrocutions.
- It is important to note that technology changes over time, and as it changes, the hazards to which a worker can be exposed can also change.
- *NFPA 70E* Section 110.6 addresses training requirements in five major categories: Safety training, type of training, emergency procedures training, qualified and unqualified person training, and documentation of training.
- Section 110.6(A) points out that safety training requirements are applicable to people subjected to the hazards of electric shock, arc flash burn, thermal burn, or blast from contact or equipment failure when those hazards exceed an acceptable risk level, even when the equipment was installed in accordance with relevant installation requirements such as, but not limited to, the *NEC*.

- Shock protection is required if workers are working on live parts, since they could be exposed to a shock hazard.
- Arc flash protection might be necessary if an employee is working on live parts. The FPN to the definition of arc flash hazard (*NFPA 70E* Article 100) warns that an arc flash hazard may exist for enclosed equipment when "a person is interacting with the equipment in such a manner that could cause an electric arc" and that "enclosed energized equipment that has been properly installed and maintained is not likely to pose an arc flash hazard."
- Training must also include the practices and procedures required to protect from hazards from which injury or death could result. Practices and procedures could include, but are not limited to, lockout/tagout, establishing an electrically safe work condition, and performing an electrical hazard/risk analysis (including a shock and arc flash hazard analysis that includes labeling, determination of boundaries, and PPE).
- Section 110.6(B) indicates that training should be appropriate for the task—in some cases in the classroom or on the job, and in other cases, both. The more complex the task becomes, the more thorough and detailed the training must be.
- Section 110.6(C) addresses emergency procedure requirements, such as how to safely release victims from contact with energized electrical equipment, as well as how to administer first aid and perform resuscitative measures.
- There are specific training requirements for qualified and unqualified employees. Section 110.6(D) covers these. A qualified person must receive safety training to recognize and avoid the hazards involved. Unqualified employees must be trained in and familiar with electrical safety–related practices that pertain to his or her safety.
- Section 110.6(D) also states that employees must be retrained in the safe performance of a task before being permitted to perform that task again if he or she performs that task less than once a year.

- Section 110.6(E) emphasizes the need for employers to document safety training, including details such as the employee name and date of training.

Vocabulary

<u>**qualified person**</u>* One who has skills and knowledge related to the construction and operation of the electrical equipment and installations and has received safety training to recognize and avoid the hazards involved.

<u>**unqualified person**</u>* A person who is not a qualified person.

*This is the *NFPA 70E* definition.

Photo courtesy of National Electrical Contractors Association (NECA).

OPENING CASE

A 48-year-old machine operator was killed when he was crushed inside a plastic injection molding machine. The employer of the victim was a manufacturer that produced plastic products. The employer had been in business for over 80 years and had approximately 175 employees. There were 55 employees at the facility where the incident occurred. The victim had been employed with the company for 19 years.

The employer had a written safety program with written task-specific safe work procedures for all positions in the shop. Employees held weekly tailgate safety meetings with the supervisors, and formal, monthly safety meetings were held and documented. The company's training program was usually accomplished by on-the-job-training (OJT) monitored by the supervisors. Training was usually measured by testing, demonstration, and workmanship. The victim's duties included troubleshooting the machines' problems when they developed and taking appropriate action.

The machine involved in the incident was a completely automated plastic injection-molding machine. The operating portion of the machine was completely enclosed for safety. Whenever any regular access panel or door to the machine was opened, the machine would automatically shut off. The pedestal upon which the machine sat was also completely enclosed with the exception of the area where the conveyor belt exited from underneath the machine. The guarding around the pedestal had to be mechanically removed to gain access.

On the day of the incident, the machine stopped working, and its warning lights came on. The victim was not at his work station. Coworkers contacted the supervisor, who discovered the victim inside the machine with his head trapped between the pedestal frame and the mobile plate frame. Coworkers called 911, and the responding firefighters had to unbolt the guarding around the pedestal frame to gain access to the victim and extricate him. Paramedics pronounced the victim dead after they removed him from the machine.

Investigation of the incident site revealed the victim's tools lying next to the opening in the pedestal where the conveyor belt was located and several pieces of the machine product scattered about the

The Control of Hazardous Energy

CHAPTER 9

Overview

The Occupational Safety and Health Administration's (OSHA's) requirements are performance-oriented in many cases. Protection of workers is required without necessarily spelling out how worker protection is to be accomplished. This includes requirements related to protecting workers from electrical hazards. While more and more employers are looking to *NFPA 70E* for guidance in an effort to understand how to comply with OSHA's performance-oriented requirements, OSHA's requirements for lockout/tagout and working on or near exposed deenergized parts are examples where OSHA offers details including the steps that must be followed to comply.

This chapter explores OSHA's requirements for control of hazardous energy (lockout/tagout), working on or near exposed deenergized parts, and how provisions in *NFPA 70E* Section 120.1 can supplement OSHA's requirements and procedures.

OPENING CASE, CONT.

conveyor belt. The guard or shield on the side of the conveyor belt was unbolted on one side. These factors suggest that the victim crawled into the machine from the opening of the conveyor belt not being used. Once inside, he might have attempted to adjust the shield or guard to prevent the product from bouncing off the conveyor belt. Because the area he was in was extremely restrictive, he might have tried to stand up when the machine cycled, not realizing that the mobile portion of the mold would cycle backward again. When the machine cycled back, the victim might have been unable to move out of the way, and his head got caught between the frame of the pedestal and the frame of the mold.

The victim had authority to troubleshoot basic machine and processing problems. He also had authority to apply the lockout/tagout procedure to prepare the machine for repair by the maintenance department. The lockout/tagout procedure was detailed, and the victim had had refresher training on lockout/tagout just 1 week prior to the incident. The machine was considered stationary and had operable shut-off devices that were clearly identifiable, visible, and accessible. The victim did not use these devices to shut down or lockout/tagout the machine prior to his entrance. The cause of death according to the death certificate was blunt head trauma.

Adapted from: Fatality and Assessment and Control Evaluation (FACE) Investigation #03CA006. Accessed 9/14/08 at http://www.cdc.gov/niosh/face/stateface/ca/03ca006.html.

1. What could have been done to prevent this incident?

2. What can an employer do to increase the chances that employees implement lockout/tagout procedure requirements?

Chapter Outline

- **Overview**
- **Introduction**
- **The Control of Hazardous Energy (Lockout/Tagout)**
- **Typical Minimal Lockout Procedures**
- **Achieving an Electrically Safe Work Condition**
- **Lessons Learned**
- **Current Knowledge**
- **Summary**

Learning Objectives

1. Understand the requirements and procedures of 29 Code of Federal Regulations (CFR) 1910.147 and that this establishes minimum performance requirements for the control of hazardous energy.
2. Recognize that 29 CFR 1910.333(b) applies to work on exposed parts or work that is near enough to them to expose the employee to any electrical hazard they present.
3. Understand the conditions under which lockout/tagout procedures that comply with paragraphs (c) through (f) of 29 CFR 1910.147 will also be deemed to comply with 1910.333 (b)(2).
4. Identify the six-step process required to verify an electrically safe work condition.

References

1. *NFPA 70E*®, 2009 Edition
2. OSHA 29 CFR Part 1910
3. OSHA 29 CFR Part 1926

The electrically safe work condition is a concept first introduced in *NFPA 70E*. An electrically safe work condition is achieved when performed in accordance with the procedures of *NFPA 70E* Section 120.2 and verified by the six steps outlined in Section 120.1. Achieving an electrically safe work condition is one example of how *NFPA 70E* can supplement OSHA's requirements for lockout/tagout and working on or near exposed parts.

Introduction

Electricians and other workers often believe that a circuit is safe to approach or touch if it is deenergized. The fact that injuries and fatalities occur frequently, based upon this belief, proves that additional steps are needed to protect workers.

Some people also believe that if a lock and tag are placed on a labeled disconnecting means, the equipment is safe to work on. However, other issues must be considered. For example, labels can be marked incorrectly, equipment can be supplied from more than one source, or a temporary conductor could have been installed. It also is feasible that an unrelated energized circuit conductor could contact the conductor leading to the work area.

In other instances, workers outside the area or complicated systems can affect the work area. Often it is assumed that if the contact point is tested for absence of voltage, the point is safe for executing the task. But this only proves that no voltage is present at the time of the voltage test. Voltage could be absent due to a process interlock being open, or a second source of energy could simply be turned off for the moment. Avoiding incidents and injury requires training, planning, and preparation.

Many lockout/tagout programs are based on OSHA 1910.147. These requirements define the minimum re-quirements of a lockout/tagout program and must be referred to and complied with in their entirety. Since OSHA's lockout/tagout requirements are federal law and offer details including the steps that must be followed to comply, this chapter primarily focuses on the provisions of 29 CFR 1910.147 as well as two additional related paragraphs from 29 CFR 1910.333(b)(2).

OSHA notes that lockout and tagging procedures complying with 1910.147 are also deemed to comply with 1910.333(b)(2), provided that the procedures address the electrical safety hazards covered by Subpart S of 29 CFR 1910, and that those procedures additionally incorporate the requirements of two additional paragraphs not contained in 1910.147.

The first additional paragraph from 29 CFR 1910.333(b)(2) requires that a tag used without a lock must be supplemented by at least one additional safety measure that provides a level of safety equivalent to that obtained by the use of a lock. The second additional paragraph requires that a qualified person do the following:

- Use test equipment to test the circuit elements and electrical parts of equipment to which employees will be exposed.
- Verify that the circuit elements and equipment parts are deenergized.
- Determine if any energized condition exists as a result of inadvertently induced voltage or unrelated voltage backfeed even though specific parts of the circuit have been deenergized and presumed to be safe.
- Check the test equipment for proper operation immediately after this text, if the circuit to be tested is over 600 volts, nominal.

In addition to developing lockout/tagout programs based on requirements from OSHA, a number of the requirements from *NFPA 70E* Article 120 should be considered. The provisions of *NFPA 70E* Article 120 generally meet or exceed OSHA's lockout/tagout re-

---

OSHA Tip

29 CFR 1910.269 applies to the operation and maintenance of electric power generation, control, transformation, transmission, and distribution lines and equipment for the purpose of protecting employees.

Control of hazardous energy sources used in the generation of electric energy is covered in paragraph 29 CFR 1910.269(d). 29 CFR 1910.269(m) applies to the deenergizing of transmission and distribution lines and equipment for the purpose of protecting employees.

OSHA Tip

The standards contained in 29 CFR 1926 Subpart V apply to the construction of electric transmission and distribution lines and equipment. When a worker is deenergizing lines and equipment that are operated in excess of 600 volts, and the means of disconnecting from electric energy is not visibly open or visibly locked out, the provisions of paragraphs 29 CFR 1926.950(d)(1)(i) through (vii) shall be implemented

quirements and should be examined carefully for any additional considerations for a more comprehensive lockout/tagout program. This includes, but is not limited to, the six-step process outlined in Section 120.1 to verify that an electrically safe work condition exists after the provisions contained in *NFPA 70E* Section 120.2 have been taken into account.

The Control of Hazardous Energy (Lockout/Tagout)

OSHA 29 CFR 1910.147 is entitled "the control of hazardous energy (lockout/tagout)." It is broken down into six different major headings, and each is indicated by a different letter of the alphabet.

- (a) Scope, application, and purpose
- (b) Definitions
- (c) General
- (d) Application of control
- (e) Release from lockout or tagout
- (f) Additional requirements

Scope, Application, and Purpose

OSHA 29 CFR 1910.147(a) covers the scope, application, and purpose. 1910.147(a)(1) indicates what is covered and what is not covered. 1910.147(a)(2) describes when these requirements are applicable and when they are not applicable. 1910.147(a)(3) outlines the purpose of 1910.147 and its relationship to other standards. Part 1910.147(a)(3) is especially important in light of the relationship between 1910.147 and 1910.333(b). The requirements of 1910.333(b) must be used and supplemented by the procedural and training requirements of 1910.147.

1910.147(a)(1)

Scope.

1910.147(a)(1)(i)

This standard covers the servicing and maintenance of machines and equipment in which the unexpected energization or start up of the machines or equipment, or release of stored energy could cause injury to employees. This standard establishes minimum performance requirements for the control of such hazardous energy (Figure 9.1).

1910.147(a)(1)(ii)

This standard does not cover the following:

Construction, agriculture and maritime employment;

Installations under the exclusive control of electric utilities for the purpose of power generation, transmission and distribution, including related equipment for communication or metering; and

Figure 9.1 OSHA 29 CFR 1910.147 establishes minimum performance requirements for the control of hazardous energy.

Exposure to electrical hazards from work on, near, or with conductors or equipment in electric utilization installations, which is covered by Subpart S of this part; and

1910.147(a)(1)(ii)(D)

Oil and gas well drilling and servicing.

1910.147(a)(2)

Application.

1910.147(a)(2)(i)

This standard applies to the control of energy during servicing and/or maintenance of machines and equipment (Figure 9.2).

1910.147(a)(2)(ii)

Normal production operations are not covered by this standard (See Subpart O of this Part). Servicing and/or

Figure 9.2 When production equipment undergoes service, lockout/tagout must occur.

maintenance which takes place during normal production operations is covered by this standard only if:

1910.147(a)(2)(ii)(A)

An employee is required to remove or bypass a guard or other safety device; or

1910.147(a)(2)(ii)(B)

An employee is required to place any part of his or her body into an area on a machine or piece of equipment where work is actually performed upon the material being processed (point of operation) or where an associated danger zone exists during a machine operating cycle.

Note: Exception to paragraph (a)(2)(ii): Minor tool changes and adjustments, and other minor servicing activities, which take place during normal production operations, are not covered by this standard if they are routine, repetitive, and integral to the use of the equipment for production, provided that the work is performed using alternative measures which provide effective protection (See Subpart O of this Part).

1910.147(a)(2)(iii)

This standard does not apply to the following:

1910.147(a)(2)(iii)(A)

Work on cord and plug connected electric equipment for which exposure to the hazards of unexpected energization or start up of the equipment is controlled by the unplugging of the equipment from the energy source and by the plug being under the exclusive control of the employee performing the servicing or maintenance.

1910.147(a)(2)(iii)(B)

Hot tap operations involving transmission and distribution systems for substances such as gas, steam, water or petroleum products when they are performed on pressurized pipelines, provided that the employer demonstrates that-

1910.147(a)(2)(iii)(B)(1)

continuity of service is essential;

1910.147(a)(2)(iii)(B)(2)

shutdown of the system is impractical; and

1910.147(a)(2)(iii)(B)(3)

documented procedures are followed, and special equipment is used which will provide proven effective protection for employees.

1910.147(a)(3)

Purpose.

1910.147(a)(3)(i)

This section requires employers to establish a program and utilize procedures for affixing appropriate lockout devices or tagout devices to energy isolating devices, and to otherwise disable machines or equipment to

Figure 9.3   Lockout/tagout procedures involve the application of energy control devices to create a safe working environment.

prevent unexpected energization, start up or release of stored energy in order to prevent injury to employees (**Figure 9.3**).

1910.147(a)(3)(ii)

When other standards in this part require the use of lockout or tagout, they shall be used and supplemented by the procedural and training requirements of this section.

Definitions

OSHA 29 CFR 1910.147(b) provides the definitions for 1910.147. It is important to understand each of the definitions to fully understand and apply the requirements where these defined terms are used.

1910.147(b)

Definitions applicable to this section.

Affected employee. An employee whose job requires him/her to operate or use a machine or equipment on which servicing or maintenance is being performed under lockout or tagout, or whose job requires him/her to work in an area in which such servicing or maintenance is being performed.

Authorized employee. A person who locks out or tags out machines or equipment in order to perform servicing or maintenance on that machine or equipment. An affected employee becomes an authorized employee when that employee's duties include performing servicing or maintenance covered under this section.

Capable of being locked out. An energy isolating device is capable of being locked out if it has a hasp or other means of attachment to which, or through which, a lock can be affixed, or it has a locking mechanism built into it. Other energy isolating devices are capable of being locked out, if lockout can be achieved without the need to dismantle, rebuild, or replace the energy isolating device or permanently alter its energy control capability.

Energized. Connected to an energy source or containing residual or stored energy.

Energy-isolating device. A mechanical device that physically prevents the transmission or release of energy, including but not limited to the following: A manually operated electrical circuit breaker; a disconnect switch; a manually operated switch by which the conductors of a circuit can be disconnected from all ungrounded supply conductors, and, in addition, no pole can be operated independently; a line valve; a block; and any similar device used to block or isolate energy. Push buttons, selector switches and other control circuit type devices are not energy isolating devices.

Energy source. Any source of electrical, mechanical, hydraulic, pneumatic, chemical, thermal, or other energy.

Hot tap. A procedure used in the repair, maintenance and services activities which involves welding on a piece of equipment (pipelines, vessels, or tanks) under pressure, in order to install connections or appurtenances. It is commonly used to replace or add sections of pipeline without the interruption of service for air, gas, water, steam, and petrochemical distribution systems.

Lockout. The placement of a lockout device on an energy isolating device, in accordance with an established procedure, ensuring that the energy isolating device and the equipment being controlled cannot be operated until the lockout device is removed.

Lockout device. A device that utilizes a positive means such as a lock, either key or combination type, to hold an energy isolating device in the safe position and prevent the energizing of a machine or equipment. Included are blank flanges and bolted slip blinds.

Normal production operations. The utilization of a machine or equipment to perform its intended production function.

Servicing and/or maintenance. Workplace activities such as constructing, installing, setting up, adjusting, inspecting, modifying, and maintaining and/or servicing machines or equipment. These activities include lubrication, cleaning, or unjamming of machines or equipment and making adjustments or tool changes, where the employee may be exposed to the unexpected energization or start-up of the equipment or release of hazardous energy.

Setting up. Any work performed to prepare a machine or equipment to perform its normal production operation.

Tagout. The placement of a tagout device on an energy isolating device, in accordance with an established procedure, to indicate that the energy isolating device and the equipment being controlled may not be operated until the tagout device is removed.

Tagout device. A prominent warning device, such as a tag and a means of attachment, that can be securely fastened to an energy isolating device in accordance with an established procedure, to indicate that the energy isolating device and the equipment being controlled may not be operated until the tagout device is removed.

General

OSHA 29 CFR 1910.147(c) is further broken down into nine topics. Each change in topic is indicated by a change numerically from (1) through (9).

1910.147(c)

General -

1910.147(c)(1)

Energy control program. The employer shall establish a program consisting of energy control procedures, employee training and periodic inspections to ensure that before any employee performs any servicing or maintenance on a machine or equipment where the unexpected energizing, startup or release of stored energy could occur and cause injury, the machine or equipment shall be isolated from the energy source and rendered inoperative (**Figure 9.4**).

1910.147(c)(2)

Lockout/tagout.

1910.147(c)(2)(i)

If an energy isolating device is not capable of being locked out, the employer's energy control program under paragraph (c)(1) of this section shall utilize a tagout system (**Figure 9.5**).

Figure 9.4 It is the responsibility of the employer to provide an energy control program as well as training regarding that program.

Figure 9.5 Some devices cannot accept a lock. In these cases a tag must be applied.

1910.147(c)(2)(ii)

If an energy isolating device is capable of being locked out, the employer's energy control program under paragraph (c)(1) of this section shall utilize lockout, unless the employer can demonstrate that the utilization of a tagout system will provide full employee protection as set forth in paragraph (c)(3) of this section (**Figure 9.6**).

1910.147(c)(2)(iii)

After January 2, 1990, whenever replacement or major repair, renovation or modification of a machine or equipment is performed, and whenever new machines or equipment are installed, energy isolating

Figure 9.6 Any device that can be locked out must be locked out.

devices for such machine or equipment shall be designed to accept a lockout device.

1910.147(c)(3)

Full employee protection.

1910.147(c)(3)(i)

When a tagout device is used on an energy isolating device which is capable of being locked out, the tagout device shall be attached at the same location that the lockout device would have been attached, and the employer shall demonstrate that the tagout program will provide a level of safety equivalent to that obtained by using a lockout program.

1910.147(c)(3)(ii)

In demonstrating that a level of safety is achieved in the tagout program which is equivalent to the level of safety obtained by using a lockout program, the employer shall demonstrate full compliance with all tagout-related provisions of this standard together with such additional elements as are necessary to provide the equivalent safety available from the use of a lockout device. Additional means to be considered as part of the demonstration of full employee protection shall include the implementation of additional safety measures such as the removal of an isolating circuit element, blocking of a controlling switch, opening of an extra disconnecting device, or the removal of a valve handle to reduce the likelihood of inadvertent energization.

1910.147(c)(4)

Energy control procedure.

1910.147(c)(4)(i)

Procedures shall be developed, documented and utilized for the control of potentially hazardous energy when employees are engaged in the activities covered by this section.

Note: Exception: The employer need not document the required procedure for a particular machine or equipment, when all of the following elements exist: (1) The machine or equipment has no potential for stored or residual energy or reaccumulation of stored energy after shut down which could endanger employees; (2) the machine or equipment has a single energy source which can be readily identified and isolated; (3) the isolation and locking out of that energy source will completely deenergize and deactivate the machine or equipment; (4) the machine or equipment is isolated from that energy source and locked out during servicing or maintenance; (5) a single lockout device will achieve a locked-out condition; (6) the lockout device is under the exclusive control of the authorized employee performing the servicing or maintenance; (7) the servicing or maintenance does not create hazards for other employees; and (8) the employer, in utilizing this exception, has

Figure 9.7 Locks, tags, and other hardware must be provided by the employer for isolating, securing, or blocking of machines or equipment from energy sources.

had no accidents involving the unexpected activation or reenergization of the machine or equipment during servicing or maintenance.

1910.147(c)(4)(ii)

The procedures shall clearly and specifically outline the scope, purpose, authorization, rules, and techniques to be utilized for the control of hazardous energy, and the means to enforce compliance including, but not limited to, the following:

1910.147(c)(4)(ii)(A)

A specific statement of the intended use of the procedure;

1910.147(c)(4)(ii)(B)

Specific procedural steps for shutting down, isolating, blocking and securing machines or equipment to control hazardous energy;

1910.147(c)(4)(ii)(C)

Specific procedural steps for the placement, removal and transfer of lockout devices or tagout devices and the responsibility for them; and

1910.147(c)(4)(ii)(D)

Specific requirements for testing a machine or equipment to determine and verify the effectiveness of lockout devices, tagout devices, and other energy control measures.

1910.147(c)(5)

Protective materials and hardware.

1910.147(c)(5)(i)

Locks, tags, chains, wedges, key blocks, adapter pins, self-locking fasteners, or other hardware shall

be provided by the employer for isolating, securing or blocking of machines or equipment from energy sources (**Figure 9.7**).

1910.147(c)(5)(ii)

Lockout devices and tagout devices shall be singularly identified; shall be the only devices(s) used for controlling energy; shall not be used for other purposes (**Figure 9.8**); and shall meet the following requirements:

1910.147(c)(5)(ii)(A)

Durable.

1910.147(c)(5)(ii)(A)(1)

Lockout and tagout devices shall be capable of withstanding the environment to which they are exposed for the maximum period of time that exposure is expected (**Figure 9.9**).

1910.147(c)(5)(ii)(A)(2)

Tagout devices shall be constructed and printed so that exposure to weather conditions or

Figure 9.8 Locks to be used for lockout/tagout purposes must be labeled accordingly, and used only for that purpose.

Figure 9.9 Tags must be able to withstand the environment in which they are used.

Figure 9.10 Permanent markers are recommended for use on tags to be sure that all information remains legible.

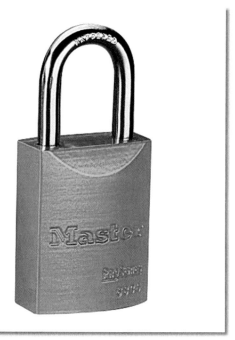

Figure 9.11 Lockout devices must be durable enough to prevent accidental removal.

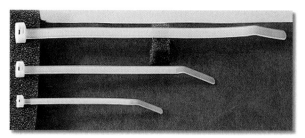

Figure 9.12 Tagout devices such as nylon cable ties must require a minimum of 50 lbs force to remove.

wet and damp locations will not cause the tag to deteriorate or the message on the tag to become illegible (**Figure 9.10**).

1910.147(c)(5)(ii)(A)(3)

Tags shall not deteriorate when used in corrosive environments such as areas where acid and alkali chemicals are handled and stored.

1910.147(c)(5)(ii)(B)

Standardized. Lockout and tagout devices shall be standardized within the facility in at least one of the following criteria: Color; shape; or size; and additionally, in the case of tagout devices, print and format shall be standardized.

1910.147(c)(5)(ii)(C)

Substantial -

1910.147(c)(5)(ii)(C)(1)

Lockout devices. Lockout devices shall be substantial enough to prevent removal without the use of excessive force or unusual techniques, such as with the use of bolt cutters or other metal cutting tools (**Figure 9.11**).

1910.147(c)(5)(ii)(C)(2)

Tagout devices. Tagout devices, including their means of attachment, shall be substantial enough to prevent inadvertent or accidental removal. Tagout device attachment means shall be of a non-reusable type, attachable by hand, self-locking, and non-releasable with a minimum unlocking strength of no less than 50 pounds and having the general design and basic characteristics of being at least equivalent to a one-piece, all environment-tolerant nylon cable tie (**Figure 9.12**).

1910.147(c)(5)(ii)(D)

Identifiable. Lockout devices and tagout devices shall indicate the identity of the employee applying the device(s).

1910.147(c)(5)(iii)

Tagout devices shall warn against hazardous conditions if the machine or equipment is energized and shall include a legend such as the following: Do Not Start. Do Not Open. Do Not Close. Do Not Energize. Do Not Operate (**Figure 9.13**).

1910.147(c)(6)

Periodic inspection.

1910.147(c)(6)(i)

The employer shall conduct a periodic inspection of the energy control procedure at least annually to

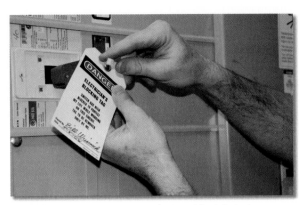

Figure 9.13 Tags must warn against accidental start-up.
Photo courtesy of National Electrical Contractors Association (NECA).

ensure that the procedure and the requirements of this standard are being followed (**Figure 9.14**).

1910.147(c)(6)(i)(A)

The periodic inspection shall be performed by an authorized employee other than the ones(s) utilizing the energy control procedure being inspected.

1910.147(c)(6)(i)(B)

The periodic inspection shall be conducted to correct any deviations or inadequacies identified.

1910.147(c)(6)(i)(C)

Where lockout is used for energy control, the periodic inspection shall include a review, between the inspector and each authorized employee, of that employee's responsibilities under the energy control procedure being inspected.

1910.147(c)(6)(i)(D)

Where tagout is used for energy control, the periodic inspection shall include a review, between the inspector and each authorized and affected employee, of that employee's responsibilities under the energy control procedure being inspected, and the elements set forth in paragraph (c)(7)(ii) of this section.

1910.147(c)(6)(ii)

The employer shall certify that the periodic inspections have been performed. The certification shall identify the machine or equipment on which the energy control procedure was being utilized, the date of the inspection, the employees included in the inspection, and the person performing the inspection.

1910.147(c)(7)

Training and communication.

1910.147(c)(7)(i)

The employer shall provide training to ensure that the purpose and function of the energy control program are understood by employees and that the knowledge and skills required for the safe application, usage, and removal of the energy controls are acquired by employees. The training shall include the following:

1910.147(c)(7)(i)(A)

Each authorized employee shall receive training in the recognition of applicable hazardous energy sources, the type and magnitude of the energy available in the workplace, and the methods and means necessary for energy isolation and control (**Figure 9.15**).

Figure 9.14 The employer must conduct a periodic inspection of the energy control procedure at least annually to ensure that the procedure and the requirements of 1910.147 are being followed.

Figure 9.15 Every authorized employee must be trained not only about all types of hazardous energy sources, but also in how to control that energy.
Photo courtesy of National Electrical Contractors Association (NECA).

Figure 9.16 All affected employees must know what lockout/tagout procedures are, and why they are important for the safety of all.
Photo courtesy of National Electrical Contractors Association (NECA).

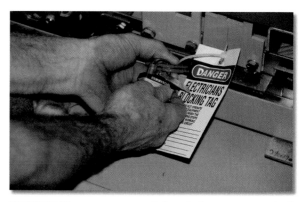

Figure 9.17 Tags are an essential part of an energy control plan, and should be removed only with the permission of the authorized person responsible.
Photo courtesy of National Electrical Contractors Association (NECA).

1910.147(c)(7)(i)(B)

Each affected employee shall be instructed in the purpose and use of the energy control procedure (**Figure 9.16**).

1910.147(c)(7)(i)(C)

All other employees whose work operations are or may be in an area where energy control procedures may be utilized, shall be instructed about the procedure, and about the prohibition relating to attempts to restart or reenergize machines or equipment which are locked out or tagged out.

1910.147(c)(7)(ii)

When tagout systems are used, employees shall also be trained in the following limitations of tags:

1910.147(c)(7)(ii)(A)

Tags are essentially warning devices affixed to energy isolating devices, and do not provide the physical restraint on those devices that is provided by a lock.

1910.147(c)(7)(ii)(B)

When a tag is attached to an energy isolating means, it is not to be removed without authorization of the authorized person responsible for it, and it is never to be bypassed, ignored, or otherwise defeated (**Figure 9.17**).

1910.147(c)(7)(ii)(C)

Tags must be legible and understandable by all authorized employees, affected employees, and all other employees whose work operations are or may be in the area, in order to be effective.

1910.147(c)(7)(ii)(D)

Tags and their means of attachment must be made of materials which will withstand the environmental conditions encountered in the workplace.

1910.147(c)(7)(ii)(E)

Tags may evoke a false sense of security, and their meaning needs to be understood as part of the overall energy control program.

1910.147(c)(7)(ii)(F)

Tags must be securely attached to energy isolating devices so that they cannot be inadvertently or accidentally detached during use.

1910.147(c)(7)(iii)

Employee retraining.

1910.147(c)(7)(iii)(A)

Retraining shall be provided for all authorized and affected employees whenever there is a change in their job assignments, a change in machines, equipment or processes that present a new hazard, or when there is a change in the energy control procedures (**Figure 9.18**).

1910.147(c)(7)(iii)(B)

Additional retraining shall also be conducted whenever a periodic inspection under paragraph (c)(6) of this section reveals, or whenever the employer has reason to believe that there are deviations from or inadequacies in the employee's knowledge or use of the energy control procedures.

1910.147(c)(7)(iii)(C)

The retraining shall reestablish employee proficiency and introduce new or revised control methods and procedures, as necessary.

1910.147(c)(7)(iv)

The employer shall certify that employee training has been accomplished and is being kept up to date.

Figure 9.18 Retraining must be provided for all authorized and affected employees whenever there is a change in equipment or processes that presents a new hazard.

Figure 9.19 Authorized employees must be familiar with the entire energy control procedure before shutting down any equipment.

The certification shall contain each employee's name and dates of training.

1910.147(c)(8)

Energy isolation. Lockout or tagout shall be performed only by the authorized employees who are performing the servicing or maintenance.

1910.147(c)(9)

Notification of employees. Affected employees shall be notified by the employer or authorized employee of the application and removal of lockout devices or tagout devices. Notification shall be given before the controls are applied, and after they are removed from the machine or equipment.

Application of Control

OSHA 29 CFR 1910.147(d) contains the elements and actions for energy control that procedures must cover and the order in which they must be executed. Application of control is broken down into six headings, which are identified by topic from (1) through (6).

1910.147(d)

Application of control. The established procedures for the application of energy control (the lockout or tagout procedures) shall cover the following elements and actions and shall be done in the following sequence:

1910.147(d)(1)

Preparation for shutdown. Before an authorized or affected employee turns off a machine or equipment, the authorized employee shall have knowledge of the type and magnitude of the energy, the hazards of the energy to be controlled, and the method or means to control the energy (**Figure 9.19**).

1910.147(d)(2)

Machine or equipment shutdown. The machine or equipment shall be turned off or shut down using the procedures established for the machine or equipment. An orderly shutdown must be utilized to avoid any additional or increased hazard(s) to employees as a result of the equipment stoppage.

1910.147(d)(3)

Machine or equipment isolation. All energy isolating devices that are needed to control the energy to the machine or equipment shall be physically located and operated in such a manner as to isolate the machine or equipment from the energy source(s) (**Figure 9.20**).

Figure 9.20 Appropriate devices must be applied to lockout or tagout equipment to properly isolate all energy sources.

1910.147(d)(4)

Lockout or tagout device application.

1910.147(d)(4)(i)

Lockout or tagout devices shall be affixed to each energy isolating device by authorized employees.

1910.147(d)(4)(ii)

Lockout devices, where used, shall be affixed in a manner that will hold the energy isolating devices in a "safe" or "off" position (**Figure 9.21**).

1910.147(d)(4)(iii)

Tagout devices, where used, shall be affixed in such a manner as will clearly indicate that the operation or movement of energy isolating devices from the "safe" or "off" position is prohibited.

1910.147(d)(4)(iii)(A)

Where tagout devices are used with energy isolating devices designed with the capability of being locked, the tag attachment shall be fastened at the same point at which the lock would have been attached.

1910.147(d)(4)(iii)(B)

Where a tag cannot be affixed directly to the energy isolating device, the tag shall be located as close as safely possible to the device, in a position that will be immediately obvious to anyone attempting to operate the device.

1910.147(d)(5)

Stored energy.

1910.147(d)(5)(i)

Following the application of lockout or tagout devices to energy isolating devices, all potentially hazardous stored or residual energy shall be relieved, disconnected, restrained, and otherwise rendered safe.

Figure 9.21 Lockout devices must be applied to safely keep energy sources isolated.

Figure 9.22 All stored energy must be released, and verified prior to service.

1910.147(d)(5)(ii)

If there is a possibility of reaccumulation of stored energy to a hazardous level, verification of isolation shall be continued until the servicing or maintenance is completed, or until the possibility of such accumulation no longer exists.

1910.147(d)(6)

Verification of isolation. Prior to starting work on machines or equipment that have been locked out or tagged out, the authorized employee shall verify that isolation and deenergization of the machine or equipment have been accomplished. (**Figure 9.22**)

Release from Lockout or Tagout

OSHA 29 CFR 1910.147(e) covers the procedures that must be followed and actions that must be taken before lockout or tagout devices are removed and energy is restored. The procedures for release from lockout or tagout are broken down into three headings.

1910.147(e)

Release from lockout or tagout. Before lockout or tagout devices are removed and energy is restored to the machine or equipment, procedures shall be followed and actions taken by the authorized employee(s) to ensure the following (**Figure 9.23**):

1910.147(e)(1)

The machine or equipment. The work area shall be inspected to ensure that nonessential items have been removed and to ensure that machine or equipment components are operationally intact.

1910.147(e)(2)

Employees.

1910.147(e)(2)(i)

The work area shall be checked to ensure that all employees have been safely positioned or removed.

Figure 9.23 Procedures must be followed and actions taken by authorized employees before lockout or tagout devices are removed and energy is restored to the machine or equipment.
Photo courtesy of National Electrical Contractors Association (NECA).

1910.147(e)(2)(ii)

After lockout or tagout devices have been removed and before a machine or equipment is started, affected employees shall be notified that the lockout or tagout device(s) have been removed.

1910.147(e)(3)

Lockout or tagout devices removal. Each lockout or tagout device shall be removed from each energy isolating device by the employee who applied the device. Exception to paragraph (e)(3): When the authorized employee who applied the lockout or tagout device is not available to remove it, that device may be removed under the direction of the employer, provided that specific procedures and training for such removal have been developed, documented and incorporated into the employer's energy control program. The employer shall demonstrate that the specific procedure provides equivalent safety to the removal of the device by the authorized employee who applied it. The specific procedure shall include at least the following elements:

1910.147(e)(3)(i)

Verification by the employer that the authorized employee who applied the device is not at the facility:

1910.147(e)(3)(ii)

Making all reasonable efforts to contact the authorized employee to inform him/her that his/her lockout or tagout device has been removed; and

1910.147(e)(3)(iii)

Ensuring that the authorized employee has this knowledge before he/she resumes work at that facility.

Additional Requirements

OSHA 29 CFR 1910.147(f) is the last of the six topic headings. The "Additional Requirements" topic is broken down into four headings as indicated in (f)(1) though (f)(4).

1910.147(f)

Additional requirements.

1910.147(f)(1)

Testing or positioning of machines, equipment or components thereof. In situations in which lockout or tagout devices must be temporarily removed from the energy isolating device and the machine or equipment energized to test or position the machine, equipment or component thereof, the following sequence of actions shall be followed:

1910.147(f)(1)(i)

Clear the machine or equipment of tools and materials in accordance with paragraph (e)(1) of this section;

1910.147(f)(1)(ii)

Remove employees from the machine or equipment area in accordance with paragraph (e)(2) of this section;

1910.147(f)(1)(iii)

Remove the lockout or tagout devices as specified in paragraph (e)(3) of this section;

1910.147(f)(1)(iv)

Energize and proceed with testing or positioning;

1910.147(f)(1)(v)

Deenergize all systems and reapply energy control measures in accordance with paragraph (d) of this section to continue the servicing and/or maintenance.

1910.147(f)(2)

Outside personnel (contractors, etc.).

1910.147(f)(2)(i)

Whenever outside servicing personnel are to be engaged in activities covered by the scope and application of this standard, the on-site employer and the outside employer shall inform each other of their respective lockout or tagout procedures (**Figure 9.24**).

1910.147(f)(2)(ii)

The on-site employer shall ensure that his/her employees understand and comply with the restrictions and prohibitions of the outside employer's energy control program (**Figure 9.25**).

1910.147(f)(3)

Group lockout or tagout.

1910.147(f)(3)(i)

When servicing and/or maintenance is performed by a crew, craft, department or other group, they shall utilize a procedure which affords the employees a level of protection equivalent to that provided by

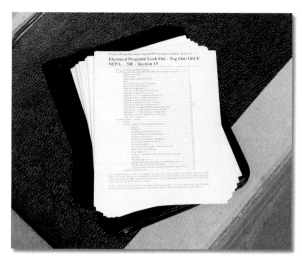

Figure 9.24 The on-site employer and the outside employer shall inform each other of their respective lockout or tagout procedures.
Photo courtesy of National Electrical Contractors Association (NECA).

Figure 9.25 The on-site employer shall ensure that his/her employees understand and comply with the outside employer's energy control program.
Photo courtesy of National Electrical Contractors Association (NECA).

the implementation of a personal lockout or tagout device.

1910.147(f)(3)(ii)

Group lockout or tagout devices shall be used in accordance with the procedures required by paragraph (c)(4) of this section including, but not necessarily limited to, the following specific requirements (**Figure 9.26**):

1910.147(f)(3)(ii)(A)

Primary responsibility is vested in an authorized employee for a set number of employees working under the protection of a group lockout or tagout device (such as an operations lock);

Figure 9.26 A group lockout/tagout procedure must provide the same level of protection as an individual procedure.

1910.147(f)(3)(ii)(B)

Provision for the authorized employee to ascertain the exposure status of individual group members with regard to the lockout or tagout of the machine or equipment and

1910.147(f)(3)(ii)(C)

When more than one crew, craft, department, etc. is involved, assignment of overall job-associated lockout or tagout control responsibility to an authorized employee designated to coordinate affected work forces and ensure continuity of protection; and

1910.147(f)(3)(ii)(D)

Each authorized employee shall affix a personal lockout or tagout device to the group lockout device, group lockbox, or comparable mechanism when he or she begins work, and shall remove those devices when he or she stops working on the machine or equipment being serviced or maintained.

1910.147(f)(4)

Shift or personnel changes. Specific procedures shall be utilized during shift or personnel changes to ensure the continuity of lockout or tagout protection, including provision for the orderly transfer of lockout or tagout device protection between off-going and oncoming employees, to minimize exposure to hazards from the unexpected energization or start-up of the machine or equipment, or the release of stored energy.

Applicable Part 1910 Subpart S Requirements

As mentioned previously, 1910.333(b)(2) Note 2 indicates that lockout and tagout procedures complying with 29 CFR 1910.147 are also deemed to comply with 1910.333(b)(2), provided that the procedures address the electrical safety hazards covered by Subpart S of 29 CFR 1910, and that those procedures additionally incorporate the requirements of two additional paragraphs not contained in 1910.147. Those additional procedures are as follows:

1910.333(b)(2) Note 2: Lockout and tagging procedures that comply with paragraphs (c) through (f) of 1910.147 will also be deemed to comply with paragraph (b)(2) of this section provided that: [1] The procedures address the electrical safety hazards covered by this Subpart; and [2] The procedures also incorporate the requirements of paragraphs (b)(2)(iii)(D) and (b)(2)(iv)(B) of this section.

1910.333(b)(2)(iii)(D)

A tag used without a lock, as permitted by paragraph (b)(2)(iii)(C) of this section, shall be supplemented by at least one additional safety measure that provides a level of safety equivalent to that obtained by use of a lock. Examples of additional safety measures include the removal of an isolating circuit element, blocking of a controlling switch, or opening of an extra disconnecting device.

BACKGROUND

OSHA Performance-Based Lockout/Tagout Requirements

An example of OSHA performance-based lockout/tagout requirements is found in Subpart K of 29 CFR 1926. Note the contrast with the detailed steps that must be followed to comply with the requirements in 29 CFR 1910.147.

1926.417

Lockout and tagging of circuits.

1926.417(a)

Controls. Controls that are to be deactivated during the course of work on energized or deenergized equipment or circuits shall be tagged.

1926.417(b)

Equipment and circuits. Equipment or circuits that are deenergized shall be rendered inoperative and shall have tags attached at all points where such equipment or circuits can be energized.

1926.417(c)

Tags. Tags shall be placed to identify plainly the equipment or circuits being worked on.

1910.333(b)(2)(iv)(B)

A qualified person shall use test equipment to test the circuit elements and electrical parts of equipment to which employees will be exposed and shall verify that the circuit elements and equipment parts are deenergized. The test shall also determine if any energized condition exists as a result of inadvertently induced voltage or unrelated voltage backfeed even though specific parts of the circuit have been deenergized and presumed to be safe. If the circuit to be tested is over 600 volts, nominal, the test equipment shall be checked for proper operation immediately after this test.

Chapter 3 covers the requirements in 29 CFR 1910.333(b) that are applicable to work on exposed deenergized parts or near enough to them to expose the employee to any electrical hazard they present.

Typical Minimal Lockout Procedures

Appendix A to OSHA 29 CFR 1910.147 (**Figure 9.27**) serves as a non-mandatory guideline to assist employers and employees in complying with the requirements of 1910.147, as well as to provide other helpful information. Nothing in the appendix adds to or detracts from any of the requirements of 1910.147. It is a simple lockout procedure to assist employers in developing their procedures so they meet the requirements of this standard. For more complex systems, more comprehensive procedures may need to be developed, documented, and utilized.

Achieving an Electrically Safe Work Condition

As discussed previously, many lockout/tagout programs are based on OSHA 1910.147, since OSHA's lockout/tagout requirements are federal law. However, in addition to developing lockout/tagout programs and procedures for working on or near exposed deenergized parts, based on 29 CFR 1910.147 and 1910.333(b), the process of achieving an electrically safe work condition from *NFPA 70E* Article 120 should be considered, including, but not limited to, *NFPA 70E* Section 120.1. Achieving an electrically safe work condition is one example of how *NFPA 70E* can supplement OSHA's requirements for lockout/tagout and working on or near exposed parts.

The Process of Achieving an Electrically Safe Work Condition

An electrically safe work condition is a concept first introduced in *NFPA 70E*. Achieving an electrically safe

General

The following simple lockout procedure is provided to assist employers in developing their procedures so they meet the requirements of this standard. When the energy isolating devices are not lockable, tagout may be used, provided the employer complies with the provisions of the standard which require additional training and more rigorous periodic inspections. When tagout is used and the energy isolating devices are lockable, the employer must provide full employee protection (see paragraph (c)(3)) and additional training and more rigorous periodic inspections are required. For more complex systems, more comprehensive procedures may need to be developed, documented, and utilized.

Lockout Procedure

Lockout Procedure for

_____

(Name of Company for single procedure or identification of equipment if multiple procedures are used).

Purpose

This procedure establishes the minimum requirements for the lockout of energy isolating devices whenever maintenance or servicing is done on machines or equipment. It shall be used to ensure that the machine or equipment is stopped, isolated from all potentially hazardous energy sources and locked out before employees perform any servicing or maintenance where the unexpected energization or start-up of the machine or equipment or release of stored energy could cause injury.

Compliance With This Program

All employees are required to comply with the restrictions and limitations imposed upon them during the use of lockout. The authorized employees are required to perform the lockout in accordance with this procedure. All employees, upon observing a machine or piece of equipment which is locked out to perform servicing or maintenance shall not attempt to start, energize, or use that machine or equipment.

_____

(Type of compliance enforcement to be taken for violation of the above.)

Sequence of Lockout

(1) Notify all affected employees that servicing or maintenance is required on a machine or equipment and that the machine or equipment must be shut down and locked out to perform the servicing or maintenance.

_____

(Name(s)/Job Title(s) of affected employees and how to notify.)

(2) The authorized employee shall refer to the company procedure to identify the type and magnitude of the energy that the machine or equipment utilizes, shall understand the hazards of the energy, and shall know the methods to control the energy.

_____

(Type(s) and magnitude(s) of energy, its hazards and the methods to control the energy.)

(3) If the machine or equipment is operating, shut it down by the normal stopping procedure (depress the stop button, open switch, close valve, etc.).

_____

(Type(s) and location(s) of machine or equipment operating controls.)

(4) De-activate the energy isolating device(s) so that the machine or equipment is isolated from the energy source(s).

_____

(Type(s) and location(s) of energy isolating devices.)

(5) Lock out the energy isolating device(s) with assigned individual lock(s).

(6) Stored or residual energy (such as that in capacitors, springs, elevated machine members, rotating flywheels, hydraulic systems, and air, gas, steam, or water pressure, etc.) must be dissipated or restrained by methods such as grounding, repositioning, blocking, bleeding down, etc.

_____

(Type(s) of stored energy—methods to dissipate or restrain.)

(7) Ensure that the equipment is disconnected from the energy source(s) by first checking that no personnel are exposed, then verify the isolation of the equipment by operating the push button or other normal operating control(s) or by testing to make certain the equipment will not operate.

Caution: Return operating control(s) to neutral or "off" position after verifying the isolation of the equipment.

_____

(Method of verifying the isolation of the equipment.)

(8) The machine or equipment is now locked out.

"Restoring Equipment to Service." When the servicing or maintenance is completed and the machine or equipment is ready to return to normal operating condition, the following steps shall be taken.

(1) Check the machine or equipment and the immediate area around the machine to ensure that nonessential items have been removed and that the machine or equipment components are operationally intact.

(2) Check the work area to ensure that all employees have been safely positioned or removed from the area.

(3) Verify that the controls are in neutral.

(4) Remove the lockout devices and reenergize the machine or equipment. Note: The removal of some forms of blocking may require reenergization of the machine before safe removal.

(5) Notify affected employees that the servicing or maintenance is completed and the machine or equipment is ready for used.

[54 FR 36687, Sept. 1, 1989 as amended at 54 FR 42498, Oct. 17, 1989; 55 FR 38685, Sept. 20, 1990; 61 FR 5507, Feb. 13, 1996]

Figure 9.27   Example of a simple lockout procedure from OSHA 29 CFR 1910.147 Appendix A.

work condition is one example of how *NFPA 70E* can supplement OSHA lockout/tagout requirements. This includes, but is not limited to, the six-step process outlined in Section 120.1 to verify that an electrically safe work condition exists after the provisions contained in *NFPA 70E* Section 120.2 have been taken into account. An electrically safe work condition is achieved when performed in accordance with the procedures of *NFPA 70E* Section 120.2 and verified by the following six steps outlined in Section 120.1.

(1) Determine all possible sources of electrical supply to the specific equipment. Check applicable up-to-date drawings, diagrams, and identification tags.

(2) After properly interrupting the load current, open the disconnecting device(s) for each source.

(3) Wherever possible, visually verify that all blades of the disconnecting devices are fully open or that drawout-type circuit breakers are withdrawn to the fully disconnected position.

(4) Apply lockout/tagout devices in accordance with a documented and established policy.

(5) Use an adequately rated voltage detector to test each phase conductor or circuit part to verify they are deenergized. Test each phase conductor or circuit part both phase-to-phase and phase-to-ground. Before and after each test, determine that the voltage detector is operating satisfactorily.

(6) Where the possibility of induced voltages or stored electrical energy exists, ground the phase conductors or circuit parts before touching them. Where it could be reasonably anticipated that the conductors or circuit parts being deenergized could contact other exposed energized conductors or circuit parts, apply ground connecting devices rated for the available fault duty.

The provisions of *NFPA 70E* Article 120 generally meet or exceed OSHA's lockout/tagout requirements

CAUTION

Consult American National Standards Institute/Instrumentation, Systems, and Automation Society (ANSI/ISA)-61010-1 (82.02.01)/Underwriters Laboratories Inc. (UL) 61010-1 for rating and design requirements for voltage measurement and test instruments intended for use on electrical systems 1000 volts and below [derived from *NFPA 70E* Section 120.1(5) Fine Print Note (FPN)].

and should be examined carefully for considerations for a more comprehensive and complete lockout/tagout program.

Summary

As discussed, OSHA's steps for lockout/tagout and detailed requirements for working on or near exposed deenergized parts must be followed. But even when a lock and tag are placed, one must also consider other factors such as incorrectly marked labels and additional sources of current. Training, planning, and preparation are critical to avoiding incidents and injury.

OSHA 29 CFR 1910.147 and OSHA CFR 1910.333 (b)(2) cover minimum steps for lockout/tagout, as well as procedures to follow when lockout or tagout devices are removed and energy is restored.

Finally, consider *NFPA 70E* Article 120 as a supplement to OSHA's requirements and procedures. Following the procedures in *NFPA 70E* Section 120.1 for verification that an electrically safe work condition has been achieved will allow for a more comprehensive lockout/tagout program.

Lessons *learned*

Electrician Electrocuted When He Contacts Energized Conductor in a Manhole

A 24-year-old electrician was electrocuted when he inadvertently contacted a 2300-volt circuit. The incident occurred while the victim was working inside a manhole splicing a conductor. The employer in this incident is an electrical contractor, engaged primarily in commercial and industrial electrical construction. The company has been in operation for 22 years and employs 97 workers, including 51 electricians. The company has a written safety policy and safety rules that are administered by the loss control/personnel manager. In addition, the employer holds weekly safety toolbox meetings and uses a safety incentive program. The employer also has a stepped (graduated) disciplinary system that consists of the following:

1. First incident, verbal counseling
2. Second incident, a written warning
3. Third incident, discharge

The victim worked for this employer for 3 years and 9 months prior to the incident.

The company had been contracted to install a lighting system for the taxiway and runway at a local airport. Preformed concrete manholes, which provided access to the underground circuitry, had been previously installed. Work was in progress to complete the wiring for permanent taxiway lights, and temporary work area lighting had already been installed. The victim and a coworker were part of a six-person crew assigned to install a new lighting system.

On the evening of the incident, a crew of six employees (one equipment operator, two apprentice electricians, two Journeyman electricians, and one electrician/foreman) arrived at the incident site to continue work on the lighting systems. The victim and a coworker were assigned the task of splicing the temporary taxiway lighting circuit conductors into the new conductor for the permanent taxiway lighting circuit and making the appropriate connections. The system consisted of three circuits:

1. An energized 2300-volt runway lighting circuit
2. An energized 700-volt temporary taxiway lighting circuit
3. A deenergized taxiway lighting circuit.

All the conductors were underground, and the manholes provided access to the conductor junctions.

Standard company procedure involved testing each circuit in the manhole by using an amp probe (a device used to detect current in a conductor) prior to working on that circuit, identifying the energized runway and temporary taxiway circuits, cutting the deenergized circuit (permanent taxiway circuit), and splicing together the appropriate conductors. Prior to the incident, the victim and coworker had completed connections for the permanent taxiway lights in four separate manholes. The victim entered the fifth manhole via a 24-inch diameter manway, descended a metal ladder attached to the inside of the manhole, and positioned himself on the ladder facing the circuit conductors. He removed a pair of insulated side (wire) cutters from his tool belt and, without using the amp probe to test for current in the conductors, cut a hanging conductor. The conductor, which was part of the energized runway lighting circuit, came in contact with the back of the victim's right hand after being cut in half. Current passed through the victim's right hand and exited his right thigh at the point of contact with the grounded ladder.

The coworker was standing near the top of the manhole observing the victim. After realizing what had occurred, he knocked the victim off the ladder away from the energized conductor. He entered the manhole and carried the victim out. The coworker then notified the electrician/foreman, who was in the area but working on a separate task. The foreman summoned airport emergency rescue personnel, who arrived within 3 minutes after being contacted. The rescue squad provided advanced cardiac life support and transported the victim to the local hospital, where he was pronounced dead 45 minutes after the incident occurred. The medical examiner listed the cause of death as electrocution.

Recommendations/Discussion

Recommendation 1: Employers must establish required procedures for the protection of employees exposed to electrical hazards and provide worker training in the recognition and avoidance of such hazards.

Discussion: Employers must comply with OSHA construction safety standard 29 CFR 1926.416(a)(1) by prohibiting employees from working in close proximity to energized electrical circuits where the employee could make contact in the course of work, unless the employee is protected against electric shock by deenergizing and grounding the circuit and/or by effective guarding. Employers should provide worker training in recognizing electrical hazards and in safe work procedures, including identifying circuits, testing circuits, deenergizing circuits, locking/tagging deenergized circuits, and verifying deenergization.

Recommendation 2: Employers should conduct initial job site surveys to identify all hazards associated with each specific job site and develop specific methods of controlling the identified hazards.

Discussion: Employers must comply with OSHA construction safety standard 29 CFR 1926.416(a)(3) by conducting initial job site surveys prior to the start of any work to identify potential situations for employee contact with energized electrical circuits, and by providing subsequent employee notification about protective measures (e.g., identification, testing, deenergization, locking/tagging of energized conductors, verification, and sufficient work area lighting) to be implemented to control the hazards.

Adapted from: FACE 90-32. Accessed 9/14/08 at http://www.cdc.gov/niosh/face/In-house/full9032.html.

Questions

1. How does measuring current indicate that a circuit is deenergized?
   A. Measuring current and measuring voltage is the same thing.
   B. An amp probe converts current measurement into voltage.
   C. The color of a conductor is an accurate indication of its purpose.
   D. It doesn't.

2. The incident report stated that the victim used insulated side cutters to cut the wire. That statement must have been made in error; we can tell that the statement is incorrect. Which of the following statements is not a valid indication of an insulated tool?
   A. The term *insulated* suggests that the side cutters had a material applied that would eliminate the risk of electrocution. If the side cutters were insulated, then the tool was voltage rated and would not have been in the worker's tool belt.
   B. Insulated hand tools are rated at 1000 volts, and the electrician was using the side cutters on a 2300-volt circuit.
   C. The victim did not verify the integrity of the "insulated" tool.
   D. The color of the comfort handles on the tool was blue.

3. The victim apparently entered the manhole without installing emergency extraction equipment as required. How did the coworker get the victim out of the manhole?
   A. He called emergency medical technicians for help.
   B. The equipment operator and the apprentice ran to assist him to lift the electrician.
   C. The coworker entered the manhole and lifted the victim through the 24-inch opening by himself.
   D. The fire department was adjacent to the site and responded with emergency extraction equipment.

4. The incident report suggests that the worker was in contact with a metal ladder. Why is that a bad idea?
   A. Metal ladders are normally rated for 100 pounds or less.
   B. ASTM D120 does not accept metal ladders.
   C. OSHA forbids the use of metal ladders in the vicinity of an electrical circuit.
   D. Metal ladders are much heavier and more difficult to maneuver than wood or fiberglass ladders.

CURRENT Knowledge

Ready for Review

- OSHA's requirements are written as performance-oriented requirements in many cases. Lockout/tagout is an area in which OSHA offers details including the steps that must be followed to comply in many cases.
- Provisions in *NFPA 70E* can supplement OSHA-based lockout/tagout programs.
- Electricians and other workers often believe that a circuit is safe to approach or touch if it is deenergized; however, additional steps are needed.
- Placing a lock and tag on a labeled disconnecting means does not ensure that the equipment is safe to work on. Incorrectly marked labels, additional energy sources, or temporary conductor could allow an unrelated energized circuit conductor to contact the conductor leading to the work area.
- Workers often assume that if the contact point is tested for absence of voltage, the point is safe for executing the task. However, this action only proves that no voltage is present at the time of the voltage test.
- Avoiding incidents and injury requires training, planning, and preparation.
- OSHA 29 CFR 1910.147 defines minimum requirements of a lockout/tagout program. These requirements are federal law and must be adhered to.
- Additional related components from OSHA include 29 CFR 1910.333(b)(2).
- A summary of OSHA lockout/tagout requirements follows:
  - A tag used without a lock must be supplemented by at least one additional safety measure that provides a level of safety equivalent to that obtained by the use of a lock.
  - A qualified person must use test equipment to test the circuit elements and electrical parts of equipment to which employees will be exposed, verify that the circuit elements and equipment parts are deenergized, and determine if any energized condition exists as a result of inad-

vertently induced voltage or unrelated voltage backfeed, even though specific parts of the circuit have been deenergized and presumed to be safe. If the circuit to be tested is over 600 volts, nominal, the test equipment shall be checked for proper operation immediately after this test.
- Lockout/tagout programs must be developed based on requirements from OSHA, but a number of the requirements from *NFPA 70E* Article 120 should also be considered, such as the six-step process outlined in Section 120.1.
- OSHA 29 CFR 1910.147 is entitled "the control of hazardous energy (lockout/tagout)," and its scope includes:
  - The servicing and maintenance of machines and equipment, in which the unexpected energization or start-up of the machines or equipment, or release of stored energy, could cause injury to employees.
  - Establishing minimum performance requirements for the control of such hazardous energy.
- OSHA 29 CFR 1910.147 requires employers to establish a program and utilize procedures for affixing appropriate lockout devices or tagout devices to energy-isolating devices and to otherwise disable machines or equipment to prevent unexpected energization, start-up, or release of stored energy.
- OSHA 29 CFR 1910.147(b) provides specific definitions relevant to 1910.147. A worker must be familiar with the meanings of these definitions, as they contain important nuances that relate to the requirements.
- OSHA 29 CFR 1910.147(c)(1) provides further requirements for the employer's energy control procedures, employee training, and periodic inspections.
- OSHA 29 CFR 1910.147(c)(2) specifies procedures for lockout/tagout. Lockout should be used wherever possible. Tagout may be used if the employer can demonstrate that the utiliza-

tion of a tagout system can provide full employee protection as set forth in paragraph (c)(3). When a tagout device is used on an energy-isolating device that is capable of being locked out, the tagout device shall be attached at the same location that the lockout device would have been attached.

- OSHA 29 CFR 1910.147(c)(3)(ii) requires employers to demonstrate full compliance with all tagout-related provisions of the standard together with such additional elements as are necessary to provide the equivalent safety available from the use of a lockout device. These additional means can include safety measures, such as the removal of an isolating circuit element, blocking of a controlling switch, opening of an extra disconnecting device, or the removal of a valve handle to reduce the likelihood of inadvertent energization.

- OSHA 29 CFR 1910.147(c)(4) requires that procedures be developed, documented, and utilized for the control of potentially hazardous energy when employees are engaged in the activities covered by this section. The procedures shall clearly and specifically outline the scope, purpose, authorization, rules, and techniques to be utilized for the control of hazardous energy and the means to enforce compliance.

- OSHA 29 CFR 1910.147(c)(5) identifies protective materials and hardware that the employer must provide for isolating, securing, or blocking machines or equipment from energy sources. It also lists the requirements that these protective materials must meet.

- OSHA 29 CFR 1910.147(c)(6) requires the employer to conduct a periodic inspection of the energy control procedure to ensure that it is being followed and that the specific requirements are being met.

- OSHA 29 CFR 1910.147(c)(7) covers training and communication, e.g., the employer must provide training to ensure that employees understand the purpose and function of the energy control program and that employees have the necessary knowledge and skills.

- OSHA 29 CFR 1910.147(c)(8) indicates that lockout or tagout shall be performed only by the authorized employees who are performing the servicing or maintenance.

- OSHA 29 CFR 1910.147(c)(9) requires that the employer notify affected employees of application and removal of lockout devices or tagout devices and specifies when notification should occur.

- OSHA 29 CFR 1910.147(d) covers application of control, broken down into six headings identifying the elements and actions.

- OSHA 29 CFR 1910.147(d)(1) lists the first step as preparation for shutdown. The authorized employee shall have knowledge of the type and magnitude of the energy, the hazards of the energy to be controlled, and the method or means to control the energy.

- OSHA 29 CFR 1910.147(d)(2) requires that the machine or equipment be turned off or shut down, using the procedures established for the machine or equipment.

- OSHA 29 CFR 1910.147(d)(3) requires machine or equipment isolation from the energy source(s).

- OSHA 29 CFR 1910.147(d)(4) outlines lockout or tagout device application as follows:
  - Lockout or tagout devices shall be affixed to each energy-isolating device by authorized employees.
  - Lockout devices, where used, shall be affixed in a manner that will hold the energy-isolating devices in a "safe" or "off" position.
  - Tagout devices, where used, shall be affixed in such a manner as will clearly indicate that the operation or movement of energy-isolating devices from the "safe" or "off" position is prohibited.
  - Where tagout devices are used with energy-isolating devices designed with the capability of being locked, the tag attachment shall be fastened at the same point at which the lock would have been attached.

- Where a tag cannot be affixed directly to the energy-isolating device, the tag shall be located as close as safely possible to the device, in a position that will be immediately obvious to anyone attempting to operate the device.
- OSHA 29 CFR 1910.147(d)(5) requires that potentially hazardous stored or residual energy be relieved, disconnected, restrained, and otherwise rendered safe.
- OSHA 29 CFR 1910.147(d)(6) requires the authorized employee to verify that isolation and deenergization of the machine or equipment have been accomplished prior to starting work.
- OSHA 29 CFR 1910.147(e) covers the procedures and actions required before lockout or tagout devices are removed and energy is restored. The work area must be inspected to ensure removal or safety of all employees, removal of nonessential items, and to ensure that machine or equipment components are operationally intact. After lockout or tagout devices have been removed and before a machine or equipment is started, affected employees shall be notified.
- OSHA 29 CFR 1910.147(e)(3) outlines the procedure for removal of lockout or tagout devices. Each lockout or tagout device shall be removed from each energy-isolating device by the employee who applied the device. This section outlines exceptions to this rule as well.
- OSHA 29 CFR 1910.147(f) lists additional requirements, such as temporary removal of lockout/tagout devices while testing the machine or equipment, working with outside personnel, and group lockout/tagout.
- OSHA 29 CFR 1910.147(f)(4) covers procedures to ensure the continuity of lockout or tagout protection during shift or personnel changes.
- OSHA 29 CFR 1910.333(b)(2) indicates additional requirements relating to lockout and tagging procedures:
  - OSHA 29 CFR 1910.333(b)(2)(iii)(D) indicates that a tag used without a lock shall be supplemented by at least one additional safety measure that provides a level of safety equivalent to that obtained by use of a lock.
  - OSHA 29 CFR 1910.333(b)(2)(iv)(B) relates to equipment testing and states that a qualified person shall use test equipment to test the circuit elements and electrical parts of equipment to which employees will be exposed and shall verify that the circuit elements and equipment parts are deenergized.
- OSHA 29 CFR 1910.147 App A provides a typical minimal lockout procedure as an example of practical implementation of 1910.147. Workers should become familiar with this procedure in an effort to understand how OSHA requirements translate into the actual procedure.
- *NFPA 70E* Section 120.2 outlines procedures for achieving an electrically safe work condition. Section 120.1 outlines the following six steps to verify that this has been achieved.

  (1) Determine all possible sources of electrical supply to the specific equipment. Check applicable up-to-date drawings, diagrams, and identification tags.

  (2) After properly interrupting the load current, open the disconnecting device(s) for each source.

  (3) Wherever possible, visually verify that all blades of the disconnecting devices are fully open or that drawout-type circuit breakers are withdrawn to the fully disconnected position.

  (4) Apply lockout/tagout devices in accordance with a documented and established policy.

  (5) Use an adequately rated voltage detector to test each phase conductor or circuit part to verify they are deenergized. Test each phase conductor or circuit part both phase-to-phase and phase-to-ground. Before and after each test, determine that the voltage detector is operating satisfactorily.

  (6) Where the possibility of induced voltages or stored electrical energy exists, ground the phase conductors or circuit parts before touching them. Where it could be reasonably anticipated that the conductors or circuit parts being deenergized could contact other exposed energized conductors or circuit parts, apply ground connecting devices rated for the available fault duty.

Vocabulary

affected employee§ An employee whose job requires him/her to operate or use a machine or equipment on which servicing or maintenance is being performed under lockout or tagout, or whose job requires him/her to work in an area in which such servicing or maintenance is being performed.

authorized employee§ A person who locks out or tags out machines or equipment in order to perform servicing or maintenance on that machine or equipment. An affected employee becomes an authorized employee when that employee's duties include performing servicing or maintenance covered under this section.

capable of being locked out§ An energy isolating device is capable of being locked out if it has a hasp or other means of attachment to which, or through which, a lock can be affixed, or it has a locking mechanism built into it. Other energy isolating devices are capable of being locked out, if lockout can be achieved without the need to dismantle, rebuild, or replace the energy isolating device or permanently alter its energy control capability.

energized§ Electrically connected to, or is, a source of voltage.

energy-isolating device§ A mechanical device that physically prevents the transmission or release of energy, including but not limited to the following: A manually operated electrical circuit breaker; a disconnect switch; a manually operated switch by which the conductors of a circuit can be disconnected from all ungrounded supply conductors, and, in addition, no pole can be operated independently; a line valve; a block; and any similar device used to block or isolate energy. Push buttons, selector switches and other control circuit type devices are not energy isolating devices.

energy source§ Any source of electrical, mechanical, hydraulic, pneumatic, chemical, thermal, or other energy.

lockout§ The placement of a lockout device on an energy isolating device, in accordance with an established procedure, ensuring that the energy isolating device and the equipment being controlled cannot be operated until the lockout device is removed.

lockout device§ A device that utilizes a positive means such as a lock, either key or combination type, to hold an energy isolating device in the safe position and prevent the energizing of a machine or equipment. Included are blank flanges and bolted slip blinds.

tagout§ The placement of a tagout device on an energy isolating device, in accordance with an established procedure, to indicate that the energy isolating device and the equipment being controlled may not be operated until the tagout device is removed.

tagout device§ A prominent warning device, such as a tag and a means of attachment, which can be securely fastened to an energy isolating device in accordance with an established procedure, to indicate that the energy isolating device and the equipment being controlled may not be operated until the tagout device is removed.

§This is the OSHA definition.

⚠ WARNING

Arc Flash and Shock Hazard

Appropriate PPE Required

10 inch	Flash Hazard Boundary
0.45 cal/cm^2	Flash Hazard at 18 inches
Category 0	Untreated Cotton
480 VAC	Shock Hazard when cover is removed
00	Glove Class
42 inch	Limited Approach
12 inch	Restricted Approach
1 inch	Prohibited Approach

Location: **ATS-ESB-L**

OPENING CASE

Photo courtesy of National Electrical Contractors Association (NECA).

A 39-year-old Journeyman electrician was removing a metal fish tape from the access hole at the base of a metal light pole. The fish tape became energized, electrocuting him. It was raining at the time that the incident occurred.

The victim had been on the project for 1 day and had 16 years experience at this type of work. There was no training and education provided by the employer, and it was determined that an inadequate safety and health program was in effect.

As a result of its inspection, the Occupational Safety and Health Administration (OSHA) issued a citation for three serious violations of the agency's construction standards. Had requirements for deenergizing energy sources been followed, the electrocution might have been prevented.

Adapted from: Accident Summary No. 30, *Fatal Facts Accident Report*. Occupational Safety and Health Administration. Accessed 8/27/ 08 at http://osha.gov/OshDoc/data_FatalFacts/f-facts30.html.

1. What procedures should have been followed that could have prevented this incident?
2. Was the energized work justified in this case?

Justification, Assessment, and Documentation

CHAPTER 10

Chapter Outline

- **Introduction**
- **Justification**
- **Energized Electrical Work Permit**
- **Approach Boundaries**
- **Arc Flash Hazard Analysis**
- **Summary**
- **Lessons Learned**
- **Current Knowledge**

Introduction

"Work Involving Electrical Hazards" is the title of *NFPA 70E* Article 130. This chapter will begin the discussion of Article 130 by examining the first three sections of that article:

- The justification for energized work and the energized electrical work permit requirements of Section 130.1
- The shock hazard analysis and approach boundary requirements of Section 130.2
- The arc flash hazard analysis requirements of Section 130.3

These topics detail a prescriptive-type solution to OSHA's performance-based requirements for a written hazard assessment to determine and select personal protective equipment (PPE) based on the hazard assessment.

Justification

NFPA 70E Section 130.1(A) requires that energized electrical equipment be placed into an electrically safe work condition if a worker is to cross the limited approach boundary (LAB) associated with the energized electrical equipment. For example, a worker cannot get closer than 42 in. to an exposed energized fixed circuit part operating at 480 volts, according to Table 130.2(C), unless the equipment is in an electrically safe work condition. An exception to this is if the employer can demonstrate that deenergizing the circuit introduces additional or increased hazards or is infeasible due to equipment design or operational limitations.

Two fine print notes (FPNs) to Section 130.1(A) provide examples of infeasibility and increased or additional hazards that would permit energized work. These examples are similar to those that OSHA offers in 29 CFR 1910.333(a)(1).

> 1910.333(a)(1) Live parts to which an employee may be exposed shall be deenergized before the employee works on or near them, unless the employer can demonstrate that deenergizing introduces additional or increased hazards or is infeasible due to equipment design or operational limitations. Live parts that operate at less than 50 volts to ground need not be deenergized if there will be no increased exposure to electrical burns or to explosion due to electric arcs.
>
> Note 1: Examples of increased or additional hazards include interruption of life support equipment, deactivation of emergency alarm systems, shutdown of hazardous location ventilation equipment, or removal of illumination for an area.
>
> Note 2: Examples of work that may be performed on or near energized circuit parts because of infeasibility due to equipment design or operational limitations include testing of electric circuits that can only be performed with the circuit energized and work on circuits that form an integral part of a continuous industrial process in a chemical plant that would otherwise need to be completely shut down in order to permit work on one circuit or piece of equipment.

Examples of increased or additional hazards according to *NFPA 70E* include, but are not limited to:

- Interruption of life support equipment
- Deactivation of emergency alarm systems
- Shutdown of hazardous location ventilation equipment

Learning Objectives

1. Describe when energized work is permitted.
2. Understand the elements that are required to be in an energized electrical work permit and when they are required to be implemented.
3. Understand approach boundaries to energized electrical conductors or circuit parts and how they apply to qualified and unqualified persons.
4. Describe the three components necessary to fulfill what is required as part of an arc flash hazard analysis.

References

1. *NFPA 70E®*, 2009 Edition
2. OSHA 29 CFR Part 1910

Examples of work that might be infeasible to perform while deenergized due to equipment design or operational limitations according to *NFPA 70E* include:

- Diagnostics and testing of circuits that can only be performed while energized
- Work on circuits that form an integral part of a continuous process that would otherwise need to be completely shut down in order to permit work on one circuit or piece of equipment

It's not safe to assume that energized electrical equipment operating at less than 50 volts is free of hazard. *NFPA 70E* Section 130.1(A)(3) requires that the capacity of the source and any overcurrent protection between the energy source and the worker be evaluated to determine the need to deenergize.

A third FPN to Section 130.1(A) discusses arcing faults within enclosures, how occurrences of arcing faults differ from those of bolted faults, and the equipment and design practices available to mitigate related hazards. These topics are covered in detail in Chapters 3, 12, and 15 of this textbook.

Safe work practices, including personal and other protective equipment, are required even if the employer has demonstrated that deenergizing introduces additional or increased hazards or is infeasible due to equipment design or operational limitations.

Consider the two photos that follow. **Figure 10.1** illustrates a task performed inside equipment with live parts. Whether or not this task met the criteria for energized work (greater hazard or infeasibility demonstrated), the results of what can and does happen are shown in **Figure 10.2** . This illustrates the need for safe work practices, PPE, and insulated tools. Both OSHA and *NFPA 70E* require the use of insulated (voltage-rated) tools or insulating tools (such as a hot stick) for energized work.

Figure 10.1   Example illustrating a task to be performed energized.

Figure 10.2   Example illustrating the result of working energized without the use of an insulated tool.

Working in an electrically safe work condition, or using a voltage-rated tool where energized work is warranted, could have prevented the incident in Figure 10.2.

Energized Electrical Work Permit

OSHA requires a written hazard assessment in 29 CFR 1910.132. The performance language requires an employer to assess the workplace to determine if hazards are present, or are likely to be present, that would necessitate the use of PPE. The employer must provide the affected worker with the appropriate PPE for the hazards identified in the assessment. *NFPA 70E* also requires a written hazard assessment. However, *NFPA 70E* Section 130.1(B)(2) provides specific minimum requirements as a means for employers to assess the task to be performed and determine appropriate PPE to protect against electrical hazards.

Energized work is generally to be performed in accordance with the energized electrical work permit requirements of Section 130.1(B) if the equipment is not in an electrically safe work condition. *NFPA 70E* Annex J.1 offers a sample permit (**Figure 10.3**).

Annex J.1, like all *NFPA* annexes, is nonmandatory. *NFPA 70E* Section 130.1(B)(2), however, details the minimum information that must be documented in an energized electrical work permit.

- Description and location of the equipment and circuit
- Justification for the need to work on energized circuits
- A description of the safe work practices to be used
- The voltage that workers will be exposed to (results of the shock hazard analysis)

ENERGIZED ELECTRICAL WORK PERMIT

PART I: TO BE COMPLETED BY THE REQUESTER:

Job/Work Order Number _____

(1) Description of circuit/equipment/job location: _____

(2) Description of work to be done: _____

(3) Justification of why the circuit/equipment cannot be de-energized or the work deferred until the next scheduled outage: _____

_____

Requester/Title   Date

PART II: TO BE COMPLETED BY THE ELECTRICALLY QUALIFIED PERSONS *DOING* THE WORK:

Check when complete

(1) Detailed job description procedure to be used in performing the above detailed work: _____ ☐

(2) Description of the Safe Work Practices to be employed: _____ ☐

(3) Results of the Shock Hazard Analysis: _____ ☐

(4) Determination of Shock Protection Boundaries: _____ ☐

(5) Results of the Arc Flash Hazard Analysis: _____ ☐

(6) Determination of the Arc Flash Protection Boundary: _____ ☐

(7) Necessary personal protective equipment to safely perform the assigned task: _____ ☐

(8) Means employed to restrict the access of unqualified persons from the work area: _____ ☐

(9) Evidence of completion of a Job Briefing including discussion of any job-related hazards: _____ ☐

(10) Do you agree the above described work can be done safely? ☐ Yes ☐ No (If *no*, return to requester)

Electrically Qualified Person(s)   Date

Electrically Qualified Person(s)   Date

PART III: APPROVAL(S) TO PERFORM THE WORK WHILE ELECTRICALLY ENERGIZED:

Manufacturing Manager   Maintenance/Engineering Manager

Safety Manager   Electrically Knowledgeable Person

General Manager   Date

Note: Once the work is complete, forward this form to the site Safety Department for review and retention.

© 2008 National Fire Protection Association   NFPA 70E

Figure 10.3 An energized electrical work permit is generally required for energized electrical equipment that is not in an electrically safe work condition.

- The shock protection boundaries
- Results of the arc flash hazard analysis
- The arc flash protection boundary (AFPB)
- The PPE required for shock protection
- The PPE required for arc flash
- How unqualified persons will be kept from the work area
- Evidence that a job briefing was completed, including job-specific hazards
- Signature of the person authorizing the approval of energized work

NFPA 70E Section 130.1(B)(3) recognizes a limited number of examples where qualified persons can perform energized work inside the LAB without an energized electrical work permit as long as appropriate work practices and PPE are used. These include, but are not limited to:

- Testing
- Troubleshooting
- Voltage measuring
- Visual inspection where the restricted approach boundary will not be crossed

Tasks that are exempt from the energized electrical work permit requirements are not exempt from the need to select, provide, and use adequate personal and other protective equipment in accordance with *NFPA 70E* Chapter 1. A job briefing would still be required in accordance with *NFPA 70E* Section 110.7, including tasks qualifying under the exemption. Using the first 10 steps of *NFPA 70E* Section 130.3(B)(2), even for tasks exempt from the permit, would provide a consistent means to determine the required PPE and the boundaries where PPE should be used, for example. Information on the permit would also be useful in the job briefing to cover the hazards, procedures, precautions, energy source controls, and PPE associated with the task to be performed.

■ Approach Boundaries

NFPA 70E Section 130.2 discusses shock protection through four major headings:

- Shock hazard analysis
- Shock protection boundaries
- Approach to exposed energized electrical conductors or circuit parts operating at 50 volts or more
- Approach by unqualified persons

Shock Hazard Analysis

NFPA 70E Section 130.2(A) requires a shock hazard analysis. **Shock hazard** is defined in *NFPA 70E* Article 100:

Shock Hazard. A dangerous condition associated with the possible release of energy caused by contact or approach to energized electrical conductors or circuit parts.

The shock hazard analysis must determine:

- The voltage that workers will be exposed to
- The shock protection boundaries
- The PPE required for shock protection

The results of the shock hazard analysis are part of the information required to be documented on the energized electrical work permit to meet the requirements of *NFPA 70E* Section 130.1(B)(2).

Shock Protection Boundaries

There are three shock protection boundaries defined in *NFPA 70E* Article 100:

- **Limited approach boundary** (LAB)
- **Restricted approach boundary**
- **Prohibited approach boundary**

Boundary, Limited Approach. An approach limit at a distance from an exposed energized electrical conductor or circuit part within which a shock hazard exists.

Boundary, Restricted Approach. An approach limit at a distance from an exposed energized electrical conductor or circuit part within which there is an increased risk of shock, due to electrical arc over combined with inadvertent movement, for personnel working in close proximity to the energized electrical conductor or circuit part.

Boundary, Prohibited Approach. An approach limit at a distance from an exposed energized electrical conductor or circuit part within which work is considered the same as making contact with the electrical conductor or circuit part.

These shock protection boundaries apply when workers are exposed to energized electrical conductors or circuit parts. Table 130.2(C) establishes the distances associated within a range of nominal system voltages (as illustrated in Table 8.1 in Chapter 8). The three approach boundaries and associated distances for each are located in this table. **Nominal voltage** is defined by *NFPA 70E* as follows:

Voltage, Nominal. A nominal value assigned to a circuit or system for the purpose of conveniently designating its voltage class (e.g., 120/240 volts, 480Y/277 volts, 600 volts). The actual voltage at which a circuit operates can vary from the nominal within a range that permits satisfactory operation of equipment.

The table heading clarifies that the approach boundaries are for shock protection from energized electrical equipment and that all dimensions indicated are the distance from an energized part to the worker. The table is broken down into five vertical columns as follows:

- Column 1 contains the range of phase-to-phase voltage.
- Column 2 contains dimensions for a LAB for an exposed movable conductor. *Note*: The third footnote to Table 130.2(C) describes the term **exposed movable conductor**:

  > A condition in which the distance between the conductor and a person is not under the control of the person. The term is normally applied to overhead line conductors supported by poles.

- Column 3 contains dimensions for a LAB for an exposed fixed circuit part.
- Column 4 contains dimensions for a restricted approach boundary including inadvertent movement adder. (Commentary in the 2009 edition of the *NFPA 70E Handbook* advises that the restricted approach boundary factors in the potential for unintended movement by a worker who might move his or her hand unintentionally.)
- Column 5 contains the dimensions for a prohibited approach boundary.

NFPA Annex C elaborates on Table 130.2(C). It discusses the reasons for the content of the five columns in the Table, as well as some insight into the requirements of Section 130.2. Annex C should be read thoroughly to gain a better understanding of the requirements of Section 130.2.

The LAB is the first shock protection boundary that a worker encounters as he or she approaches and is exposed to energized electrical equipment. However, the LAB might not be the first boundary that a worker encounters as he or she approaches electrical equipment. The FPN to Section 130.2(B) warns that an AFPB might be a greater distance from the exposed energized electrical conductors or circuit parts than the limited approach boundary.

The shock protection boundaries are based on nominal system voltage, while the AFPB is conditional. The AFPB is determined in accordance with Section 130.1(A). The boundary is determined by Section 130.1(A)(1) for voltage levels between 50 and 600:

130.3(A)(1) Voltage Levels Between 50 and 600 Volts.

In those cases where detailed arc flash hazard analysis calculations are not performed for systems that are between 50 and 600 volts, the Arc Flash Protection Boundary shall be 4.0 ft, based on the product of clearing time of 2 cycles (0.033 second) and the available bolted fault current of 50 kA or any combination not exceeding 100 kA cycles (1667 ampere seconds). When the product of clearing times and bolted fault current exceeds 100 kA cycles, the Arc Flash Protection Boundary shall be calculated.

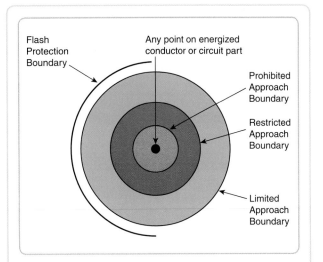

Figure 10.4 An AFPB might be a greater distance from the exposed energized electrical conductors or circuit parts than the limited approach boundary.

For example, according to Table 130.2(C), the LAB for an exposed energized fixed circuit part operating at 480 volts is 42 in. However, the AFPB would be 48 in. based on the product of clearing time and the available bolted short-circuit (fault) current of any combination not exceeding 100-kiloampere cycles. In this example, a worker would reach the boundary requiring arc flash protection before he or she would reach the outermost shock protection boundary.

The AFPB and the three shock protection boundaries that must be observed per *NFPA 70E* are depicted to provide an example of how the relationship between these boundaries might look (**Figure 10.4**). The *NFPA 70E* prescribed distance for each boundary could be in all directions from the exposed parts, which creates a protection boundary sphere, or it could be semicircular or linear, where access to the equipment is from only one direction (**Figure 10.5**), such as at a panelboard or MCC bucket.

Approach by Qualified Persons

NFPA 70E Section 130.2(C) describes approach to exposed energized electrical equipment operating at or above 50 volts by a qualified person. Workers are prohibited from approach closer than the restricted approach boundary, including with conductive objects, unless the following certain conditions are met:

- The worker is insulated or guarded from the energized electrical equipment, and no uninsulated part of the body crosses the prohibited approach boundary. A combination of the techniques described in this and the following two points must

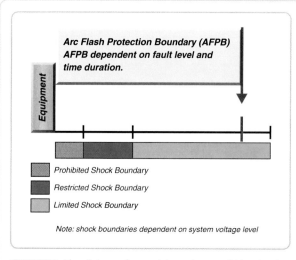

Arc Flash Protection Boundary (AFPB)
AFPB dependent on fault level and
time duration.

Equipment

☐ Prohibited Shock Boundary

☐ Restricted Shock Boundary

☐ Limited Shock Boundary

Note: shock boundaries dependent on system voltage level

Figure 10.5 The distance for each boundary could be circular, semicircular, or linear, where access to the equipment is from only one direction such as at a panelboard.

be used if an uninsulated part of the worker will cross the prohibited approach boundary.

- The energized electrical equipment is insulated from the worker and from any other conductive object at a different potential.
- The worker is insulated from any other conductive object.

Approach by Unqualified Persons

- Unqualified persons are prohibited from entering areas required to be accessible only to qualified persons unless the equipment is in an electrically safe work condition (**Figure 10.6**).
- Unqualified persons must be made aware of the electrical hazards they are or might be exposed

Figure 10.6 Unqualified persons are prohibited from entering areas accessible only to qualified persons unless the equipment is in an electrically safe work condition.

to when in the vicinity of the limited approach boundary.

- Unqualified persons must not cross the LAB unless escorted by a qualified person and made aware of the electrical hazards they are or might be exposed to.
- Unqualified persons are never permitted to cross the restricted approach boundary.

Arc Flash Hazard Analysis

NFPA 70E Section 130.3 requires an arc flash hazard analysis to determine the AFPB and the PPE necessary for workers that cross that boundary. **Arc flash hazard analysis** is defined in *NFPA 70E* Article 100 as follows:

Arc Flash Hazard Analysis. A study investigating a worker's potential exposure to arc flash energy, conducted for the purpose of injury prevention and the determination of safe work practices, flash protection boundary, and the appropriate levels of PPE.

The analysis consists of three components:

- Determination of the AFPB
- Determination of protective clothing and other PPE
- Field labeling of equipment

Recognizing that electrical systems might change over their years of service, the *NFPA 70E* Technical Committee added new conditions to Section 130.3 for the 2009 edition of *NFPA 70E* to account for changes in the electrical distribution system that could affect the results of the arc flash hazard analysis:

- The AFPB is required to be updated when a major modification or renovation takes place.
- The AFPB must be reviewed periodically. The review must take place at least every 5 years.

Additional new language was added into Section 130.3 to recognize the important role that the design and maintenance of overcurrent devices play in worker protection. As stated by Dennis Neitzel in his substantiation for Proposal 70E-204, published in Volume 1 of the National Fire Protection Association (NFPA) 2008, *Annual Revision Cycle Report on Proposals (ROP)*, "If maintenance is not performed, or if it is performed inadequately, extended clearing times are very likely to occur, which will increase the amount of incident energy in an arc flash situation." The arc flash hazard analysis must take into consideration:

- The design of the overcurrent protective device (OCPD)
- The opening time of the OCPD including its condition of maintenance

Comment 70E-444, published in the NFPA 2008, *Annual Revision Cycle Report on Comments*, contends that an arc flash hazard analysis is not always necessary. The submitter of the comment reports that "the Institute of Electrical and Electronics Engineers (IEEE) 1584 committee has analyzed the exposure in incident energy for circuits fed by transformers less than 125 kilovolt-amperes and less than 240 volts, and found them to be less than 1.2 cal/cm²." Accordingly, an exception was added recognizing that an arc flash hazard analysis is not required when all of the following apply:

- The circuit is rated 240 volts or less.
- The circuit is supplied by one transformer.
- The transformer supplying the circuit is rated less than 125 kilovolt-amperes.

Finally, two FPNs were added to *NFPA 70E* Section 130.3 reinforcing the importance of overcurrent device maintenance. The first reiterates the importance of appropriate and adequate OCPD maintenance. Without it, an increased opening time of the OCPD and an increase in incident energy would be expected. The second FPN is a reminder that information on performing maintenance on OCPDs can be found in Chapter 2 of *NFPA 70E*.

Arc Flash Protection Boundary

The requirements for determining the AFPB are broken down into two voltage level categories: Between 50 and 600 volts and above 600 volts. **Arc flash protection boundary** is defined by *NFPA 70E* as follows:

> **Boundary, Arc Flash Protection.** When an arc flash hazard exists, an approach limit at a distance from a prospective arc source within which a person could receive a second degree burn if an electrical arc flash were to occur.

NFPA 70E Section 130.3(A)(1) covers voltage levels between 50 and 600 volts. The boundary at this voltage range can either be calculated or determined based on the product of clearing time and the available bolted short-circuit current.

If the product of clearing time and bolted short-circuit current exceeds 100-kiloampere cycles, the AFPB must be calculated. A formula to calculate the AFPB has been relocated into *NFPA 70E* Annex D for the 2009 edition. Accordingly, there is no longer one formula that must be used, since *NFPA* Annex information is not mandatory.

While it is often referred to as a "default" boundary, the 4 ft boundary is really a conditional one. Where detailed arc flash hazard analysis calculations are not performed, the AFPB is 4 ft for the product of clearing time of two cycles and an available bolted short-circuit

current of 50 kiloamperes (or any combination not exceeding 100 kiloamperes).

NFPA 70E Section 130.3(A)(2) covers voltage levels above 600 volts. The AFPB must be calculated for voltage levels above 600 volts. There is no conditional fixed distance option at this voltage level.

No regulations offer guidance to prevent a first-degree burn, similar to a sunburn. The arc flash protection boundary is defined as the point where a second-degree burn injury is possible. FR protection is necessary to prevent a second-degree burn or worse. The requirements for determining boundary are broken down into the following two incident energy categories:

- The AFPB is the generally the distance where the incident energy is 1.2 cal/cm².
- The distance where the incident energy level equals 1.5 cal/cm² if the fault-clearing time is less than or equal to 0.1 seconds.

This topic, including sample calculations, is covered in greater detail in Chapter 12.

Personal Protective Equipment

Section 130.3(B) describes the second of three steps to fulfill what is required as part of an arc flash hazard analysis. While Section 130.3(A) established the AFPB, Section 130.3(B) will determine the PPE required once the arc flash boundary is crossed. Two options are available to determine the appropriate PPE for the work to be performed inside the AFPB determined by Section 130.3(A).

Incident Energy Analysis

NFPA 70E Section 130.3(B)(1) details one option permitted to determine the required PPE to cross the arc flash boundary; namely, incident energy analysis. This analysis determines the incident energy exposure of the worker in cal/cm². This exposure must be documented by the employer. The level of incident energy is required to be included on the energized electrical work permit to meet the requirements of Section 130.1(B)(2) and on a field-installed equipment label to meet the requirements of Section 130.3(C).

PPE must be selected and used based on the incident energy analysis. The incident energy level is determined by where the worker is positioned in relationship to the electrical equipment.

The currently accepted methods of calculating incident energy assume that a worker's chest is at a predetermined distance from a potential fault. In some instances, that distance is assumed to be 18 inches, and in other instances, the distance is assumed to be 36 inches. The amount of incident energy is determined by the distance that the worker's face or chest (whichever is

Figure 10.7 Equipment must be field marked with a label as part of an arc flash hazard analysis.

closer) will be from the potential electric arc. No distance is mandated in *NFPA 70E*. However, the distance used to perform the incident energy analysis must reflect the closest distance of the affected worker's face or chest so that the calculation will be valid. Additional PPE is required for any part of a worker's body that is closer to the potential arc source than the distance used in the incident energy analysis.

Annex D contains nonmandatory guidance on estimating incident energy. Typical working distances used for the incident energy calculations in Annex D.5 are offered in *NFPA 70E* Annex D.5(3).

> *NFPA 70E* Annex D.3, in part
>
> Typical working distances used for incident energy calculations are given below.
>
> • Low voltage (600V and below) MCC and panelboards – 455 mm [18 in.]
>
> • Low voltage (600V and below) switchgear – 610 mm [24 in.]
>
> • Medium voltage (above 600V) switchgear – 910 mm [36 in.]

This topic, including sample calculations, is covered in greater detail in Chapter 12.

Hazard Risk Categories

NFPA 70E Section 130.3(B)(2) details a second option permitted for the determination, selection, and use of the personal and other protective equipment required for a worker to be protected once he or she crosses the arc flash boundary. The requirements of *NFPA 70E* Sections 130.7(C)(9), 130.7(C)(10), and 130.7(C)(11) are permitted as an alternative to conducting an incident energy analysis.

The hazard risk category associated with the task to be performed is determined from *NFPA 70E* Table 130.7(C)(9), including notes. This table must be used in accordance with *NFPA 70E* Section 130.7(C)(9). Tables 130.7(C)(10) and 130.7(C)(11) are used once the level of PPE [the hazard/risk category (HRC)] is determined from Table 130.7(C)(9). The necessary level of PPE must be included on a field-installed equipment label to meet the requirements of Section 130.3(C) and on the energized electrical work permit to meet the requirements of Section 130.1(B)(2).

Equipment Labeling

NFPA 70E Section 130.3(C) details the third component of an arc flash hazard analysis. It is the requirement for equipment labeling. Electrical equipment must have a label affixed that documents either the results of the incident energy analysis (in cal/cm^2) or the HRC for the task to be performed (**Figure 10.7**). This information is among that required on the energized electrical work permit.

A label is not required to be placed on electrical equipment when the transformer supplying the circuit is rated less than 125 kilovolt-amperes, the circuit is rated 240 volts or less, and the circuit is supplied by one transformer.

■ Summary

This chapter focused on the first three concepts addressed in *NFPA 70E* Article 130:

• Justification for energized work and the energized electrical work permit
• Approach boundaries to energized electrical conductors or circuit parts
• Arc flash hazard analysis

These concepts offer prescriptive-type requirements as a solution to comply with performance-based requirements. Energized work is generally required to be performed by authorization of an energized electrical work permit once energized work is justified. The shock hazard analysis and arc flash hazard analysis determine the boundaries and PPE required and protect against those hazards. The results of the arc flash hazard analysis are required to be documented on a field-installed label, while the results of both the shock hazard analysis and arc flash hazard analysis are required to be documented on the energized electrical work permit.

Lessons *learned*

Electrician Electrocuted After Contacting an Energized 480-Volt Busbar

A 30-year-old electrician was electrocuted when he inadvertently touched an energized 480-volt busbar in a main service disconnect breaker panel. The company, which employs 2000 employees (carpenters, electricians, plumbers, and laborers), has no written safety program or safety director. The foreman is responsible for job site safety. The victim in this incident, who had worked for this employer for 7 months, was a licensed electrician with 6 years of experience.

On the day of the incident, a crew of 15 contract employees was performing various tasks. The victim and a coworker had been assigned the task of identifying and eliminating a ground fault that repeatedly tripped a circuit breaker in the breaker panel. The two employees proceeded to the maintenance room where the panel was located. The panel contained six separate circuits (three 480-volt and three 277-volt circuits) with accompanying breaker switches. The problem was determined to be a 400-amp breaker in one of the 480-volt circuits.

The two employees switched all of the breakers, including the main breaker, to the off position, and then removed the breaker panel covers. The next step was to check the continuity of the conductors leading to a branch circuit panel located in another room. As the coworker turned away from the victim to place one of the panel covers on the floor, he heard what sounded like the click of a breaker being switched on/off. He then heard a spitting sound and turned around to see the victim's right hand in contact with a conductor and his left hand in contact with a busbar. The coworker kicked the victim, causing him to fall and break contact.

The foreman, who had been on his way to the maintenance room to check on the workers' progress, heard the commotion and rushed to the area. After realizing what had happened, the foreman instructed another worker to call the local police, fire, and emergency medical service (EMS) departments. Approximately 2 minutes after the incident, the victim stopped breathing, and the foreman began cardiopulmonary resuscitation (CPR), continuing it until the police arrived. The police arrived approximately 4 minutes after the incident and continued CPR. The EMS arrived 55 minutes after being called, provided advanced cardiac life support, and transported the victim to the hospital, where he was pronounced dead approximately 1 hour later. The medical examiner's office reported the cause of death as electrocution.

Recommendations/Discussion

Recommendation 1: Employers should develop, implement, and enforce a comprehensive safety program, which includes worker training in recognizing and avoiding hazards, especially electrical hazards.

Discussion: In this incident, the victim switched the main breaker from an on position to an off position, removed a panel box cover, and for unknown reasons, switched the main breaker back to the on position. This procedure exposed an energized busbar, which the victim contacted. Employers should evaluate the tasks performed by workers, identify all potential hazards, and then develop, implement, and enforce a comprehensive safety program addressing these issues as required by OSHA standard 29 CFR 1926.21 (Office of the Federal Register: Code of Federal Regulations, Labor 29 Part 1926. p. 20. July 1, 1989). This safety program should include, but not be limited to, worker training in electrical hazard recognition.

Recommendation 2: An electrical system should be deenergized, and tested to verify that it has been deenergized, prior to any work being performed on it.

Discussion: The breaker panel was not deenergized before the repair work was attempted. The circuitry might have been left energized so that other repair workers would not be inconvenienced. A job of this type should be scheduled at a time (a weekend or before or after hours) when the incoming power could be deenergized without disrupting operations. Employers should develop specific job procedures for tasks that are performed by employees, including deenergizing electrical circuits before beginning to work on them and verifying that the system has been deenergized. These procedures should detail the various safety hazards associated with each task. Once these specific procedures have been developed, employers should ensure that they are implemented and enforced by a qualified person at each job site. Additionally, when employees must work away from the control point (i.e., the breaker panel in this incident), lockout and tagout procedures should be implemented (Federal Register: Part IV, Department of Labor, 29 CFR Part 1910. pp. 36644-36696. September 1, 1989).

Adapted from: Fatality and Assessment and Control Evaluation (FACE) 90-22. Accessed 8/27/08 at http://www.cdc.gov/niosh/face/In-house/full9022.html.

Questions

1. How might following 29 CFR 1910.333 have changed the outcome of this incident?
   A. The main disconnecting means would have been switched off, and both workers would have attached a lockout device on the disconnecting means.
   B. 29 CFR 1910.333 does not contain lockout requirements.
   C. The workers were construction electricians, so general industry standards do not apply.
   D. OSHA does not address this work task.

2. How would following *NFPA 70E* might have changed outcome of this incident?
   A. The outcome would have remained the same because the worker had no volt-meter.
   B. The coworker would also have been electrocuted by kicking the victim.
   C. Article 120 of *NFPA 70E* requires that an electrically safe work condition exist before a worker removes any cover or exposes a potentially energized electrical conductor. An electrically safe work condition would have prevented the electrocution.
   D. *NFPA 70E* does not address this work task.

3. What is the impact of the emergency medical response taking 55 minutes?
   A. No impact, because foremen are required to be trained to provide emergency medical assistance.
   B. OSHA requires that emergency medical assistance be available in 4 minutes.
   C. Emergency medical personnel must be a part of the work crew.
   D. The worker died instantaneously, so the extended EMS response time had no bearing.

Ready for Review

- *NFPA 70E* Article 130 covers justification for energized work and the energized electrical work permit requirements, shock hazard analysis and approach boundary requirements, and arc flash hazard analysis requirements.
- Section 130.1(A) requires that energized electrical equipment be placed into an electrically safe work condition if a worker is to cross the LAB associated with the energized electrical equipment.
- Circumstances that would permit energized work include infeasibility and increased or additional hazards resulting from deenergization.
- Do not assume that energized electrical equipment operating at less than 50 volts is free of hazard. The capacity of the source and any overcurrent protection between the energy source and the worker must be evaluated to determine the need to deenergize.
- Safe work practices are required even if the employer has demonstrated that deenergizing introduces additional or increased hazards or is infeasible due to equipment design or operational limitations.
- OSHA and *NFPA 70E* require a written hazard assessment; an employer must assess the workplace and determine if hazards are or could be present, then must provide appropriate PPE.
- *NFPA 70E* Section 130.1(B) provides energized electrical work permit requirements for working on equipment that is not in an electrically safe work condition. Annex J is an example of an energized electrical work permit.
- *NFPA 70E* Section 130.2(A) requires a shock hazard analysis to determine the voltage that workers will be exposed to, shock protection boundaries, and PPE for shock.
- Three shock protection boundaries are established by *NFPA 70E*: limited, restricted, and prohibited. These apply when workers are exposed to energized electrical conductors or circuit parts.
- *NFPA 70E* Table 130.2(C) establishes the distances associated within a range of nominal system voltages. The three approach boundaries and associated distances for each are located in this table.

- The LAB is the first shock protection boundary that a worker would encounter as he or she approaches exposed energized electrical equipment, but it might not be the first boundary that a worker would encounter as he or she approaches electrical equipment. The AFPB might be a greater distance from the exposed energized electrical conductors or circuit parts than the LAB.
- For a qualified person's approach to exposed energized electrical equipment operating at or above 50 volts, *NFPA 70E* Section 130.2(C) prohibits approach from closer than the restricted approach boundary unless certain conditions are met. Additional requirements are listed for unqualified persons.
- *NFPA 70E* Section 130.3 requires an arc flash hazard analysis to determine the AFPB and the PPE necessary for workers who cross that boundary. The analysis consists of determining the AFPB, determining protective clothing and other PPE, and field labeling of equipment.
- To account for changes in the electrical distribution system, *NFPA 70E* requires that the AFPB be updated when a major modification or renovation takes place and that the AFPB be reviewed at least every 5 years.
- The arc flash hazard analysis must also consider the design of the OCPD and the opening time of the OCPD, including its condition of maintenance.
- An arc flash hazard analysis is not required when all of the following apply:
  1. The circuit is rated at 240 volts or less.
  2. The circuit is supplied by one transformer.
  3. The transformer supplying the circuit is rated at less than 125 kilovolt-amperes.
- The requirements for determining the AFPB are broken down into two voltage level categories: between 50 and 600 volts and above 600 volts.
- *NFPA 70E* Section 130.3(A)(1) covers voltage levels between 50 and 600 volts. The boundary at this voltage range can either be calculated or determined based on the product of clearing time and the available bolted short-circuit current.
- If the product of clearing time and bolted short-circuit current exceeds 100-kiloampere cycles, the

AFPB must be calculated. A formula to calculate the AFPB has been relocated to *NFPA 70E* Annex D, 2009 Edition, and is not mandatory.

- Where detailed arc flash hazard analysis calculations are not performed, the conditional AFPB is 4 ft. This conditional boundary applies only when the product of the clearing time of the overcurrent device is two cycles and the available bolted short-circuit current of 50 kiloamperes, or any combination thereof, in which their product does not exceed 100 kiloamperes.
- *NFPA 70E* Section 130.3(A)(2) covers voltage levels above 600 volts. The AFPB must be calculated for voltage levels above 600 volts.
- Section 130.3(B) covers two options for determining the appropriate PPE for the work to be performed inside the AFPB: The incident energy analysis and hazard risk categories.
- The incident energy analysis determines the incident energy exposure of the worker in cal/cm², which is required to be documented by the employer. It must be included on the energized electrical work permit and on a field-installed equipment label.
- The distance used to perform the incident energy analysis must reflect the closest distance that the affected worker's face or chest will be located. Additional PPE is required for any part of a worker's body that is closer to the potential arc source than the distance used in the incident energy analysis.
- An alternative to conducting an incident energy analysis is to base PPE on hazard risk categories associated with the task to be performed, as determined from *NFPA 70E* Table 130.7(C)(9).
- *NFPA 70E* Section 130.3(C) requires that electrical equipment have an affixed label that documents either the results of the incident energy analysis (in cal/cm²) or the HRC for the task to be performed. A label is not required to be placed on electrical equipment when the transformer supplying the circuit is rated at less than 125 kiloamperes, the circuit is rated at 240 volts or less, and the circuit is supplied by one transformer.

Vocabulary

arc flash hazard analysis* A study investigating a worker's potential exposure to arc-flash energy, conducted for the purpose of injury prevention and the determination of safe work practices, arc flash protection boundary, and the appropriate levels of PPE.

arc flash protection boundary (AFPB)* When an arc flash hazard exists, an approach limit at a distance from a prospective arc source within which a person could receive a second degree burn if an electrical arc flash were to occur.

exposed movable conductor A condition in which the distance between the conductor and a person is not under the control of the person. The term is normally applied to overhead line conductors supported by poles.

limited approach boundary* An approach limit at a distance from an exposed energized electrical conductor or circuit part within which a shock hazard exists.

nominal voltage* A nominal value assigned to a circuit or system for the purpose of conveniently designating its voltage class (e.g., 120/240 volts, 480Y/277 volts, 600 volts). The actual voltage at which a circuit operates can vary from the nominal within a range that permits satisfactory operation of equipment.

prohibited approach boundary* An approach limit at a distance from an exposed energized electrical conductor or circuit part within which work is considered the same as making contact with the electrical conductor or circuit part.

restricted approach boundary* An approach limit at a distance from an exposed energized electrical conductor or circuit part within which there is an increased risk of shock, due to electrical arc over combined with inadvertent movement, for personnel working in close proximity to the energized electrical conductor or circuit part.

shock hazard* A dangerous condition associated with the possible release of energy caused by contact or approach to energized electrical conductors or circuit parts.

*This is the *NFPA 70E* definition.

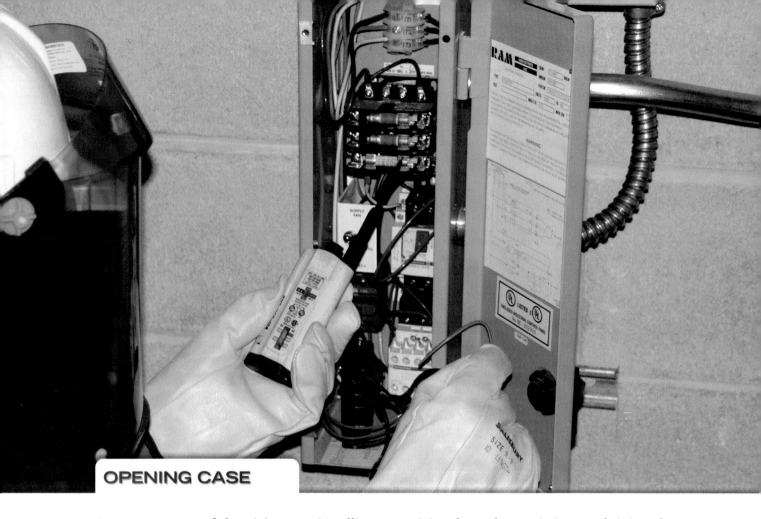

A two-man crew of electricians was installing new wiring through an existing conduit in a department store ceiling. While using a personnel lift to work overhead, one of the workers made contact with an energized 277-volt electrical circuit.

The employer is an electrical contractor who has been in business for 40 years. The contractor employs 12 workers, 11 of whom are electricians or electrical helpers. The employer was cited by Occupational Safety and Health Administration (OSHA) in 1989 because of the fatal electrocution of another employee. After the 1989 fatality, employees were required to sign a statement acknowledging that "deenergizing electrical circuits is required when work is performed and should be tagged or locked out." It was signed by all electrical workers including the victim and his coworker. The victim had worked for the company for 5 years and had learned his trade through on-the-job training.

The company was contracted to install extra wiring in the store. Some of the previous work had been done by the victim. The victim and his coworker were using a spare wire to pull two new wires through an existing ceiling conduit. They were not aware that an existing 277-volt line had been spliced and had a potential of 277 volts to ground. It is unclear what sequence of events lead to the fatal contact with the energized circuit, since the coworker's reports have been inconsistent.

The electrical inspector's report noted that an open junction box with several circuits passing through it was directly above the lift on which the victim had been working. Energized conductors were hanging out of the box with no splice connector installed. A connector was found laying on the lift platform. The spliced conductors were energized and had a potential of 277 volts to ground. The building steel, metal conduits and boxes, flexible metal conduit, and the personnel lift were tested and found to be properly grounded. No electrical code violations were noted.

The victim was in the lift, elevated to working height, when he made contact with the energized circuit, either by touching it with his hand or a pair of pliers. He was electrocuted with 277 volts, lost

Calculation of Short-Circuit Currents

CHAPTER 11

Introduction

The purpose of a **short-circuit study** is to determine the available **short-circuit current** at various points in an electrical system in order to assure compliance with requirements of the *NEC* and *NFPA 70E*. To comply with the NEC requirements, the short-circuit current must be known to verify that the **interrupting rating (IR)** of overcurrent devices is adequate. In addition, the short-circuit current must be known to verify that the **short-circuit current rating** (SCCR) of equipment is adequate.

To comply with *NFPA 70E* requirements, the short-circuit current must be known in order to determine the associated arcing current and opening time of the overcurrent device. Once these variables are known, the arc flash hazard analysis can be performed in order to determine and document the **arc flash protection boundary** (AFPB) and required protective clothing and personal protective equipment (PPE).

OPENING CASE, CONT.

consciousness, and fell out of the lift to the floor, a distance of approximately 16 feet (the approximate height of the lift platform). In order to enter the lift it is necessary to stoop and step under the guardrail on the side of entrance. A worker using the lift is protected from a fall by the guardrail on four sides and midrail on three sides. On the entry side a chain, the two ends of which must be fastened, substitutes for the midrail. The victim did not fasten the two ends of the chain. When he lost consciousness he became limp and fell out of the lift.

Emergency services were summoned and a store customer who was a paramedic started cardiopulmonary resuscitation (CPR) less than a minute after the incident. She was joined by a policeman who arrived first at the scene. Paramedics arrived about 15 minutes after being called, followed by the local rescue squad. The employee was transported to the local emergency room where he died. The medical examiner's report listed the cause of death as electrocution.

Adapted from: New Jersey Case Report 90NJ006. Accessed 5/28/08 at http://www.cdc.gov/niosh/face/stateface/nj/90nj006.html.

1. Do you think the electrician understood the acknowledgement statement when he signed it?

2. Why do you think the electrician did not fasten the midrail chain after he entered the lift?

3. Do you think it is all right to install a new circuit in a conduit that contains an existing energized circuit? Do you believe that additional hazards might be present?

Chapter Outline

- **Introduction**
- **Short-Circuit Calculation Requirements**
- **Effect of Short-Circuit Current on Arc Flash Hazards**
- **Short-Circuit Calculation Basics**
- **Procedures and Methods**
- **Point-to-Point Method**
- **Short-Circuit Calculation Software Programs**
- **Summary**
- **Lessons Learned**
- **Current Knowledge**

Learning Objectives

1. Understand why short-circuit studies are required and when they are needed.
2. Understand the effect of short-circuit current on arc flash hazards.
3. Understand the procedures and calculation methods used to perform a short-circuit study.
4. Understand the point-to-point method and examples for calculating the available short-circuit current.

References

1. *National Electrical Code®* (*NEC®*), 2008 Edition
2. *NFPA 70E®*, 2009 Edition
3. Cooper Bussmann® SPD (Selecting Protective Devices)
4. Cooper Bussmann Short-Circuit Calculator

70E Highlights

NFPA 70E uses the terms *short-circuit current* and *fault current* interchangeably; the meanings are essentially the same.

The short-circuit study starts from the service point and extends to the branch-circuit devices. The study determines the short-circuit current at all critical points in the system. Equipment, such as service equipment, transfer switches, switchboards, panelboards, motor starters, and motor control centers are analyzed in this study. The level of short-circuit current is dependent upon many factors such as the utility, generator, and motor contribution. In general, the short-circuit current decreases as distance from the source increases due to the increase of impedance that is added by transformers, conductors, and busway. The available short-circuit current is the output of a short-circuit study.

There are procedures and different calculation methods that can be used to perform a short-circuit study. The point-to-point calculation method will be covered in this chapter to determine the available short-circuit current at various locations. Short-circuit calculation software programs are available from several sources. The Cooper Bussmann Short-Circuit Calculator program (available as a free download) based on the point-to-point method will be illustrated in this chapter.

Short-Circuit Calculation Requirements

Knowledge of and the ability to calculate the short-circuit current at various points in an electrical system is required to comply with various sections of the *NEC* and *NFPA 70E*. The list below details a number of those requirements and why the short-circuit current is needed for compliance.

- *NEC* Section 110.9 and *NFPA 70E* Section 210.5—Overcurrent devices must have adequate interrupting ratings for the short-circuit current available.
- *NEC* Section 110.10—System components must have adequate short-circuit current ratings for the short-circuit current available.
- *NEC* Section 110.16—Field-installed warning label is required to be installed on certain equipment to warn of an arc flash hazard.
- *NFPA 70E* Section 110.8(B), Section 130.3, Table 130.7(C)(9), and Annex D—An arc flash hazard analysis is required to be conducted before work involving electrical hazards is performed. An arc flash hazard analysis is required to determine the AFPB and required PPE. PPE must be selected based on the documented **incident energy** level, as calculated per Annex D. The Table notes detail the maximum permitted level of short-circuit current if Table 130.7(C)(9) is used to determine the hazard/risk category to select PPE in lieu of calculating the incident energy.
- *NFPA 70E* Section 130.1(B)(2)—The AFPB and necessary PPE (based on the incident energy or required level of PPE) must be documented on the energized work permit.

Effect of Short-Circuit Current on Arc Flash Hazards

The short-circuit current is a key factor that determines the extent of the arc flash hazard. An integral part of protecting workers from an arc flash hazard includes, but is not limited to, conducting an arc flash hazard analysis. An arc flash hazard analysis will determine, among other things, the AFPB and incident energy or required level of PPE.

70E Highlights

NFPA 70E Section 130.1(B)(2)

Elements of Work Permit. The energized electrical work permit shall include, but not be limited to, the following items:

(6) Results of the arc flash hazard analysis (130.3)

(7) The arc flash protection boundary [130.3(A)]

(8) The necessary personal protective equipment to safely perform the assigned task [130.3(B), 130.7(C)(9), and Table 130.7(C)(9)]

70E Highlights

The first paragraph of *NFPA 70E* Section 130.3, Arc Flash Hazard Analysis, states:

An arc flash hazard analysis shall determine the arc flash protection boundary and the personal protective equipment that people within the arc flash protection boundary shall use.

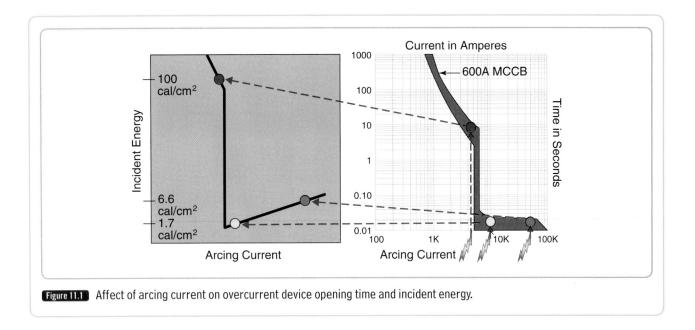

Figure 11.1 Affect of arcing current on overcurrent device opening time and incident energy.

It is important to determine both the low and high values of short-circuit current. This is because in some cases, a lower value of short-circuit current (and resulting arcing current) can actually increase the clearing time of the overcurrent device and result in an increase of the incident energy, AFPB, and level of PPE. **Figure 11.1** illustrates how lower arcing short-circuit currents can result in increased opening times of overcurrent devices and higher incident energy compared to higher arcing short-circuit currents, which can result in decreased overcurrent device opening time and lower incident energies. However, as the arcing short-circuit current increases, the incident energy increases for circuit breakers.

In addition, it is important to realize that short-circuit currents can change over time due to changes, including but not limited to, the utility, generators, motors, transformers, or the building distribution system. Because of this, *NFPA 70E* Section 130.3 requires the arc flash hazard analysis to be updated when a major modification or renovation takes place. It is to be reviewed periodically, but no longer than 5 years, to account for changes that could affect the results of the arc flash hazard analysis.

Short-Circuit Calculation Basics

Normally, short-circuit studies involve calculating a bolted 3-phase short-circuit condition. This can be characterized as all three phases "bolted" together to create a zero **impedance** connection. This establishes a condition that results in maximum thermal and me-

CAUTION

Short-circuit currents can change over time due to system changes. To assure protection of people and equipment, short-circuit studies and arc flash hazard analysis studies must be updated as necessary.

chanical stress in the system. From this calculation, other types of fault conditions, such as bolted phase-to-phase fault, bolted phase-to-ground fault, arcing phase-to-phase fault, and arcing phase-to-ground fault can be obtained.

Note: Bolted short-circuit current determines the highest amount of current that the electrical system can deliver during a short-circuit condition.

Short-circuit calculations should be completed for all critical points in the system (**Figure 11.2**). These would include, but are not limited to, the following:

- Switchboards
- Panelboards
- Motor control centers (MCC)
- Motor starters
- Disconnect switches
- Transfer switches

Sources of Short-Circuit Current

Sources of short-circuit current that are normally taken under consideration include: utility generation, local generation, synchronous motors, and induction motors. When electrical systems are supplied from a

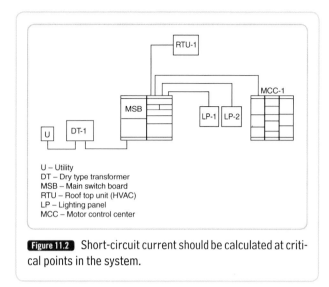

Figure 11.2 Short-circuit current should be calculated at critical points in the system.

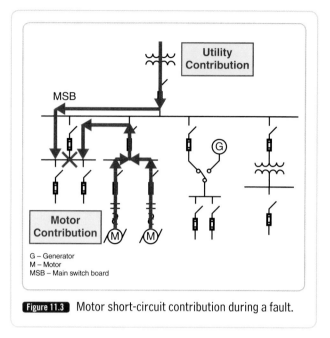

Figure 11.3 Motor short-circuit contribution during a fault.

utility or customer owned transformer, the amount of short-circuit current depends upon the size (kVA) and impedance (%Z) of the transformer. The larger the size and/or the smaller the impedance, the higher the short-circuit current.

Small residential building systems (100- to 200-ampere service) typically have short-circuit currents of 10,000 to 15,000 amperes or less. Small commercial building systems (400- to 800-ampere service) typically have short-circuit currents of 20,000 to 30,000 amperes. Larger commercial and manufacturing building systems (2000- to 3000-ampere service) typically have short-circuit currents of around 50,000 to 65,000 amperes. These short-circuit current values can be much higher where low impedance transformers are used to increase efficiency. When commercial buildings are directly connected to the utility "grid system," such as in major metropolitan cities (New York, Chicago, Dallas, etc.) the short-circuit currents can exceed 200,000 amperes. It important to realize that the available short-circuit current at the service may change, and often will increase, due to changes in utility contribution and/or transformer size and impedance.

Generators and induction motors can increase the short-circuit current from the utility. Typically the worst-case short-circuit current from a generator is around 10 times the full-load **ampacity** of the genera-

tor. If the generator is paralleled with the utility during transfer, the short-circuit current of the generator is added to the short-circuit current of the utility. If not, the generator short-circuit current is typically less than the utility and not used unless looking at arc flash hazards when under generator power. During a fault, a motor acts likes a generator and will contribute to the short-circuit current as shown in **Figure 11.3**. The motor contribution is typically four to six times the full-load ampacity of the motor. The additional motor contribution must be added to the utility contribution to determine the required ratings of overcurrent devices and equipment.

Short-Circuit Current Factors

The amount of short-circuit current depends upon many factors, including the short-circuit current contribution of the utility, local generation, and motors as well as the size and impedance of the transformer, the size and length of the wire, and more (**Figure 11.4**). Generators and motors also affect the amount of short-circuit current.

The short-circuit current is typically the highest at the service point. Conductors and transformers will decrease the available short-circuit current in downstream equipment as shown in **Figure 11.5**.

CAUTION

Short-circuit currents can be very high (65,000 to more than 200,000 amperes). Verify short-circuit current levels when applying overcurrent devices and equipment.

CAUTION

Each worker must understand his or her qualifications or lack of qualifications.

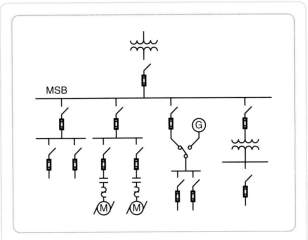

Figure 11.4 Factors that affect short-circuit current include utility, transformer (kVA and %Z), generators, motors, voltage, and conductor size and length.

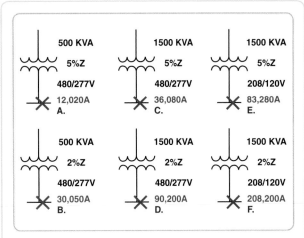

Figure 11.6 Effect of transformer size, impedance, and secondary voltage on short-circuit current.

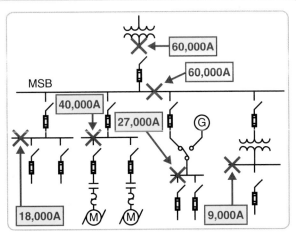

Figure 11.5 Short-circuit current decreases downstream due to conductors and transformers.

Effect of Transformers on Short-Circuit Current

How the transformer size (kVA), impedance (%Z), and secondary voltage affect short-circuit current is illustrated in **Figure 11.6**. Assuming an "infinite" primary current, the short-circuit current at point A for a 5% impedance transformer with a secondary voltage of 480/277 is equal to the full-load ampacity of the transformer (601 amperes) times 1/Z (1/0.05 = 20), which equals 12,020 amperes. This value may be 10% higher due to the tolerance of the transformer impedance.

If the impedance is reduced to 2% (to increase transformer efficiency), as shown at point B), the short-circuit current increases by 2.5 times (1/Z increases from 20 to 50) to 30,050 amperes.

If the kVA is increased to 1500 kVA (1804 amperes full-load amps), as shown at points C and D, the short-circuit current triples to 36,080 and 90,200 amperes, respectively.

If the voltage is decreased to 208/120, as shown in points E and F, the transformer full-load ampacity increases approximately 2.3 times (480 divided by 208 equals approximately 2.3) to 4164 amperes, and the short-circuit current increases approximately 2.3 times to 83,280 and 208,200 amperes, respectively.

Effect of Conductors on Short-Circuit Current

How the conductor (or bus duct) size (ampacity), length and voltage affect short-circuit current is illustrated in **Figure 11.7**. Assuming a short-circuit current at point A of 40,000 amperes [at the beginning of 50-foot run of a 1 American Wire Gauge (AWG) conductor] at 208/120 volts, the short-circuit current at the end of the conductor at point B would be 12,412 amperes. If the voltage was increased to 480/277 as shown at point C, with the same size and length conductor, the short-circuit current at point D increases to 20,376 amperes due to the increased voltage potential. If the size of the conductor is then increased to 250 kcmil as shown at

CAUTION

Short-circuit currents can greatly increase when transformer impedance and voltage is decreased or when transformer kVA increases. When service transformers are changed, short-circuit studies and arc flash studies should be updated.

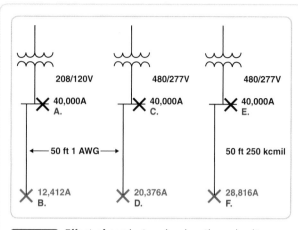

Figure 11.7 Effect of conductor size, length, and voltage on short-circuit current.

point E, the short-circuit current at point F in increases to 28,816 amperes due to the decrease in impedance of the larger conductor. Although not shown, if the length of the conductor is increased, the short-circuit current would decrease due to the increase of impedance of the longer conductor.

Procedures and Methods

To determine the short-circuit current at any point in the system, first draw a one-line diagram showing all of the sources of short-circuit current. Then include the system components, such as conductors and busway (sizes and lengths), transformers (sizes, voltages, and impedances), and overcurrent protective devices (OCPDs) (ampere, voltage, and interrupting ratings).

It must be understood that short-circuit calculations are performed without OCPDs in the system. Calculations are done as though these devices are replaced with copper bars, to determine the maximum "available" short-circuit current. After the calculation is completed throughout the system, current-limiting devices can be used to show reduction of the available short-circuit current at a single location only. Current-limiting devices do not operate in series to produce a "compounding" current-limiting effect.

Various methods have been developed to actually calculate the available short-circuit current, but all are based on Ohm's law. These methods include the ohmic method, per-unit method, point-to-point method, and computer-based versions of all three.

The application of the point-to-point method permits the quick determination of available short-circuit currents with a reasonable degree of accuracy at various points for either 3-phase or single-phase electrical distribution systems.

Point-to-Point Method

The point-to-point method uses formulas to calculate the short-circuit current at the secondary of a transformer assuming infinite primary short-circuit current available, fault at the end of a run of conductor or busway, and fault at the secondary of a transformer with primary short-circuit current known.

Begin by reviewing steps 1 to 6 of the *Basic Short-Circuit Calculation Procedure* section by studying both the procedure and the formulas for each step, including the associated notes. Then examine how each of the steps of the procedure and associated formulas are used to determine the short-circuit current available at Fault #1 as shown in **Figure 11.8** by reviewing the section on *Calculation Example* for Fault #1.

Note that Fault #1 (main service panel) is located 20 feet away from the secondary side of the service transformer. It is first necessary to determine the short-circuit current at the service transformer secondary using the procedures and formulas in steps 1 to 3, including the associated notes. These formulas calculate the short-circuit current at the secondary of a transformer assuming infinite primary short-circuit current available, which is often assumed for service transformers. Be sure to review steps 1 to 3 of the *Basic Short-Circuit Calculation Procedure* again if there are any steps that you do not understand.

Once the short-circuit current at the transformer secondary has been determined, use the procedures and formulas in steps 4 to 6, including the associated notes,

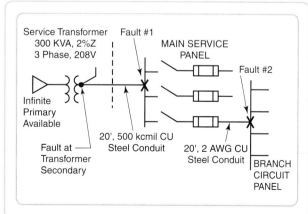

Figure 11.8 A sample diagram for short-circuit current calculations at various points.

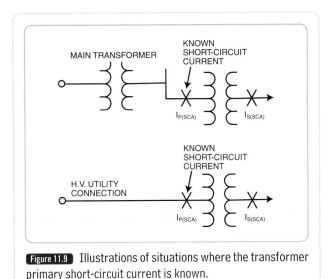

Figure 11.9 Illustrations of situations where the transformer primary short-circuit current is known.

to determine the short-circuit current at the location of Fault #1 (main service panel). Note that the short-circuit current at the service transformer secondary (step 3) is used to complete steps 4 and 6. Again, be sure to review steps 4 to 6 of the *Basic Short-Circuit Calculation Procedure* if there are any steps that you do not understand.

Now determine the available short-circuit current for Fault #2 (branch circuit panel) as shown in Figure 11.8 by reviewing the section *Calculation Example* for Fault #2. Follow the procedures and formulas in steps 4 to 6 using the short-circuit current at Fault #1 (main service panel) to complete steps 4 and 6.

Next, review the section *Calculation of Short-Circuit Current at Transformer Secondary with Primary Short-Circuit Current Known*. Then review the procedures and formulas in steps A through C and the associated example based on **Figure 11.9**.

Finally, review Table 11.1 through Table 11.7 along with their associated notes to become familiar with how this information is applicable in performing short-circuit current calculations. Tables are provided in one section later in this chapter.

Note the layout of **Table 11.1** and **Table 11.2** for the full-load amperes of transformers based on voltage. These values can be used in lieu of the formulas shown in step 1 of the *Basic Short-Circuit Calculation Procedure*. Table 11.1 is for 3-phase transformers, and Table 11.2 is for single-phase transformers.

Note the layout of **Table 11.3** and **Table 11.4** for the "C" values for conductors and **Table 11.5** for the "C" values for Busway (Bus Duct). This is where the constant ("C" value) is obtained to use in the formula in step 4 of the

Basic Short-Circuit Calculation Procedure. Table 11.3 is for copper conductors, and Table 11.4 is for aluminum conductors.

For Table 11.3 and Table 11.4, the conductor size is in the far left column. The left side of the table heading indicates that it is used to determine the constant for three single conductors. The right side of the table heading indicates that it is used to determine the constant for a three-conductor cable. Note that the left and right side of these tables are further broken down into steel conduit and nonmagnetic conduit. Various voltage levels are indicated from left to right on both the left and right side of these tables.

Note that **Table 11.6** can be used to determine the "M" Multiplier value in step 5 or step B after the "f" value has been calculated in step 4 or step A.

Table 11.7 can be used to find the worst-case short-circuit current for transformers. This table assumes the lowest impedance of transformers based on actual installations, typical worst-case or infinite primary short-circuit current available, as well as worst-case adjustment values for transformer impedance tolerance and system voltage fluctuations. Actual short-circuit current for installations may be lower than what this table indicates.

Table 11.8 can be used to find typical values of short-circuit currents as a percentage of the calculated 3-phase bolted short-circuit current for other types of bolted and arcing faults.

Taken together, all of the material contained in the *Basic Short-Circuit Calculation Procedure* section and Table 11.1 through Table 11.8 will provide you with a good introduction to the procedures and formulas necessary to meet the learning objectives for this material.

Basic Short-Circuit Calculation Procedure

Use the following procedure to determine the short-circuit current (I_{SCA}) at the transformer secondary terminals assuming infinite primary short-circuit current available:

Step 1: Determine the transformer full-load amperes (I_{FLA}):
   a) Name plate
   b) Table 11.1 and Table 11.2
   c) Formula:

$$\text{3-phase (3\o) transformer: } I_{FLA} = \frac{kVA \times 1000}{E_{L-L} \times 1.732}$$

$$\text{1-phase (1\o) transformer: } I_{FLA} = \frac{kVA \times 1000}{E_{L-L}}$$

Note 1: E_{L-L} = line-to-line voltage

Step 2: Determine the transformer multiplier:

$$\text{Multiplier} = \frac{100}{\%Z}$$

Note 2: The marked transformer impedance (% Z) value may vary ±10% from the actual values determined by the American National Standards Institute/Institute of Electrical and Electronics Engineers (ANSI/IEEE) test. See UL Standard 1561. For worst-case conditions, multiply transformer %Z by 0.9.

Step 3: Determine the short-circuit current (ISCA) at the transformer secondary:

  a) Table 11.7

  b) Formula:

      3ø transformer (see Notes 3 through 5):

$$I_{SCA\,(L\text{-}L\text{-}L)} = I_{FLA} \times \text{Multiplier}$$

      1ø transformer (see Notes 3 and 4):

$$I_{SCA\,(L\text{-}L)} = I_{FLA} \times \text{Multiplier}$$

Note 3: Utility voltage may vary ±10% for power, and ±5.8% for 120-volt lighting services. For worst-case conditions, multiply values calculated in step 3 by 1.1 and/or 1.058, respectively.

Note 4: Motor short-circuit contribution, if significant, may be determined at all fault locations throughout the system. A practical estimate of motor short-circuit contribution is to multiply the total load current by 4 or 6. For worst-case, calculate total motor load current, multiply by 4 and add to step 3.

Note 5: For 3-phase systems, line-to-line-to-line (L-L-L), line-to-line (L-L), line-to-ground (L-G), and line-to-neutral (L-N) bolted faults and arcing faults can be found in Table 11.8 based on the calculated 3-phase line-to-line-to-line (L-L-L) bolted fault.

Use the following procedure to determine the short-circuit current (I_{SCA}) at the end of a run of conductor or busway:

Step 4: Calculate the "f" factor:

  a) 3ø line-to-line-to-line (L-L-L) fault (see Note 6):

$$f = \frac{1.732 \times L \times I_{SCA(L\text{-}L\text{-}L)}}{C \times E_{L\text{-}L}}$$

  b) 1ø line-to-line (L-L) fault (see Notes 5 and 6):

$$f = \frac{2 \times L \times I_{SCA(L\text{-}L)}}{C \times E_{L\text{-}L}}$$

  c) 1ø line-to-neutral (L-N) fault (see Notes 5, 6, and 7):

$$f = \frac{2 \times L \times I_{SCA(L\text{-}N)}}{C \times E_{L\text{-}N}}$$

Note 6: See below for an explanation of variables:

L = length (feet) of conduit to the fault.

C = constant from Table 11.3, Table 11.4, or Table 11.5. For parallel runs, multiply C values by the number of conductors per phase.

I_{SCA} = available short-circuit current in amperes (A) at beginning of circuit.

$E_{L\text{-}N}$ = line-to-neutral voltage

Note 7: The L-N short-circuit current is higher than the L-L short-circuit current at the secondary terminals of a single-phase center-tapped transformer. The short-circuit current available ($I_{L\text{-}N}$) at the transformer terminals is typically adjusted as follows: $I_{L\text{-}N} = 1.5 \times I_{L\text{-}L}$ at Transformer Terminals

The 1.5 multiplier is an approximation and will theoretically vary from 1.33 to 1.67. At some distance from the terminals, depending upon wire size, the L-N short-circuit current is lower than the L-L short-circuit current.

Step 5: Determine the "M" (multiplier):

  a) Table 11.6

  b) Formula: $M = \dfrac{1}{1+f}$

Step 6: Calculate the available short-circuit current (I_{SCA}) at end of circuit:

3ø line-to-line-to-line (L-L-L) fault:

$$\underset{\text{AT END OF CIRCUIT}}{I_{SCA\,(L\text{-}L\text{-}L)}} = \underset{\text{AT BEGINNING OF CIRCUIT}}{I_{SCA\,(L\text{-}L\text{-}L)}} \times M$$

Note 8: For 3-phase systems, line-to-line-to-line (L-L-L), line-to-line (L-L), line-to-ground (L-G), and line-to-neutral (L-N) bolted faults and arcing faults can be found in Table 11.8 based on the calculated 3-phase line-to-line-to-line (L-L-L) bolted fault.

1ø line-to-line (L-L) fault:

$$\underset{\text{AT END OF CIRCUIT}}{I_{SCA\,(L\text{-}L)}} = \underset{\text{AT BEGINNING OF CIRCUIT}}{I_{SCA\,(L\text{-}L)}} \times M$$

1ø line-to-neutral (L-N) fault:

$$\underset{\text{AT END OF CIRCUIT}}{I_{SCA\,(L\text{-}N)}} = \underset{\text{AT BEGINNING OF CIRCUIT}}{I_{SCA\,(L\text{-}N)}} \times M$$

Calculation Example for Fault #1

Below is an example of the calculation for Fault #1, shown in Figure 11.8.

Step 1: Determine the transformer full-load amperes (I_{FLA}):

$$I_{FLA} = \frac{kVA \times 1000}{E_{L-L} \times 1.732} = \frac{300 \times 1000}{208 \times 1.732} = 833A$$

Step 2: Find the transformer multiplier:

$$Multiplier = \frac{100}{*0.9 \times Transf. \%Z} = \frac{100}{1.8} = 55.55$$

*Transformer Z is multiplied by 0.9 to establish a worst-case condition. See Note 2 in the section on *Basic Short-Circuit Calculation Procedure*.

Step 3: Determine the short-circuit current (I_{SCA}) at the transformer secondary:

$$**I_{SCA (L-L-L)} = I_{FLA} \times M = 833 \times 55.55 = 46,273A$$

**For simplicity, the motor contribution and voltage variance was not included. See Notes 3 and 4 in the section on *Basic Short-Circuit Calculation Procedure*. See Note 5 for other types of short-circuit current values.

Step 4: Calculate the "f" factor:

$$f = \frac{1.732 \times L \times I_{SCA(L-L-L)}}{C \times E_{L-L}} = \frac{1.732 \times 20 \times 46,273}{22,185 \times 208} = 0.35$$

Step 5: Determine the "M" (multiplier):

$$***M = \frac{1}{1+f} = \frac{1}{1+0.35} = 0.74$$

***Table 11.6 could be used to determine "M."

Step 6: Calculate the available short-circuit current (I_{SCA}) at Fault #1:

$$I_{SCA (L-L-L) \text{ AT END OF CIRCUIT}} = I_{SCA (L-L-L) \text{ AT BEGINNING OF CIRCUIT}} \times M$$

$$= 46,273 \times 0.74 = 34,242A$$

Calculation Example for Fault #2

Below is an example of a calculation for Fault #2 in Table 11.8. (Use $I_{SCA (L-L-L)}$ at Fault #1 in steps 4 and 6.)

Step 4: Calculate the "f" factor:

$$f = \frac{1.732 \times 20 \times 34,242}{5,907 \times 208} = 0.97$$

Step 5: Determine the "M" (multiplier):

$$M = \frac{1}{1+0.97} = 0.51$$

Step 6: Calculate the available short-circuit current (I_{SCA}) at Fault #2:

$$I_{SCA (L-L-L) \text{ AT FAULT #2}} = 34,242 \times 0.51 = 17,463A$$

Calculation of Short-Circuit Currents on Transformer Secondary with Primary Short-Circuit Current Known

When calculating the short-circuit current at the transformer secondary, it is not always necessary to assume infinite primary short-circuit current. Figure 11.9 illustrates where the primary short-circuit current at a transformer is known, either through a previous calculation (if it is the second transformer in the system) or if the short-circuit current of the utility connection is provided. In these cases, a more accurate calculation can be performed. Use the following procedure to calculate the short-circuit current at the secondary of a second, when the short-circuit current at the transformer primary is known.

Procedure for Calculation of Short-Circuit Current on Transformer Secondary with Primary Short-Circuit Current Known

The following steps show the procedure.

Step A: Calculate the "f" factor for the transformer:

$$3\emptyset \text{ transformer}: \quad f = \frac{I_{P(SCA)L-L-L} \times V_P \times 1.732 \times \%Z}{100,000 \times kVA}$$

$$1\emptyset \text{ transformer}: \quad f = \frac{I_{P(SCA)L-L} \times V_P \times \%Z}{100,000 \times kVA}$$

Note 9: See below for explanation of variables

$I_{P(SCA)}$ = primary short-circuit current

V_P = primary voltage (L-L)

Note 10: For simplicity, the transformer tolerance was not included. See Note 2 in the section on *Basic Short-Circuit Calculation Procedure*.

Step B: Determine the multiplier "M" for the transformer:

$$M = \frac{1}{1+f}$$

Note 11: Table 11.6 could be used to determine "M."

Step C: Determine short-circuit current at transformer secondary:

$$I_{S(SCA)} = \frac{V_P}{V_S} \times M \times I_{P(SCA)}$$

Note 12: See below for an explanation of variables:
$I_{S(SCA)}$ = secondary short-circuit current
V_S = secondary voltage (L-L)

Note 13: For simplicity, the motor contribution and voltage variance was not included. See Notes 3 and 4 in *Basic Short-Circuit Calculation Procedure*. See Note 5 for other types of short-circuit current values.

Note 14: To calculate the short-circuit current at the end of run of conductor or busway, follow steps 4, 5, and 6 in the *Basic Short-Circuit Calculation Procedure*.

Calculation Example

The following is an example of the procedure for calculation of short-circuit current on a transformer secondary with the primary short-circuit current known. This example uses information from **Figure 11.10**.

Step A: Calculate the "f" factor for the transformer:

$$f = \frac{30,059 \times 480 \times 1.732 \times 1.2^*}{100,000 \times 225} = 1.33$$

*For simplicity, the transformer tolerance was not included. See Note 2 in *Basic Short-Circuit Calculation Procedure*.

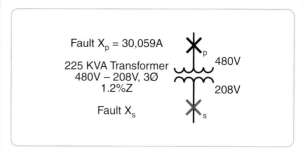

Fault X_p = 30,059A
225 KVA Transformer
480V – 208V, 3Ø
1.2%Z
480V
208V
Fault X_s

Figure 11.10   A sample diagram for calculating the transformer secondary short-circuit current with the primary short-circuit current known.

Step B: Determine the multiplier "M" for the transformer:

$$M = \frac{1}{1+1.33} = 0.43$$

Step C: Determine the short-circuit current at the transformer secondary:

$$I_{S(SCA)} = \frac{480}{208} \times 0.43 \times 30,059 = 29,828\,A^{**}$$

**For simplicity, the motor contribution and voltage variance was not included. See Notes 3 and 4 in *Basic Short-Circuit Calculation Procedure*. See Note 5 for other types of short-circuit current values.

Table 11.1	3-Phase Transformer—Full Load Current Rating (in Amperes)								
Voltage (Line-to-Line)					**Transformer kVA Rating**				
	150	167	225	300	500	750	1000	1500	2000
208	417	464	625	833	1388	2080	2776	4164	5552
220	394	439	592	788	1315	1970	2630	3940	5260
240	362	402	542	722	1203	1804	2406	3609	4812
440	197	219	296	394	657	985	1315	1970	2630
460	189	209	284	378	630	945	1260	1890	2520
480	181	201	271	361	601	902	1203	1804	2406
600	144	161	216	289	481	722	962	1444	1924

| Table 11.2 | Single-Phase Transformer—Full Load Current Rating (in Amperes) | | | | | | | | | |

Voltage	Transformer kVA Rating									
	25	50	75	100	150	167	200	250	333	500
115/230	109	217	326	435	652	726	870	1087	1448	2174
120/240	104	208	313	416	625	696	833	1042	1388	2083
230/460	54	109	163	217	326	363	435	544	724	1087
240/480	52	104	156	208	313	348	416	521	694	1042

| Table 11.3 | "C" Values for Conductors—Copper |

AWG or kcmil	Three Single Conductors Conduit						Three Conductor Cable Conduit					
	Steel			Nonmagnetic			Steel			Nonmagnetic		
	600 V	5 kV	15 kV	600 V	5 kV	15 kV	600 V	5 kV	15 kV	600 V	5 kV	15 kV
14	389	—	—	389	—	—	389	—	—	389	—	—
12	617	—	—	617	—	—	617	—	—	617	—	—
10	981	—	—	982	—	—	982	—	—	982	—	—
8	1557	1551	—	1559	1555	—	1559	1557	—	1560	1558	—
6	2425	2406	2389	2430	2418	2407	2431	2425	2415	2433	2428	2421
4	3806	3751	3696	3826	3789	3753	3830	3812	3779	3838	3823	3798
3	4774	4674	4577	4811	4745	4679	4820	4785	4726	4833	4803	4762
2	5907	5736	5574	6044	5926	5809	5989	5930	5828	6087	6023	5958
1	7293	7029	6759	7493	7307	7109	7454	7365	7189	7579	7507	7364
1/0	8925	8544	7973	9317	9034	8590	9210	9086	8708	9473	9373	9053
2/0	10,755	10,062	9390	11,424	10,878	10,319	11,245	11,045	10,500	11,703	11,529	11,053
3/0	12,844	11,804	11,022	13,923	13,048	12,360	13,656	13,333	12,613	14,410	14,119	13,462
4/0	15,082	13,606	12,543	16,673	15,351	14,347	16,392	15,890	14,813	17,483	17,020	16,013
250	16,483	14,925	13,644	18,594	17,121	15,866	18,311	17,851	16,466	19,779	19,352	18,001
300	18,177	16,293	14,769	20,868	18,975	17,409	20,617	20,052	18,319	22,525	21,938	20,163
350	19,704	17,385	15,678	22,737	20,526	18,672	22,646	21,914	19,821	24,904	24,126	21,982
400	20,566	18,235	16,366	24,297	21,786	19,731	24,253	23,372	21,042	26,916	26,044	23,518
500	22,185	19,172	17,492	26,706	23,277	21,330	26,980	25,449	23,126	30,096	28,712	25,916
600	22,965	20,567	17,962	28,033	25,204	22,097	28,752	27,975	24,897	32,154	31,258	27,766
750	24,137	21,387	18,889	29,735	26,453	23,408	31,051	30,024	26,933	34,605	33,315	29,735
1000	25,278	22,539	19,923	31,491	28,083	24,887	33,864	32,689	29,320	37,917	35,749	31,959

Table 11.4   "C" Values for Conductors—Aluminum

AWG or kcmil	Three Single Conductors Conduit						Three Conductor Cable Conduit					
	Steel			Nonmagnetic			Steel			Nonmagnetic		
	600 V	5 kV	15 kV	600 V	5 kV	15 kV	600 V	5 kV	15 kV	600 V	5 kV	15k V
14	237	—	—	237	—	—	237	—	—	237		
12	376	—	—	376	—	—	376	—	—	376		
10	599	—	—	599	—	—	599	—	—	599		
8	951	950	—	952	951	—	952	951	—	952	952	
6	1481	1476	1472	1482	1479	1476	1482	1480	1478	1482	1481	1479
4	2346	2333	2319	2350	2342	2333	2351	2347	2339	2353	2350	2344
3	2952	2928	2904	2961	2945	2929	2963	2955	2941	2966	2959	2949
2	3713	3670	3626	3730	3702	3673	3734	3719	3693	3740	3725	3709
1	4645	4575	4498	4678	4632	4580	4686	4664	4618	4699	4682	4646
1/0	5777	5670	5493	5838	5766	5646	5852	5820	5717	5876	5852	5771
2/0	7187	6968	6733	7301	7153	6986	7327	7271	7109	7373	7329	7202
3/0	8826	8467	8163	9110	8851	8627	9077	8981	8751	9243	9164	8977
4/0	10,741	10,167	9700	11,174	10,749	10,387	11,185	11,022	10,642	11,409	11,277	10,969
250	12,122	11,460	10,849	12,862	12,343	11,847	12,797	12,636	12,115	13,236	13,106	12,661
300	13,910	13,009	12,193	14,923	14,183	13,492	14,917	14,698	13,973	15,495	15,300	14,659
350	15,484	14,280	13,288	16,813	15,858	14,955	16,795	16,490	15,541	17,635	17,352	16,501
400	16,671	15,355	14,188	18,506	17,321	16,234	18,462	18,064	16,921	19,588	19,244	18,154
500	18,756	16,828	15,657	21,391	19,503	18,315	21,395	20,607	19,314	23,018	22,381	20,978
600	20,093	18,428	16,484	23,451	21,718	19,635	23,633	23,196	21,349	25,708	25,244	23,295
750	21,766	19,685	17,686	25,976	23,702	21,437	26,432	25,790	23,750	29,036	28,262	25,976
1000	23,478	21,235	19,006	28,779	26,109	23,482	29,865	29,049	26,608	32,938	31,920	29,135

Note: The values for Tables 11.3 and 11.4 are equal to one over the impedance per 1000 feet and based upon resistance and reactance values found in IEEE Std 241-1990 (Gray Book), *IEEE Recommended Practice for Electric Power Systems in Commercial Buildings* and IEEE Std 242-1986 (Buff Book), *IEEE Recommended Practice for Protection and Coordination of Industrial and Commercial Power Systems*. Where resistance and reactance values differ or are not available, the Buff Book values have been used. The values for reactance in determining the C Value at 5 and 15 kV are from the Gray Book only (Values for 14-10 AWG at 5 kV and 14-8 AWG at 15 kV are not available, and values for 3 AWG have been approximated).

Table 11.5   "C" Values for Busway

Ampacity	Plug-In	Feeder		High Impedance	
	Copper	Aluminum	Copper	Aluminum	Copper
225	28,700	23,000	18,700	12,000	—
400	38,900	34,700	23,900	21,300	—
600	41,000	38,300	36,500	31,300	—
800	46,100	57,500	49,300	44,100	—
1000	69,400	89,300	62,900	56,200	15,600
1200	94,300	97,100	76,900	69,900	16,100
1350	119,000	104,200	90,100	84,000	17,500
1600	129,900	120,500	101,000	90,900	19,200
2000	142,900	135,100	134,200	125,000	20,400
2500	143,800	156,300	180,500	166,700	21,700
3000	144,900	175,400	204,100	188,700	23,800
4000	—	—	277,800	256,400	—

Note: These values are equal to one over the impedance per foot for impedance in a survey of industry.

Table 11.6	"M" (Multiplier)*						
f	**M**	**f**	**M**	**f**	**M**	**f**	**M**
0.01	0.99	0.25	0.8	1.75	0.36	10	0.09
0.02	0.98	0.3	0.77	2	0.33	15	0.06
0.03	0.97	0.35	0.74	2.5	0.29	20	0.05
0.04	0.96	0.4	0.71	3	0.25	30	0.03
0.05	0.95	0.5	0.67	3.5	0.22	40	0.02
0.06	0.94	0.6	0.63	4	0.2	50	0.02
0.07	0.93	0.7	0.59	5	0.17	60	0.02
0.08	0.93	0.8	0.55	6	0.14	70	0.01
0.09	0.92	0.9	0.53	7	0.13	80	0.01
0.1	0.91	1.0	0.5	8	0.11	90	0.01
0.15	0.87	1.2	0.45	9	0.1	100	0.01
0.2	0.83	1.5	0.4				

*M = 1/(1+f)

Table 11.7	Short-Circuit Currents Available from Various Size Transformers								
Voltage and Phase	**kVA**	**Full Load Amps**	**% Impedance†† (Nameplate)**	**Short-Circuit Amps†**	**Voltage and Phase**	**kVA**	**Full Load Amps**	**% Impedance†† (Nameplate)**	**Short-Circuit Amps†**
120/240	25	104	1.58	11,574		25	69	1.6	4,791
1-phase*	37.5	156	1.56	17,351		50	139	1.6	9,652
	50	209	1.54	23,122		75	208	1.11	20,821
	75	313	1.6	32,637		112.5	278	1.11	27,828
	100	417	1.6	42,478		150	416	1.07	43,198
	167	695	1.8	60,255		225	625	1.12	62,004
277/480	112.5	135	1	15,000		300	833	1.11	83,383
3-phase**	150	181	1.2	16,759	120/208	500	1388	1.24	124,373
	225	271	1.2	25,082	3ph.**	750	2082	3.5	66,095
	300	361	1.2	33,426		1000	2776	3.5	88,167
	500	601	1.3	51,362		1500	4164	3.5	132,190
	750	902	3.5	28,410		2000	5552	5	123,377
	1000	1203	3.5	38,180		2500	6950	5	154,444
	1500	1804	3.5	57,261		–	–	–	–
	2000	2406	5	53,461		–	–	–	–
	2500	3007	5	66,822		–	–	–	–

*Single-phase values are L-N values at transformer terminals. These figures are based on change in turns ratio between primary and secondary, 100,000 kVA primary, zero feet from terminals of transformer, 1.2 (%X) and 1.5 (%R) multipliers for L-N versus L-L reactance and resistance values and transformer X/R ratio = 3.
**3-phase short-circuit currents based on "infinite" primary.
††U.L., listed transformers 25 kVA or greater have a ±10% impedance tolerance. Short-circuit amps reflect a "worst case" condition.
†Fluctuations in system voltage will affect the available short-circuit current. For example, a 10% increase in system voltage will result in a 10% increase in the available short-circuit currents shown in the table.

Table 11.8	Various Types of Short-Circuit Currents as a Percentage of 3-Phase Bolted Fault (Typical)
3-Phase (L-L-L) Bolted Fault	100%
Line-to-Line (L-L) Bolted Fault	87%
Line-to-Ground (L-G) Bolted Fault	25%–125%* (Use 100% at transformer, 50% elsewhere)
Line-to-Neutral (L-N) Bolted Fault	25%–125%* (Use 100% at transformer, 50% elsewhere)
3-Phase (L-L-L) Arcing Fault (480V)	89%** (maximum)
Line-to-Line (L-L) Arcing Fault (480V)	74%** (maximum)
Line-to-Ground (L-G) Arcing Fault (480V)	38%** (minimum)

*Typically much lower but can actually exceed the 3-phase bolted fault if it is near the transformer terminals. Will normally be between 25% to 125% of the 3-phase bolted fault value.
**See IEEE 1584, *Guide for Performing Arc-Flash Hazard Calculations*, for formulas to calculate the arcing current when performing an arc flash hazard analysis.
Data source: IEEE Standard 241-1990, *Recommended Practice for Electric Power Systems in Commercial Buildings*, Table 63.

Short-Circuit Calculation Software Programs

There are several short-circuit calculation software programs available. Some of the programs not only calculate the available short-circuit current but also include capabilities to draw one-line diagrams, perform coordination studies, and determine the arc flash hazard. These programs are available for purchase, and the price of the software program can vary greatly based on the capabilities of the software. The Cooper Bussmann Short-Circuit Calculator, shown in **Figure 11.11**, is a basic program that is available free for download, but only calculates short-circuit currents on the load side of transformers or at the end of a conductor or busway.

The first step in using the software program is to select the system type and type of calculation as shown in **Figure 11.12**. First, select the system type, either 3-phase

or single-phase. Next select the type of calculation to be performed. There are three options:

- Fault on the load side of transformer secondary with infinite primary available short-circuit current
- Fault on the load side of transformer secondary with primary available short-circuit current known
- Fault at the end of a run of conductor or busway

Figure 11.13 shows the first calculation option—fault on the load side of the transformer with infinite primary available short-circuit current. This is often used at the first transformer (service transformer) in the system, either utility or customer owned. For a typical 480-volt, 600-ampere service, the transformer selected would be 500 kVA. Assuming the transformer nameplate impedance was 2%, the short-circuit current is calculated at 30,071 amperes.

Figure 11.11 Cooper Bussmann Short-Circuit Current Calculator.

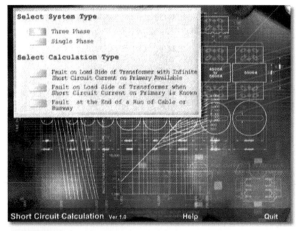

Figure 11.12 Cooper Bussmann Short-Circuit Current Calculator basic selection options.

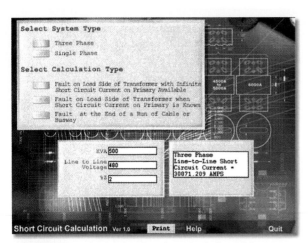

Figure 11.13 Example showing the calculation of the transformer secondary short-circuit current with infinite available primary short-circuit current.

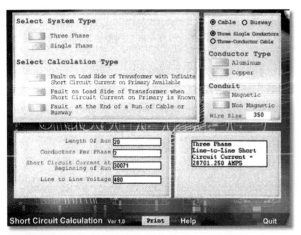

Figure 11.14 Example showing the calculation of short-circuit current at the end of a run of cable.

The next step is to determine the reduction in available short-circuit current from the transformer secondary to the service entrance switchboard as shown in **Figure 11.14**. For a 600-ampere service, the conductor typically used would be 350 kcmil (two conductors per phase). Assuming the distance from transformer to the switchboard is 20 feet and three single copper conductors in nonmagnetic conduit (PVC) are used, the short-circuit current at the service entrance switchboard is calculated to be 28,701 amperes.

Summary

Conducting a short-circuit study is crucial in increasing electrical safety and is necessary to comply with the requirements of the *NEC* and *NFPA 70E*. This helps ensure that the interrupting rating of overcurrent devices and the short-circuit current rating of the equipment is adequate. In addition, a short-circuit study must be completed before an arc flash hazard analysis can be performed, which improves workplace safety by determining the AFPB and the necessary PPE.

Short-circuit calculations help determine the short-circuit current, which can vary depending on the utility, transformers, generators, motors, voltage, conductors, and busway. Becoming familiar with the formulas, tables, and notes associated with the Point-to-Point Short-Circuit Calculation Method in this chapter will enable you to make the proper calculations and use them to assure proper protection of equipment and people.

Lessons *learned*

Maintenance Worker Dies from Injuries Received from an Electrical Flash

On September 2, 2006, at approximately 7:00 a.m., a 39-year-old hotel maintenance worker died from injuries he received on August 27, 2006, from an electrical flash burn and inhalation injuries when he attempted to change a fuse in an electrical panel.

The employer of the victim was a national hotel chain with over 575 hotels throughout the U.S. and Canada. The company had been in business for over 30 years and had approximately 9000 employees. The hotel where the victim worked had 67 employees. The victim had worked for the hotel for 3 months. He was hired as a maintenance worker, and his duties were to perform janitorial functions, minor repairs, and preventative maintenance. The victim had worked as a welder and computer programmer before taking this job. According to the hotel manager, the victim's past work experience qualified him to perform the duties he had been assigned at the hotel. The victim's job description did not include changing fuses in the electrical panel. The victim was born in Mexico and had been in the U.S. for 17 years. The victim was a high school graduate and spoke English and Spanish.

The company had a written Injury and Illness Prevention Program that was printed in English. The program had the elements required by state law. Safety meetings were held on a regular basis and were documented. The company had a documented program that provided general safety training to employees. According to the company's manager, the employee orientation and initial training program consisted of DVD training. Employees would watch DVDs that demonstrated how to complete a specific task. This was followed by a question and answer period and an employees' demonstration of what they just learned in order to determine comprehension.

The site of the incident was an enclosed electrical room in the hotel at the top floor of the garage where an electrical panel was housed. The victim had access to this electrical room but only to turn on and off the power to different systems throughout the hotel. The day of the incident was a Sunday, and the hotel had limited staff working. The power to the lights in the garage had gone out, and the assistant manager for the hotel asked the victim to "check out" the problem. The victim went to the enclosed electrical room and opened a cover on an electrical switch to expose a burned-out fuse. The victim then called the maintenance supervisor at home, and the supervisor told him not to touch the fuse. Despite the supervisor's warning, he removed the burned 30-amperage barrel-type fuse from the panel and proceeded to replace it with a blade-type fuse of different amperage. When he did this, an electrical flash occurred, burning the victim's arms and face.

Although the company had a standard safety electrical procedure for changing fuses, the victim had not been trained in the procedure and consequently did not follow it. The victim was able to exit the electrical room by himself and call for help.

When the paramedics arrived they found the victim conscious and treated his injuries. They then transported him to a local hospital, where he was examined and treated. The victim was then transferred to a burn unit, where he complained of shortness of breath and was intubated as a precautionary measure. His respiratory status remained unstable, and a bronchoscopy was performed and confirmed an inhalation injury. The victim's condition worsened over time, and he died on September 2, 2006, 5 days after the incident. The cause of death, according to the death certificate, was sequelae of electrical burns.

Recommendations/Discussion

Recommendation 1: Ensure workers only perform tasks that are part of their well-defined duties.

Discussion: Well-defined duty lists, when carefully administered, enhance worker safety by making it possible to predict the hazards workers might encounter and so to implement programs that abate or mitigate the hazards. In this incident, the victim was performing a task that was not part of his job description. The victim did not understand that he was not qualified to do this task. He certainly did not understand the hazards of changing the fuse. The maintenance supervisor properly instructed the victim not to perform the task. However, the explicit instruction by the assistant hotel manager to "check-out" the problem, and the implicit permission he had been given to enter the electrical panel room when he was given the keys to the room might have been interpreted by the victim as giving him permission or even a duty to try to change the fuse. Obviously, the assistant hotel manager was unqualified to be providing instructions associated with electrical installations. Mixed signals regarding employer intent can drastically reduce the effectiveness of any worker safety program.

Company-wide standardized programs and procedures for assigning tasks can help supervisors manage job assignments. Tasks should be discussed and planned for in advance. This decreases misassignments, and employees are less likely to attempt tasks for which they don't have adequate experience or training. Employers can enhance workers compliance with duty-restricted work using programs of task-specific training, supervision, recognition, and progressive disciplinary measures.

Adapted from: California Case Report 06CA008. Accessed 6/23/08 at http://www.cdc.gov/niosh/face/stateface/ca/06ca008.html.

Questions

1. Which of the following describes the victim?
   A. Qualified
   B. Unqualified
   C. Licensed
   D. Trained

2. How did the victim know that the fuse was blown?
 A. He measured voltage on the load side.
 B. The blown fuse indicator suggested the fuse was open.
 C. The victim did not know the fuse was blown, but he assumed that replacing the fuse would restore the circuit and bring the lights back on.
 D. The barrel of the fuse is made of paper, and the victim saw that the paper was burned.

3. The barrel-type fuse could be replaced with which of the following?
 A. A barrel-type fuse of lower amperage
 B. A barrel-type fuse with the same amperage rating and a short-circuit rating at least as high as the short-circuit rating of the panel
 C. A blade-type fuse of the same amperage
 D. A blade-type fuse with a higher short-circuit rating

4. _____ was/were the cause of death.
 A. Current flowing through the heart
 B. Damage to the medulla oblongata from current flow
 C. Burns to the inside of the lungs from inhaling superheated air and vaporized metal particles
 D. The burns on his arms and face

5. Who was responsible for causing the accident?
 A. The assistant hotel manager
 B. The victim
 C. The hotel manager
 D. The maintenance supervisor

Ready for Review

- To comply with the requirements of the *NEC* and *NFPA 70E*, a short-circuit study must be performed.
- It is necessary to ensure that the interrupting rating of overcurrent devices and the short-circuit current rating of equipment are adequate.
- An arc flash hazard analysis determines the AFPB and incident energy or hazard/risk category so that appropriate PPE can be determined.
- The short-circuit study starts from the service point and extends to the branch-circuit devices. It determines the short-circuit current at all critical points in the system and analyzes service equipment, transfer switches, switchboards, panelboards and motor control centers, motor starters, and disconnects.
- The point-to-point calculation method is a method that is used to determine the available short-circuit current at various locations.
- It is important to determine both the low and high values of short-circuit current.
- The lower value of the short-circuit current might actually increase the overcurrent device clearing time, resulting in an increase of the incident energy.
- Short-circuit currents can change over time. *NFPA 70E* requires that the arc flash hazard analysis be updated when a major modification or renovation takes place (but not to exceed 5 years).
- Short-circuit studies normally involve calculating a bolted 3-phase fault condition.
- Sources of short-circuit current include utility generation, local generation, synchronous motors, and induction motors.
- When electrical systems are supplied from a utility- or customer-owned transformer, the amount of short-circuit current depends upon the size and impedance of the transformer.
- Generators and induction motors can increase the short-circuit current from the utility.
- The amount of short-circuit current depends on factors including, but not limited to, utility, transformers, generators, motors, voltage, conductors, and busway.
- The short-circuit current is typically the highest at the service point. Conductors and transformers will decrease the available short-circuit current in downstream equipment.
- If transformer kVA and voltage are constant, but impedance is reduced, the short-circuit current increases.
- If transformer kVA increases, impedance and voltage remains constant, the short-circuit current increases.
- If transformer kVA and impedance are constant, but voltage decreases, the short-circuit current increases.
- If voltage increases, but the size and length of the conductor or busway remains constant, the short-circuit current increases.
- If the voltage and conductor or busway length remains constant, but the conductor or busway size increases, the short-circuit current increases.
- If the voltage and conductor or busway size remains constant, but the conductor or busway length increases, the short-circuit current decreases.
- To determine the short-circuit current at any point in the system, start by drawing a one-line diagram showing all of the sources of short-circuit current.
- Short-circuit calculations are performed without OCPDs in the system, to determine the maximum "available" short-circuit current.
- The point-to-point method for calculating available short-circuit current is reasonably accurate for either 3-phase or single-phase electrical distribution systems.
- Become familiar with the *Basic Short-Circuit Calculation Procedure* presented in this chapter, including the formulas, notes, and tables.

- Steps 1 through 3 are used to determine the short-circuit current available at the transformer secondary.
- Steps 4 through 6 are used to determine the short-circuit current available at the end of a run of conductor or busway.
- Steps A through C are used to determine the short-circuit current available at the transformer secondary with primary short-circuit current known.
- The Cooper Bussmann Short-Circuit Calculator program is available as a free download and is based on the point-to-point method. This program only calculates short-circuit currents on the load side of transformers or at the end of a conductor or busway.

Vocabulary

ampacity* The current, in amperes, that a conductor can carry continuously under the conditions of use without exceeding its temperature rating.

arc flash protection boundary (AFPB)* When an arc flash hazard exists, an approach limit at a distance from a prospective arc source within which a person could receive a second degree burn if an electrical arc flash were to occur.

impedance‡ Ratio of the voltage drop across a circuit element to the current flowing through the same circuit element, measured in ohms (Ω).

incident energy* The amount of energy impressed on a surface, a certain distance from the source, generated during an electrical arc event. One of the units used to measure incident energy is calories per centimeter squared (cal/cm^2).

interrupting rating (IR)† The highest current at rated voltage that a device is intended to interrupt under standard test conditions.

FPN: Equipment intended to interrupt current at other than fault levels may have its interrupting rating implied in other ratings, such as horsepower or locked rotor current.

short-circuit (fault) current‡ Overcurrent resulting from a short circuit due to a fault or an incorrect connection in an electric circuit.

short-circuit current rating† The prospective symmetrical fault current at a nominal voltage to which an apparatus or system is able to be connected without sustaining damage exceeding defined acceptance criteria.

short-circuit study An analysis of an electrical distribution system that determines the available short-circuit system; the purpose is to verify compliance with the *NEC* and *NFPA 70E* and increase electrical safety.

*This is the *NFPA 70E* definition.
†This is the *National Electrical Code*® (*NEC*®) definition.
‡This definition is from the National Fire Protection Association's (NFPA's) *Pocket Dictionary of Electrical Terms*.

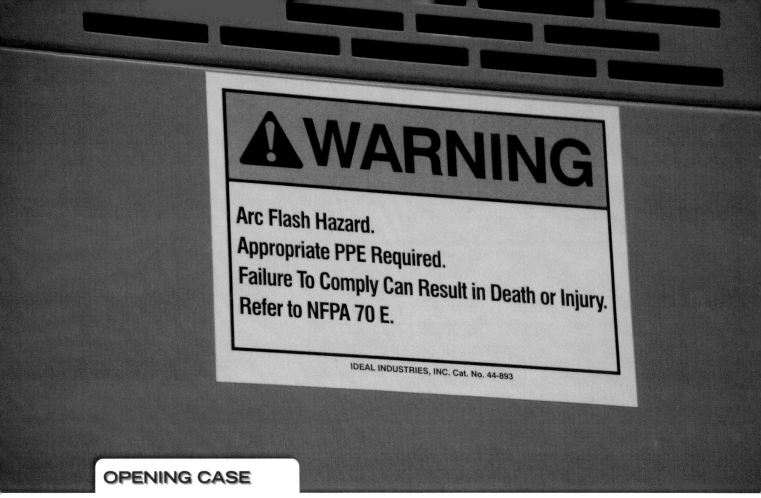

Photo courtesy of National Electrical Contractors Association (NECA).

OPENING CASE

A 53-year-old Journeyman wireman (the victim) was electrocuted when he contacted the two phases (A and C) of an energized busbar. The employer had been in business for 76 years and employed 10,000 workers nationwide, about half of whom were Journeymen wiremen. The employer was one of two contractors installing a sulfur dioxide emissions control system and had been on site for 1 year. The company maintained a full-time corporate safety director and an on-site safety coordinator. The company normally adhered to a company-wide lockout and tagout procedure; however, for this particular job, the company had adopted the utility company's procedure, a tagout-only policy. Employees of the contractor received on-the-job training in addition to safety manual and new hire orientation. Weekly safety meetings were held at the job site, and safety talks were given each time a job task was changed.

The incident occurred on the third floor of the electrical equipment building, where workers were in the final stages of installing electrical components inside a 14-compartment switchhouse. The switchhouse and associated switchgear were part of a newly installed sulfur dioxide emissions control system and had never been energized.

Employees of the contractor began work at 7 a.m. The victim and two coworkers went to the third floor of the electrical equipment building, where they had been installing electrical components inside the switchhouse. The workers were completing the installation, which included making the final connections to the ground circuit, checking electrical connections for proper tightness, and cleaning the inside of the switchhouse compartments prior to prestartup inspection by plant personnel. They were engaged in this activity during the morning and most of the afternoon. The switchgear had been isolated according to the tagout procedure in effect, but was not physically locked against energization. Workers had isolated all of the circuit breakers inside the switchhouse by placing the panel-mounted circuit breaker operating handles in the remote position and tagging them out (red tag). The circuit breakers can be opened in two ways—remotely or manually—depending on which mode of operation is selected

Arc Flash Hazard Analysis Methods

CHAPTER 12

Introduction

An electrical hazard analysis, consisting of a shock hazard analysis and an **arc flash hazard analysis**, is required if energized electrical conductors or circuit parts at 50 volts or more are not placed in an electrically safe work condition. The arc flash hazard analysis per *NFPA 70E* Section 130.3 is required to determine the **arc flash protection boundary (AFPB)** and the personal protective equipment (PPE) that is required when people are within the AFPB.

It is important to note that, in order to determine the AFPB and required PPE (based on incident energy or the *NFPA 70E* HRC tables), the available short-circuit (fault) current and the overcurrent protective device (OCPD) clearing time must be known. In order to determine the OCPD clearing time, the arcing current must be determined and compared with manufacturer's time-current curves. Therefore, these fundamental parameters must be determined before an arc flash hazard analysis can be performed.

OPENING CASE, CONT.

at the switch handle on the front panel of each compartment. In the remote position, the breakers can be operated from outside the electrical equipment room. When manual operation is selected, the circuit breakers can only be closed from the front panel. Power (6.9 kilovolts) is supplied to the north end of the switchhouse by insulated conductors that are connected to an internal busbar running the length of the switchhouse. Power is then distributed from the enclosure to various areas of the facility.

At about 3:30 p.m., without the knowledge of the workers, someone replaced the red tag posted on the circuit breaker protecting the internal busbar with a green tag, and power plant personnel closed the breaker. This energized the internal busbar of the switchhouse. The workers continued working, unaware that the switchhouse busbar was energized. At about 4:40 p.m., the victim reached inside the upper part of the second compartment from the south end of the switchhouse to wipe it down. This compartment contained three busbar terminals, A, B, and C phases, where a jumper had been connected to carry power to an adjacent switchhouse. As he reached inside the compartment, he contacted the A phase terminal with his right hand and the C phase terminal with his left hand and was electrocuted. One of the coworkers had been walking past the victim at the time of contact, was knocked down to the floor by the blast, and suffered first-degree burns on his neck and face. A second coworker, at the north end of the switchhouse, heard the explosion and went to render assistance to the victim, who had fallen to the floor and whose clothes were on fire. The second coworker suffered burns to his hands when he tried to beat out the flames. The contractor's safety

Chapter Outline

- Introduction
- Arc Flash Hazard Analysis
- Arcing Currents in the Long Time Characteristic of Overcurrent Protective Devices
- Additional Considerations
- Summary
- Lessons Learned
- Current Knowledge

Learning Objectives

1. Understand how to determine the arcing current and the clearing time of an OCPD.
2. Understand the advantages and limitations of various methods available to determine the AFPB and selection of PPE.
3. Understand how to determine the AFPB using the conditional value of 4 ft and associated system limitations or by calculation.
4. Understand how to select PPE using an incident energy analysis, available commercial tools and resources, or using the Hazard/Risk Category (HRC) tables.

References

1. *NFPA 70E®*, 2009 Edition
2. *Cooper Bussmann®* Safety BASICs™ Handbook for Electrical Safety

Due to the complexity of the arc flash hazard analysis, this chapter focuses on various methods including the calculation methods shown in *NFPA 70E* Annex D, Table D.1, to determine the AFPB and required PPE. The limitations and advantages of each method will be discussed as well.

The AFPB can be determined by using the conditional 4 ft boundary for systems 600 volts and less, provided system limitations are met, or by calculations using the formulas such as those in *NFPA 70E* Annex D. AFPB calculations are always required for systems over 600 volts. This chapter illustrates the requirements for use of the conditional 4 ft boundary and use of one calculation method to determine the AFPB.

The next step in performing the arc flash hazard analysis is to determine the PPE that is required. The incident energy can be calculated and then used to select PPE. The formulas in *NFPA 70E* Annex D are among examples of calculation methods that can be used to determine incident energy. Alternatively, if the specific task and equipment is listed in the *NFPA 70E* Table 130.7(C)(9), and system characteristic limits are met, the required PPE can be selected using the HRC tables. The system characteristic limits are described in the notes that follow the table. The incident energy or level of PPE is required to be field marked on the equipment. This chapter illustrates the use of one calculation method to determine the incident energy and the use of the HRC table method to determine required PPE.

Where work on energized equipment is justified due to either increased hazards or infeasibility, an energized electrical work permit is required per *NFPA 70E* Section 130.1(B)(1). An energized electrical work permit is not required for routine tasks such as testing or troubleshooting. However, the use of safe work practices and PPE are required in either case. In order to complete the energized electrical work permit and determine safe work practices and PPE, an arc flash hazard analysis must be completed. To complete the arc flash hazard analysis, the AFPB and incident energy or HRC must be determined.

In order to determine the AFPB and incident energy or HRC, a **short-circuit study** must be performed. Then, the arcing current must be determined and used to verify the OCPD's clearing time. After these steps are completed, the AFPB can be determined either by using the conditional 4 ft boundary or by calculating the boundary distance. For systems above 600 volts, the AFPB must be calculated. The selection of PPE can be determined by an incident energy analysis according to *NFPA 70E* Section 130.3(B)(1) or by the use of the HRC tables according to *NFPA 70E* Section 130.3(B)(2), provided the system characteristics do not exceed the specified notes to the tables. Finally, *NFPA 70E* Section 130.3(C) requires the incident energy or required level of PPE to be field marked on the equipment.

The AFPB and required PPE as part of an arc flash hazard analysis are perhaps the most important considerations when it comes to protecting employees from electrical hazards in the workplace. Because of this, this chapter has been dedicated to illustrate how to determine the AFPB and required PPE. It is important to note that additional considerations, such as equipment maintenance, can affect the AFPB and required PPE. Because of this, equipment maintenance, as covered in Chapter 15, must also be considered to ensure proper protection of employees.

OPENING CASE, CONT.

coordinator was notified by the operator of a cherry picker, who had witnessed the flash of the explosion from outside the building. The safety coordinator notified plant security to request emergency medical services (EMS) and then went to the third floor. He checked the victim for a pulse and, finding none, started cardiopulmonary resuscitation (CPR). In approximately 15 minutes, the EMS arrived and transported the victim to a local hospital emergency room where he was pronounced dead. The cause of death was established by the deputy medical examiner as electrocution.

Adapted from: Fatality and Assessment and Control Evaluation (FACE) 94-10. Accessed 9/2/08 at http://www.cdc.gov/niosh/face/In-house/full9410.html.

1. Name at least two measures that could have been taken by the employer or the electrical workers to prevent this incident.

2. Why was a tagout-only procedure used in this situation? Was the elimination of lockout justified?

Arc Flash Hazard Analysis

Fundamental Information Required

Regardless of which method is used to determine the AFPB and the required PPE, some fundamental information is required to determine both of these parameters, including the available short-circuit current and the OCPD clearing time. Chapter 11 illustrates how to calculate the available short-circuit current [3-phase bolted short-circuit current]. When performing an arc flash hazard analysis, the short-circuit current should be as exact as possible. In some cases, due to system operating variations, determining a short-circuit current range and using both the low and high value of short-circuit current is necessary to see which results in the worst case. This consideration is noted in *NFPA 70E* Section 130.7(C)(9) Fine Print Note (FPN) No. 2:

> Both larger and smaller available short-circuit currents could result in higher available arc flash energies. If the available short-circuit current increases without a decrease in the opening time of the overcurrent protective device, the arc flash energy will increase. If the available short-circuit current decreases, resulting in a longer opening time for the overcurrent protective device, arc flash energies could also increase.

Once the short-circuit current is calculated, the associated arcing current must be determined and used with the manufacturer's time-current curve to find the OCPD clearing time. Note: the short-circuit current and arcing current value must be determined at the point where the work is to be performed. The arcing current value is then compared to the upstream OCPD time-current curve to determine the device clearing time.

Arcing Current

There are two different methods that can be used to determine the arcing current, based on the available short-circuit current (bolted short-circuit current). First, the minimum and maximum arcing current can be determined using percentages based on industry research and experience. As shown in Chapter 11, Table 11.8, the typical minimum sustainable arcing fault (at 480 volts) is 38% of the 3-phase short-circuit current. It also shows that the maximum 3-phase (line-to-line-to-line) arcing current is 89% of the 3-phase short-circuit current. So in theory, the arcing current could be from 38% to 89% of the 3-phase short-circuit current.

However, a more accurate calculation is now available, based on actual arc flash testing, and found in Institute of Electrical and Electronics Engineers (IEEE) 1584 *Guide for Performing Arc-Flash Hazard Calcula-*

tions. (The IEEE 1584 equations shown in this chapter are from IEEE 1584 Guide. Copyright 2002, by IEEE. All rights reserved.) The arcing current can be calculated by the formulas in *NFPA 70E* Annex D.7.2, taken from IEEE 1584 as follows (note that the term arc-in-a-box refers to an arc that occurs within a limited amount of space, such as when working on enclosed equipment):

For systems under 1 kV [per equation D.7.2(a)]:

$$\lg I_a = K + 0.662 \lg I_{bf} + 0.0966\, V + 0.000526\, G + 0.5588\, V\, (\lg I_{bf}) - 0.00304\, G\, (\lg I_{bf})$$

where:

$\lg$ = the log10

I_a = arcing current in kA

K = −0.153 for open air arcs; −0.097 for arcs-in-a-box

I_{bf} = bolted three-phase available short-circuit current (symmetrical rms) (kA)

V = system voltage in kV

G = conductor gap (mm) (See Table D.7.2. **Table 12.1**)

For systems 1 kV or greater [per equation D.7.2(b)]:

$$\lg I_a = 0.00402 + 0.983\, (\lg I_{bf})$$

Convert from $\lg I_a$ [per equation D.7.2(c)]:

$$I_a = 10^{\lg Ia}$$

Note: if using these formulas to calculate incident energy, use I_a and 0.85 I_a arc durations and use the resulting higher incident energy. The 0.85 I_a arc duration accounts for variations in the arcing current and the time for the OCPD to open.

Using the formula above for systems less than 1 kilovolt, **Table 12.2** shows the arcing current values compared to the short-circuit current. This table is only valid for arcs-in-a-box, 480-volt systems, and applies to switchgear or motor control centers (MCCs) or panelboards as indicated in the table. Table 12.2 also indicates that the arcing current can vary from 43% to 63% of the 3-phase short-circuit current value for switchgear at 480 volts or 47% to 66% for MCC or panelboards at 480 volts. Note that the arcing current increases as the short-circuit current decreases. Now that the arcing current has been determined, this value can be used to determine the clearing time of the OCPD.

OCPD Clearing Time

The first step in determining the OCPD clearing time is to obtain the time-current curve from the OCPD manufacturer. The time-current curve shows the minimum or average opening time and/or the maximum clearing time of the OCPD, based on a range of current values. The time-current curve for circuit break-

| Table 12.1 | *NFPA 70E* Table D.7.2 Factors for Equipment and Voltage Classes |

System Voltage (kV)	Type of Equipment	Typical Conductor Gap (mm)	Distance Exponent Factor X
0.208–1	Open-air	10–40	2.000
	Switchgear	32	1.473
	MCCs and panels	25	1.641
	Cables	13	2.000
>1–5	Open-air	102	2.000
	Switchgear	13–102	0.973
	Cables	13	2.000
>5–15	Open-air	13–153	2.000
	Switchgear	153	0.973
	Cables	13	2.000

| Table 12.2 | Arcing Current Versus 3-Phase Short-Circuit Current (Arcs-in-a-Box) |

480-Volt Switchgear		480-Volt MCC/Panel	
3-Phase Short-Circuit Current (kA)	Arcing Current (kA)	3-Phase Short-Circuit Current (kA)	Arcing Current (kA)
10	6.30	10	6.56
20	11.22	20	11.85
30	15.72	30	16.76
40	19.98	40	21.43
50	20.46	50	25.93
60	28.01	60	30.30
70	31.85	70	34.57
80	35.59	80	38.74
90	39.26	90	42.84
100	42.86	100	46.88

ers can vary based on the type of circuit breaker and the manufacturer. To illustrate the differences, thermal magnetic molded case circuit breakers, electronic trip molded case circuit breakers, and low-voltage power circuit breakers time-current curves are analyzed. For current-limiting fuses, the time-current curves are typically not as complex as circuit breakers, because no setting adjustments exist. However, the concept of determining the OCPD clearing time is the same.

The following sections illustrate the use of time-current curves to determine the OCPD clearing time for various types of circuit breakers and fuses protecting a 480-volt panelboard. A 400-ampere rating has been selected for each type of OCPD, and the corresponding time-current curve is used to determine the clearing time, based on a bolted short-circuit current value of 20,000 amperes, which produces an arcing current of 11,850 amperes.

Thermal Magnetic Molded Case Circuit Breaker

Figure 12.1 shows a time-current curve for a typical 400-ampere thermal magnetic (TM) molded case circuit breaker. As mentioned earlier, the time-current curve indicates the amount of time, based upon current, required for the circuit breaker to open. Time (in seconds) is shown on the vertical axis, and current (in amps) is shown on the horizontal axis. Typically, as indicated in Figure 12.1, a minimum opening (sometimes referred to as unlatching) and a maximum clearing are shown.

Also of importance is the instantaneous trip (IT) setting and instantaneous region. In this case, the instantaneous setting is adjustable and set at 5× (5 times the ampere rating of the circuit breaker) or 2000 amperes (± operating time tolerance, approximately 1500 to 2500 amperes) for the 400-ampere circuit breaker. Note that this setting can typically be as low as 3–5×, but no more than 8–10× for a molded case circuit breaker. The instantaneous region is to the right of the IT and results in the lowest clearing time.

The time-current curve has been drawn up to the short-circuit current. (The curve could extend up to the interrupting rating of the circuit breaker, which must be greater than or equal to the short-circuit current value). The arcing current is marked with a circle on the time-current curve, and the corresponding maximum clearing (arc duration) time is found to be 0.017 seconds for this circuit breaker. Note that arcing current is in the IT region. If the arcing current is below the setting of the IT (in the overload region), the clearing time increases greatly.

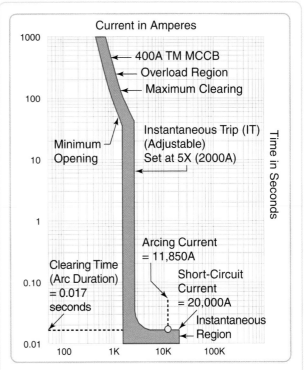

Figure 12.1 A 400-ampere thermal magnetic molded case circuit breaker—arc duration for an arcing current of 11,850 amperes is 0.017 seconds.

Electronic Trip Molded Case Circuit Breaker

With electronic trip (ET) molded case circuit breakers, as shown in **Figure 12.2**, a current transformer is used in each current phase to monitor and reduce the current to the appropriate input level required by the electronic trip unit. The electronic trip unit is generally temperature-insensitive and can offer increased accuracy, curve shaping (adjustments), programming, monitoring, communication, and testing.

Electronic trip units have two adjustable types of long-time settings. The "long-time" (LT) setting determines the ampere rating of the circuit breaker and the "long-time delay" (LTD) setting determines the overload region performance. Electronic trip units might also be equipped with three adjustable types of short-time settings as shown in Figure 12.2. The first is called the "short-time pick up" (STPU) setting, and for this, the circuit breaker's STPU is set to 5× (2000 amperes). The second is called the IT, or in some cases an "instantaneous override," and is set at the maximum setting of 8× (3200 amperes) for this circuit breaker. Note that with an electronic trip circuit breaker, more adjustments can be made, but typically the maximum IT setting cannot be more than 8–10× the ampere rating of the circuit breaker. The third setting is called the "short-time delay" (STD), which connects the STPU

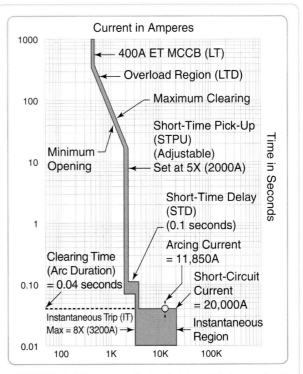

Figure 12.2 A 400-ampere electronic trip molded case circuit breaker—arc duration for an arcing current of 11,850 amperes is 0.04 seconds.

to the IT setting. For electronic trip molded case circuit breakers, this setting is typically available at 0.1 (6 cycles), 0.2 (12 cycles), or 0.3 (18 cycles). In this case, the STD is set at 0.1 seconds.

For the electronic trip molded case circuit breaker in this example, the arcing current is still in the instantaneous region but the clearing (arc duration) time is now 0.04 seconds compared with the clearing time (0.017 seconds) for the thermal magnetic molded case circuit breaker. This is not always the case. In some instances, electronic trip circuit breakers might open faster than a thermal magnetic circuit breaker.

Also note that if the arcing current is less than the IT, extended clearing time can be experienced due to the STD setting or the overload setting (LTD).

Low-Voltage Power Circuit Breaker

Low-voltage power circuit breakers are similar to electronic trip molded case circuit breakers with regards to adjustability of overcurrent protection settings. The difference is that these devices are much larger in size than molded case circuit breakers, offer draw-out capabilities (for increased maintenance capabilities), and can be provided with or without an IT (to increase selective coordination capabilities).

If equipped without an IT, the low-voltage power circuit breaker is set to hold faults for a period of time equal to the STD setting. The STD setting can be from 0.1 seconds (6 cycles) up to 0.5 seconds (30 cycles). The purpose of this is to allow the downstream OCPD to clear faults before the low-voltage power circuit breaker

trips to avoid unnecessary power loss to other loads as shown in Figure 12.3 .

This arrangement works well provided the fault is on the load side of the downstream circuit breaker. However, if the fault is on the line side of the downstream circuit breaker, extended clearing times of up to 0.5 seconds (30 cycles) during fault conditions can occur. The low-voltage power circuit breaker in Figure 12.4 has a STD setting of 0.1 seconds (6 cycles). With this setting and an arcing current of 11,850 amperes, the clearing (arc duration) time is 0.14 seconds. This increase of time can result in a dramatic increase in incident energy compared to the thermal magnetic and electronic trip molded case circuit breakers. Because of this, technology such as zone selective interlocking, arc reduction maintenance switches, and remote operation are recommended when STD settings are used on low-voltage power circuit breakers.

Current-Limiting Fuses

Current-limiting fuses do not have the adjustability of overcurrent protection settings such as molded case circuit breakers or low-voltage power circuit breakers. However, current-limiting fuses can provide reduced clearing times compared to molded case circuit breakers or low-voltage power circuit breakers provided the arcing current is within the current-limiting range. The current-limiting range of a fuse is where the current exceeds the current value where the clearing time of the fuse is less than 0.01 seconds. This is typically around 15× the ampacity rating of the current-limiting fuse, but

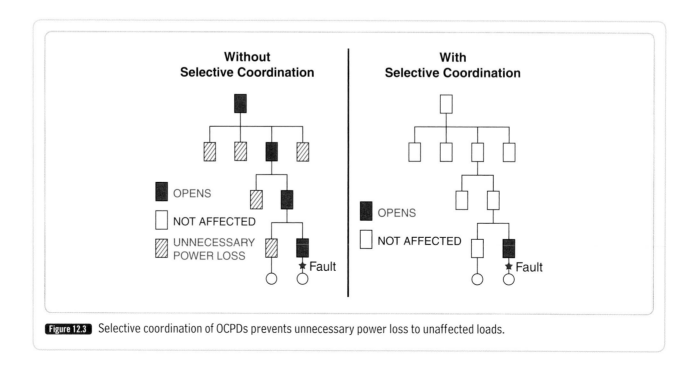

Figure 12.3 Selective coordination of OCPDs prevents unnecessary power loss to unaffected loads.

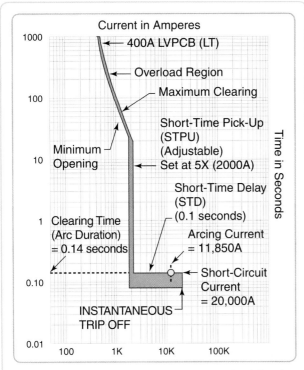

Figure 12.4 A 400-ampere low-voltage power circuit breaker without IT—arc duration for an arcing current of 11,850 amperes is 0.14 seconds.

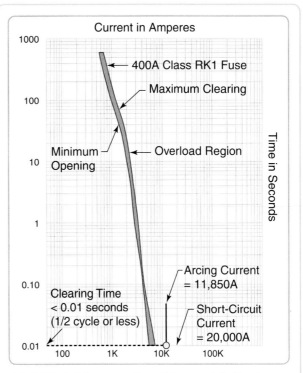

Figure 12.5 A 400-ampere Class RK1 current-limiting fuse—arc duration for an arcing current of 11,850 amperes is less than 0.01 (assumed 0.008) seconds with additionally reduced let-through arcing current.

can be verified by the fuse time-current curve. Because of this, it is important to note that current-limiting fuses are not always current-limiting. Similar to circuit breakers, current-limiting fuses have an overload region in which clearing time can be several seconds or minutes. When the arcing current is less than the current-limiting range of a fuse, extended clearing times and increased incident energy can occur.

However, if the fault is within the current-limiting range, clearing (arc duration) times of less than 0.01 seconds (assumed to be ½ cycle or 0.008 seconds) occur as shown in **Figure 12.5**. Based upon the testing conducted for IEEE 1584, the clearing time for current-limiting fuses can be assumed to be 0.004 seconds if the arcing current value is more than twice the current at 0.01 seconds (approximately 6000 amperes in Figure 12.5). In addition, the current-limiting fuse can also greatly reduce the let-through arcing current. The combination of reduced clearing times and reduced arcing current can greatly reduce the incident energy when the arcing current is within the current-limiting range of the fuse.

Once the arcing current and arc duration (OCPD clearing time) have been determined, the next step is to identify the AFPB and what PPE must be used.

Methods Available

There are several methods that can be used to identify the AFPB and PPE as required by the arc flash hazard analysis. All have advantages and disadvantages. In some cases, such as with the AFPB, a conditional value can be used, provided system characteristic limits are met. In other cases, as with the selection of PPE, tables can be used provided system characteristic limits are met. However, in many cases, both the AFPB and PPE must be determined by calculations.

NFPA 70E Annex D, Table D.1, summarizes a number of calculation methods available to determine the AFPB and incident energy. **Table 12.3** and **Table 12.4** outline the advantages and limitations for each calculation method plus additional methods to determine the AFPB and incident energy or HRC, respectively. Note that most of the methods are valid only for 3-phase alternating current (ac) systems. Methods for direct current (dc) systems are currently not available. Some methods are valid only for systems 600 volts and below or systems above 600 volts, while other methods are valid for both. In addition, some methods address only arcs in open air, while others address arcs in open air and arcs-in-a-box. In general, methods that address

Table 12.3		Methods to Determine AFPB per *NFPA 70E* (ac Systems Only)	
Value	**Method**	**Advantages**	**Limitations**

AFPB	Conditional Value [Section 130.3(A)(1)]	Easiest method	• Only applicable for
			  ○ 600 volts or less
			  ○ Combination of short-circuit current and OCPD clearing time cannot exceed 100 kiloamperes cycles
			• Does not indicate how to determine OCPD clearing time
			• Conservative compared to calculated methods

	Ralph Lee IEEE Paper (1982) [Annex D.2]	Least complicated calculated method	• Only applicable for
			  ○ 3-phase systems
			  ○ Arcs in open air
			• Conservative over 600 volts and as voltage increases
			• Does not indicate how to determine overcurrrent protective device clearing time (time of arc exposure)

	IEEE 1584 (2002) [Annex D.7.5]	• Based on the most up-to-date testing available
		• Provides formula for arcing current to assist with determining OCPD clearing time
			• Only applicable for
			  ○ 3-phase systems
			  ○ 208 to 15 kilovolts (theoretical equation above 15 kilovolts)
			  ○ Short-circuit currents between 700 amperes to 106 kiloamperes
			  ○ Arcs in open air or arcs-in-a-box
			• Most complicated formulae (more variables)
			• Only applicable for short-circuit currents of 700 amperes to 106 kiloamperes

arcs-in-a-box are more conservative (calculate higher incident energies) than methods that address arcs in open air.

Determining the AFPB

Conditional Value

NFPA 70E Section 130.3(A)(1) permits a conditional value of 4 ft to be used for the AFPB where calculations are not performed and the system characteristic limits are not exceeded. The 4 ft conditional AFPB is valid only for systems between 50 and 600 volts where:

- The clearing time is 2 cycles or less
- The available short-circuit current is 50 kiloamperes or less
- Any combination of clearing time and short-circuit current where the product does not exceed 100-kiloampere cycles

When the product of the clearing time and short-circuit current exceeds 100-kiloampere cycles, or system voltage exceeds 600 volts, the AFPB must be calculated.

Example

Referring to the system characteristics and device clearing time of Tests 4 and 3 (Figure 12.6 , Figure 12.7 , and Figure 12.8) also shown in Chapter 2, could the conditional value of 4 ft be applied?

Test 4

Bolted short-circuit current = 22.6 kiloamperes
Clearing time = 6 cycles

Note: The feeder OCPD was a 640-ampere circuit breaker with STD; the short-circuit current was permitted to flow for six cycles.

Calculation:

Combination is 22.6 kA × 6 cycles = 135.6 kA cycles

Result:

No, the 4 ft conditional AFPB cannot be used, since 135.6-kiloampere cycles exceeds 100-kiloampere cycles.

Table 12.4	Methods to Determine I_e or HRC per *NFPA 70E* (ac Systems Only)		
Value	**Method**	**Advantages**	**Limitations**

Ie or HRC	Doughty, Neal, Floyd IEEE Paper (1998) [Annex D.5, D.5.1 and Annex D.5.2]	Less complicated calculated method	• Only applicable for
			  ○ 3-phase systems
			  ○ 600 volts and below
			  ○ Short-circuit currents between 16 and 50 kiloamperes
			  ○ Arcs in open air or arcs-in-a-box
			  ○ Working distances equal to or greater than 18 in
			• Recommends calculation at the maximum short-circuit current and the minimum sustainable arcing current (38% at 480 volts)

	Ralph Lee IEEE Paper (1982) [Annex D.6]	Least complicated calculated method	• Only applicable for
			  ○ 3-phase systems
			  ○ Systems above 600 volts
			  ○ Arcs in open air
			• Conservative as voltage increases

	IEEE 1584 (2002) [Annex D.7.3 and Annex D.7.4]	• Based on the most up-to-date testing available
		• Provides formula for arcing current to assist with determining OCPD clearing time
		• Provides simplified formulas for fuses and circuit breakers
			• Only applicable for
			  ○ 3-phase systems
			  ○ 208 to 15 kilovolts (theoretical equation above 15 kilovolts)
			  ○ Short-circuit currents between 700 amperes to 106 kiloamperes
			  ○ Arcs in open air or arcs-in-a-box
			• Most complicated formulae (more variables)
			• Recommended PPE levels based on calculations are 1.2, 8, 25, and 40 to be adequate for 95% of arc flash incidents

	National Electrical Safety Code (NESC)-2007 [Annex D.8]	Uses tables and OCPD clearing time	• Only applicable for
			  ○ Arcs in open air
			  ○ Phase-to-ground arc
			  ○ 1 to 800 kilovolts only

	HRC [Tables 130.7(C)(9), 130.7(C)(10), and 130.7(C)(11)]	Easiest method	• Can be used only when the specific task is listed and the limits of the system characteristics and OCPD operating time are met as specified in the notes to the tables • Often limited to lower short-circuit currents and faster clearing times of OCPDs • Working distance is set at 18 in for systems less than 1 kilovolts or 36 in for systems 1 kilovolts and greater

Test 3

Bolted short-circuit current = 22.6 kiloamperes

Clearing time = 0.5 cycles

Note: The feeder OCPD was a 601-ampere current-limiting Class L fuse. The 601-ampere current-limiting fuse cleared the fault in ½ cycle (0.008 seconds).

Calculation:

Combination is 22.6 kA × 0.5 cycles = 11.3 kA cycles

Result:

Yes, the 4 ft conditional AFPB can be used, since 11.3-kiloampere cycles does not exceed 100-kiloampere cycles.

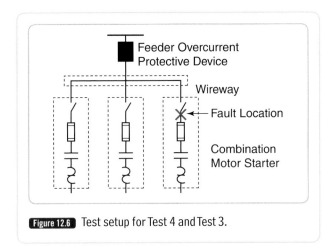

Figure 12.6 Test setup for Test 4 and Test 3.

Figure 12.7 Photo from Test 4 arc flash test.
© 1997 Institute of Electrical and Electronics Engineers. Still photos extracted from the Cooper Bussmann *Safety BASICs Handbook* with permission from Cooper Bussmann.

Figure 12.8 Photo from Test 3 arc flash test.
© 1997 Institute of Electrical and Electronics Engineers. Still photos extracted from the Cooper Bussmann *Safety BASICs Handbook* with permission from Cooper Bussmann.

Ralph Lee Method

The Ralph Lee method, detailed in *NFPA 70E* Annex D.2, is based on his paper "The Other Electrical Hazard: Electrical Arc Blast Burns," in *IEEE Trans. Industrial Applications*, vol. 1A-18. no. 3, p. 246, May/June 1982. This was the first method developed and permitted

for use by *NFPA 70E* to calculate the AFPB, but is not a preferred method because it only applies to open air arcs. The incident energy from open air arcs is less than the energy from arcs-in-a-box. Because of this, it would not be suitable to determine the AFPB for an arc-in-a-box (such as arcs in electrical equipment as seen in Test 4, Test 3, and Test 1), since this formula would yield results lower than expected.

NFPA 70E Annex D.7 and IEEE 1584 Method

The IEEE 1584 method, detailed in *NFPA 70E*, Annex D.7, is based on IEEE 1584 *Guide for Performing Arc-Flash Hazard Calculations*. IEEE 1584 used extensive testing and analysis to develop new formulas to calculate the AFPB and incident energy. Because it is the most up-to-date method, with the widest range of system characteristics, it is considered by many to be the preferred method to calculate the AFPB. In addition, the arcing current formulas in *NFPA 70E* Annex D.7.2 and IEEE 1584, discussed previously, improve the ability to accurately determine the arcing current and related OCPD clearing time. It is important to note that these formulas are in J/cm² for incident energy and millimeters for distances. To convert units: 1 cal/cm² = 4.184 J/cm² and 1 in = 25.4 mm.

The formula to calculate the AFPB per IEEE 1584 is found in *NFPA 70E* Annex D.7.5 and shown below:

For systems 208 V to 15kV [per equation D.7.5(a)]:

$$D_B = \left[4.184 C_f E_n \left(\frac{t}{0.2} \right) \left(\frac{610^x}{E_B} \right) \right]^{\frac{1}{x}}$$

where:

D_B = the distance (mm) of the arc flash protection boundary from the arcing point

C_f = a calculation factor
= 1.0 for voltages above 1 kV
= 1.5 for voltages at or below 1 kV

E_n = incident energy normalized

E_B = incident energy in J/cm² at the distance of the arc flash protection boundary

t = time (seconds)

X = the distance exponent from Table D.7.2 (Table 12.1)

As indicated by the formula, in order to calculate the AFPB, "E_n" and "t" must be determined. In order to determine "E_n" and "t", the arcing current must first be found by the formulas previously discussed. The AFPB is the distance where a person can receive a second-degree burn. A second-degree burn is where the incident energy is 1.2 cal/cm² or 5.0 J/cm² (1.2 × 4.184). Therefore, E_B is

5.0 J/cm². The value for "t" can be found from the time-current curve of the OCPD as discussed previously.

Note: when determining the AFPB, use I_a and 0.85 I_a arc durations to find the corresponding clearing time for the OCPD. The 0.85 I_a arc duration accounts for variations in the arcing current and the time for the OCPD to open. In some cases, the lower value of I_a may result in longer clearing times and a larger AFPB.

To find E_n the following formulas can be used [per equation D.7.3(a)]:

$$\lg E_n = k_1 + k_2 + 1.081(\lg I_a) + 0.0011(G)$$

where:

E_n = incident energy (J/cm²) normalized for time and distance

k_1 = −0.792 for open air arcs; −0.555 for arcs-in-a-box

k_2 = 0 for ungrounded and high-resistance grounded systems

    = −0.113 for grounded systems

I_a = arcing current in kA

G = the conductor gap (mm) (See Table D.7.2.) (Table 12.1)

Then, calculate E_n as follows per equation D.7.3(b):

$$E_n = 10^{\lg En}$$

For systems over 15 kilovolts, the IEEE 1584 formula cannot be used as the system limits are valid only up to 15 kilovolts. However, the *NFPA 70E* Annex D.7.5(b) and IEEE 1584 Standard references the Ralph Lee Equation shown in *NFPA 70E* Annex D.6. For voltages over 15 kilovolts, the short-circuit current is considered equal to the arcing current. The equation shown in below from *NFPA 70E* Annex D.7.5(b) has been modified from the formula shown in Annex D.6 to have distance in millimeters and incident energy in J/cm²:

Per equation D.7.5(b):

$$D_B = \sqrt{2.142 \times 10^6 \, VI_{bf}\left(\frac{t}{E_B}\right)}$$

where:

D_B = the distance (mm) of the arc flash protection boundary from the arcing point

V = system voltage in kV

I_{bf} = bolted 3-phase available short-circuit current in kA

t = time (seconds)

E_B = incident energy in J/cm² at the distance of the arc flash protection boundary

Note: Per *NFPA 70E* Section 130.3(A)(2), if the voltage is over 600 volts, and the clearing time is 0.1

seconds or less, the incident energy (E_B) must be adjusted from 1.2 cal/cm² to 1.5 cal/cm². Multiply these values by 4.184 to determine value in J/cm².

Example

Per *NFPA 70E* Annex D.7 and IEEE 1584, what would the calculated AFPBs be for Test 4 and Test 3?

Test 4

Bolted short-circuit current = 22.6 kilovolt-amperes, 480 volts

Clearing time = 6 cycles (0.1 seconds, assumed to be same at I_a and 0.85 I_a)

Assume arc-in-a-box (assume MCCs and panels) and ungrounded system.

Step 1. Find I_a (arcing current).

Formula:

$$\lg I_a = K + 0.662 \lg I_{bf} + 0.0966 V + 0.000526 G + 0.5588 V (\lg I_{bf}) - 0.00304 G (\lg I_{bf})$$

where:

$\lg$ = the log10

I_a = arcing current in kA

K = −0.153 for open air arcs; −0.097 for arcs-in-a-box

I_{bf} = bolted three-phase available short-circuit current (symmetrical rms) (kA)

V = system voltage in kV

G = conductor gap (mm) (See Table D.7.2.) (Table 12.1)

Calculation:

$\lg I_a = -0.097 + 0.662 (\lg 22.6) + 0.0966 (0.48) + 0.000526 (25) + 0.5588 (0.48) (\lg 22.6) - 0.00304 (25) (\lg 22.6)$

$\lg I_a = 1.119$

Then:

$I_a = 10^{\lg Ia}$

$I_a = 10^{1.119}$

$I_a = 13.15$ kA

Step 2. Find E_n (normalized incident energy).

Formula:

$$\lg E_n = k_1 + k_2 + 1.081(\lg I_a) + 0.0011(G)$$

where:

E_n = incident energy (J/cm²) normalized for time and distance

k_1 = −0.792 for open air arcs; −0.555 for arcs-in-a-box

k_2 = 0 for ungrounded and high-resistance grounded systems

    = −0.113 for grounded systems

I_a = arcing current in kA

G = the conductor gap (mm) (See Table D.7.2.) (Table 12.1)

Calculation:

$\lg E_n = -0.555 + 0 + 1.081 (\lg 13.15) + 0.0011 (25)$

$\lg E_n = 0.682$

Then:

$E_n = 10^{\lg En}$

$E_n = 10^{0.682}$

$E_n = 4.81$

Step 3. Calculate D_B (AFPB).

Formula:

Per equation D.7.5(a):

$$D_B = \left[4.184 C_f E_n \left(\frac{t}{0.2} \right) \left(\frac{610^x}{E_B} \right) \right]^{\frac{1}{x}}$$

where:

    D_B = the distance (mm) of the arc flash protection boundary from the arcing point

    C_f = a calculation factor
        = 1.0 for voltages above 1 kV
        = 1.5 for voltages at or below 1 kV

    E_n = incident energy normalized

    E_B = incident energy in J/cm² at the distance of the arc flash protection boundary

    t = time (seconds)

    x = the distance exponent from Table D.7.2 (Table 12.1)

Note: $E_B = 1.2$ cal/cm² × 4.184 = 5.0 J/cm²

Calculation:

$D_B = [4.184(1.5)(4.81)(0.1/0.2)(610^{1.641})/(5.0)]^{1/1.641}$

$D_B = 1196$ mm (1 in/25.4 mm)

$D_B = 47$ in

Test 3

    Bolted short-circuit current = 22.6 kiloamperes (formula does not account for current-limitation of OCPD, reducing short-circuit current and resulting arcing current), 480 volts

    Clearing time = ½ cycle (0.008 seconds, assumed to be same at I_a and 0.85 I_a)

    Assume arc-in-a-box (assume MCCs and panels) and ungrounded system.

Step 1. Find I_a (arcing current).

    $I_a = 13.15$ kA (same as Test 4—no change in variables)

Step 2. Find E_n (normalized incident energy),

    $E_n = 4.81$ (same as Test 4—no change in variables)

Step 3. Calculate the AFPB:

Per equation D.7.5(a):

$$D_B = \left[4.184 C_f E_n \left(\frac{t}{0.2} \right) \left(\frac{610^x}{E_B} \right) \right]^{\frac{1}{x}}$$

where:

    D_B = the distance (mm) of the arc flash protection boundary from the arcing point

    C_f = a calculation factor
        = 1.0 for voltages above 1 kV
        = 1.5 for voltages at or below 1 kV

    E_n = incident energy normalized

    E_B = incident energy in J/cm² at the distance of the arc flash protection boundary

    t = time (seconds)

    X = the distance exponent from Table D.7.2 (Table 12.1)

Calculation:

$D_B = [4.184(1.5)(4.81)(0.008/0.2)(610^{1.641})/(5.0)]^{1/1.641}$

$D_B = 257$ mm (1 in/25.4 mm)

$D_B = 10$ in

Determining the Incident Energy or PPE

Doughty/Neal/Floyd Method

The Doughty/Neal/Floyd method, detailed in *NFPA 70E* Annex D.5, D.5.1 and D.5.2, is based on the paper noted in Annex D.5.3 by R.L. Doughty, T.E. Neal, and H.L. Floyd II. The paper is entitled "Predicting Incident Energy to Better Manage the Electric Arc Hazard on 600V Power Distribution Systems," and is part of the Record of the Conference Papers IEEE IAS 45th Annual Petroleum and Chemical Industry Conference, September 28–30, 1998. The range of short-circuit current values, voltage, and working distances is limited compared to the IEEE 1584 method, discussed later.

Ralph Lee Method

This method, detailed in *NFPA 70E* Annex D.6, is based on a modification of the AFPB calculation found in Annex D.2. The equation has been modified to include incident energy as a variable in the formula in lieu of an assumed value of 1.2 cal/cm². This method can be used to determine the incident energy, but applies only to open air arcs. The energy from open air arcs is less than the energy from arcs-in-a-box. Because of this, it would not be suitable to determine the incident energy for arcs-in-a-box (such as arcs in electrical equipment as seen in Test 4, Test 3, and Test 1) since this formula would yield results lower than expected.

NFPA 70E Annex D.7 and IEEE 1584 Method

This method, detailed *NFPA 70E*, Annex D.7, is based on IEEE 1584 *Guide for Performing Arc-Flash Hazard Calculations*. IEEE 1584 used extensive testing and analysis to develop new formulas to calculate the AFPB and incident energy. It is the most up-to-date method, with the widest range of system characteristics, to calculate the incident energy. In addition, the arcing current formulas in IEEE 1584, discussed previously, improve the ability to accurately determine the arcing current and related OCPD clearing time. It is important to note that these formulas are in J/cm^2 for incident energy and millimeters for distances. To convert units: $1\ cal/cm^2 = 4.184\ J/cm^2$ and $1\ in = 25.4\ mm$.

Use of NFPA 70E Annex D.7 and IEEE 1584 Basic Formulas

The basic formulas to calculate the incident energy per IEEE 1584 are found in *NFPA 70E* Annex D.7.3 and D.7.4 and shown below:

For system voltages ranging from 208 V to 15 kV [per equation D.7.3(c)]:

$$E = 4.184 C_f E_n \left(\frac{t}{0.2} \right) \left(\frac{610^x}{D^x} \right)$$

where:

E = incident energy in J/cm^2

C_f = a calculation factor
   = 1.0 for voltages above 1 kV
   = 1.5 for voltages at or below 1 kV

E_n = incident energy normalized

t = arcing time (seconds)

D = distance (mm) from the arc to the person (see **Table 12.5**)

X = the distance exponent from Table D.7.2 (Table 12.1)

Note: To calculate E_n and t (and I_a needed to determine E_n and t), use the formulas in the previous section on calculating the AFPB per the IEEE 1584 Method.

Note: To achieve a 95% confidence level that the selected PPE will be adequate, IEEE 1584 uses the calculation factor (C_f) as shown in the equation and PPE Levels of 1.2, 8, 25, 40, and 100* cal/cm^2. If the calculated incident energy falls between these values, PPE with at least the next higher standard **arc rating** must be used; for example, for a calculated incident energy of 4 cal/cm^2, PPE with at least an 8-cal/cm^2 arc rating must be worn. If the PPE is not based on these PPE levels,

Table 12.5	*NFPA 70E* Table D.7.4 Typical Working Distances
Classes of Equipment	**Typical Working Distance* (mm)**
15kV switchgear	910
5kV switchgear	910
Low-voltage switchgear	610
Low-voltage MCCs and panelboards	455
Cable	455
Other	To be determined in field

*Typical working distance is the sum of the distance between the worker and the front of the equipment and the distance from the front of the equipment to the potential arc source inside the equipment.

there is a lower probability that the PPE selected will be adequate for the arc flash incident energy.

*In *NFPA 70E* Annex D.7, 100 cal/cm^2 is not shown. Instead, a FPN is added:

> FPN: When incident energy exceeds 40 cal/cm^2 at the working distance, greater emphasis than normal should be placed on deenergizing before working on or near the exposed electrical conductors or circuit parts.

A similar FPN is shown in *NFPA 70E* Section 130.7(A) FPN No. 2:

> When incident energy exceeds 40 cal/cm^2 at the working distance, greater emphasis may be necessary with respect to deenergizing before working within the Limited Approach Boundary of the exposed electrical conductors or circuit parts.

For systems over 15 kilovolts, the IEEE 1584 formula cannot be used, as the system limits are valid only up to 15 kilovolts. However, the *NFPA 70E* Annex D.7.4 and IEEE 1584 Standard references the Ralph Lee Equation discussed previously and shown in *NFPA 70E* Annex D.6. For voltages over 15 kilovolts, the short-circuit current is considered equal to the arcing current. The equation below has been modified from the formula shown in Annex D.6 to have distance in millimeters and incident energy in J/cm^2:

Per equation D.7.4:

$$E = 2.142 \times 10^6\, V I_{bf} \left(\frac{t}{D^2} \right)$$

where:

E = incident energy (J/cm^2)

V = system voltage (kV)

t = arcing time (seconds)

D = distance (mm) from the arc to the person (working distance) (see Table 12.5)

I_{bf} = available three-phase bolted-fault current

Example

Per *NFPA 70E* Annex D.7 and IEEE 1584, what would the calculated incident energy be for Test 4 and Test 3?

Test 4

Bolted short-circuit current = 22.6 kiloamperes, 480 volts

Clearing time = 6 cycles (0.1 seconds, assumed to be same at I_a and 0.85 I_a)

Assume arc-in-a-box (assume MCCs and panels) and ungrounded system.

Step 1. Find I_a (arcing current).

I_a = 13.15 kA (from *NFPA 70E* Annex D.7 and IEEE 1584 calculation example to determine the AFPB)

Step 2. Find E_n (normalized incident energy).

E_n = 4.81 (from *NFPA 70E* Annex D.7 and IEEE 1584 calculation example to determine the AFPB)

Step 3. Find E (incident energy).

Formula:

Per equation D.7.3(c):

$$E = 4.184 C_f E_n \left(\frac{t}{0.2} \right) \left(\frac{610^x}{D^x} \right)$$

where:

E   = incident energy in J/cm²
C_f = a calculation factor
    = 1.0 for voltages above 1 kV
    = 1.5 for voltages at or below 1 kV
E_n = incident energy normalized
t   = arcing time (seconds)
D   = distance (mm) from the arc to the person (see Table 12.5)
X   = the distance exponent from Table D.7.2 (see Table 12.1)

Calculation:

E = 4.184(1.5)(4.81)(0.1/0.2)($610^{1.641}$)/($455^{1.641}$)

E = 24.4 J/cm²

Convert:

E = 24.4 J/cm² ÷ 4.184 J/cal

E = 5.84 cal/cm² at 18 in [AFPB = 47 in per *NFPA 70E* Annex D.7 and IEEE 1584]

Note: Equipment must be marked with the incident energy per *NFPA 70E* Section 130.3(C). Per *NFPA 70E* Annex D.7 and IEEE 1584, to achieve a 95% confidence level the selected PPE will be adequate, the equipment must be marked with an incident energy of 8 cal/cm².

Select PPE using *NFPA 70E* Sections 130.7(C)(1) through (C)(16).

Test 3

Bolted short-circuit current = 22.6 kiloamperes (formula does not account for current-limitation of OCPD, reducing short-circuit current, and resulting arcing current), 480 volts

Clearing time = ½ cycle (0.008 seconds, assumed to be same at I_a and 0.85 I_a), Assume arc-in-a-box (assume MCCs and panels) and ungrounded system.

Step 1. Find I_a (arcing current).

I_a = 13.15 kA (from *NFPA 70E* Annex D.7 and IEEE 1584 calculation example to determine the AFPB)

Step 2. Find E_n (normalized incident energy).

E_n = 4.81 (from *NFPA 70E* Annex D.7 and IEEE 1584 calculation example to determine the AFPB)

Step 3. Find E (incident energy).

Formula:

Per equation D.7.3(c):

$$E = 4.184 C_f E_n \left(\frac{t}{0.2} \right) \left(\frac{610^x}{D^x} \right)$$

where:

E   = incident energy in J/cm²
C_f = a calculation factor
    = 1.0 for voltages above 1 kV
    = 1.5 for voltages at or below 1 kV
E_n = incident energy normalized
t   = arcing time (seconds)
D   = distance (mm) from the arc to the person (see Table 12.5)
X   = the distance exponent from Table D.7.2 (Table 12.1)

Calculation:

E = 4.184(1.5)(4.81)(0.008/0.2)($610^{1.641}$)/($455^{1.641}$)

E = 1.95 J/cm²

Convert:

E = 1.95 J/cm² ÷ 4.184 J/cal

E = 0.47 cal/cm² at 18 in [AFPB = 10 in per *NFPA 70E* Annex D.7 and IEEE 1584]

Note: Equipment must be marked with the incident energy per *NFPA 70E* Section 130.3(C). Per *NFPA 70E* Annex D.7 and IEEE 1584, to achieve a 95% confidence level the selected PPE will be adequate, the equipment must be marked with an incident energy of 1.2 cal/cm^2.

Select PPE using *NFPA 70E* Sections 130.7(C)(1) through (C)(16).

Use of NFPA 70E Annex D.7 and IEEE 1584 Simplified Equations

To simplify the *NFPA 70E* Annex D.7 and IEEE 1584 calculation method, fuse and circuit breaker manufacturers have developed "simplified formulas." These equations allow the incident energy to be calculated directly from the available 3-phase short-circuit current on a low-voltage system if the type and ampere rating of OCPD is known. This simplified method does not require the availability of time current curves for the devices.

The fuse formulas, shown in *NFPA 70E* Annex D.7.6 and IEEE 1584, are based on actual arc flash test data using Cooper Bussmann Low-Peak® Class RK-1 and Class L fuses. Because they are based on actual testing, the equations can be based solely on short-circuit current and do not require the arcing current to be calculated, making the use of the equations much easier than other methods. However, the equations are valid only for Cooper Bussmann Low-Peak Class RK-1 and Class L fuses. They can be used for other types and manufacturers of fuses only if the other type of fuse provides equal or faster clearing times for all overcurrent conditions. For fuses that fall outside of this range, the basic *NFPA 70E* Annex D.7 and IEEE 1584 equations must be used.

The circuit breaker formulas, shown in *NFPA 70E* Annex D.7.7 and IEEE 1584, were developed by analyzing typical circuit breaker operation and calculated arcing currents. Although the formulas are simplified, they require a "qualifying" calculation. This qualifying calculation determines if the short-circuit current is within the range of use of the formula. To use the simplified formulas, the short-circuit current must not exceed the interrupting rating of the circuit breaker and be large enough to result in an arcing current that causes instantaneous tripping to occur or short-time tripping for circuit breakers with no IT. If the short-circuit current is outside of this range, the basic *NFPA 70E* Annex D.7 and IEEE 1584 equations must be used. Where equations are used for low-voltage power circuit breakers with STD, the maximum setting (30 cycles) is assumed. Because there are several equations dependent on the type and size of the OCPD and range of short-circuit currents, the equations are not reprinted here.

Use of IEEE 1584 Resources and Tools

When IEEE 1584 is purchased, several useful Excel spreadsheets are provided with the actual document including:

- Actual test data
- Bolted short-circuit current calculator
- Arc flash hazard calculator
- Current-limiting fuse test data
- Simplified circuit breaker method

NFPA 70E Annex D.7 and IEEE 1584 Resources and Tools from Manufacturers

Other resources and tools are available from manufacturers, such as Cooper Bussmann. Cooper Bussmann has developed an online Arc Flash Incident Energy Calculator that makes it simple to find the incident energy for low-voltage systems based on the available 3-phase short-circuit current and the type of OCPD. The calculated incident energy is valid for Cooper Bussmann Low-Peak Class RK-1 and Class L fuses and specific types of circuit breakers, based on the simplified formulas shown in *NFPA 70E* Annex D.7 and IEEE 1584. This information is available using an interactive online calculator at http://www.cooperbussmann.com. For convenience, it is also available in an easy-to-use tabular format in Annex F of the *Cooper Bussmann Safety BASICs Handbook for Electrical Safety*.

The notes for the Arc Flash Incident Energy Calculator must be read before using either the online or the tabular version of the calculator. They can be found online or in *Cooper Bussmann Safety BASICs Handbook for Electrical Safety*. The calculations use a calculation factor so that the PPE selected from the calculated incident energy are adequate for 98% of arc flash incidents (compared to 95% for the calculation factor shown in the basic *NFPA 70E* Annex D.7 and IEEE 1584 formulas). In up to 2% of incidents, the level of PPE may be one level too low. For IEEE 1584, the set of PPE arc ratings was chosen as 1.2, 8, 25, 40, and 100 cal/cm^2. For incident energy results that fall between these values, PPE with the next higher standard arc rating must be used to achieve a 98% confidence level that the selected PPE will be adequate. For instance, if the incident energy is calculated to be 11 cal/cm^2, the PPE used must have a 25 cal/cm^2 arc rating. PPE with intermediate arc ratings can be utilized per this method, but at the next lower arc rating. If the PPE is not based on the PPE levels of 1.2, 8, 25, 40, and 100 cal/cm^2, there is a lower probability that the PPE selected is adequate for the arc flash incident energy. **Figure 12.9** and **Figure 12.10** show the

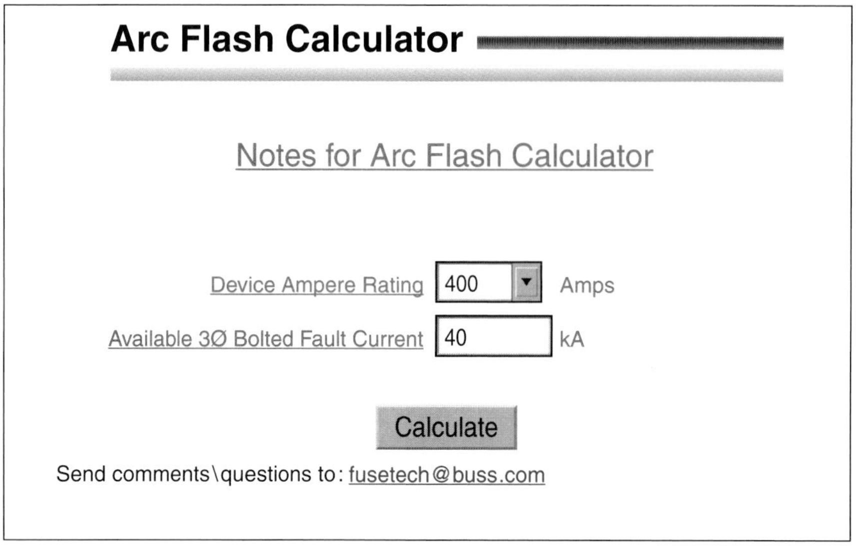

Arc Flash Calculator

Notes for Arc Flash Calculator

Device Ampere Rating   400 ▼   Amps

Available 3Ø Bolted Fault Current   40   kA

Calculate

Send comments\questions to: fusetech@buss.com

Figure 12.9   Cooper Bussmann online Arc Flash Incident Energy Calculator.

Arc Flash Calculator

Notes for Arc Flash Calculator

Device Ampere Rating   400 ▼   Amps

Available 3Ø Bolted Fault Current   40   kA

Calculate

Send comments\questions to: fusetech@buss.com

Available 3Ø Bolted Fault Current: 40kA

Overcurrent Protection Device	Incident Energy Exposure @ 18 inches (cal/cm²)	Level of PPE (cal/cm²)	Flash Protection Boundary (inches)
LPS-RK-400SP Fuse Click for the datasheet	0.25	1.2	6
400 Amp Molded Case Circuit Breaker w/ TM setting	3.08	8	34.13

Calculations generated on: 5/13/03

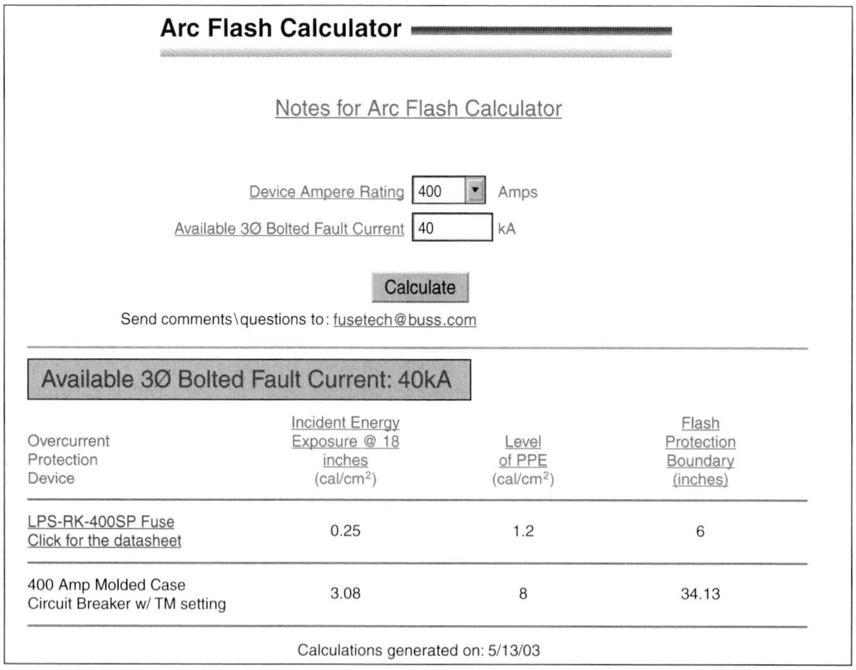

Figure 12.10   Results of Cooper Bussmann online Arc Flash Incident Energy Calculator.

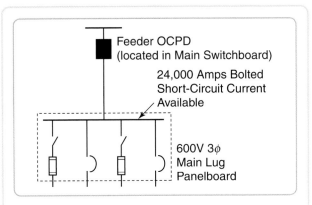

Figure 12.11 The incident energy at the panelboard is dependent upon the available short-circuit current at the panelboard and the clearing time of the upstream OCPD in the main switchboard.

Cooper Bussmann online Arc Flash Incident Energy Calculator and sample results, respectively.

Example of NFPA 70E Annex D.7 and IEEE 1584 Resources and Tools from Manufacturers
The following are examples of using *NFPA 70E* Annex D.7 and IEEE 1584 resources and tools from manufacturers. Two examples will be discussed based on using the *Cooper Bussmann Safety BASICs Handbook for Electrical Safety*.

For this example, using the tables from *Cooper Bussmann Safety BASICs Handbook for Electrical Safety* determine the incident energy for each of the following 480-volt circuits with an available short-circuit current of 24,000 amperes at the main lugs of a panelboard (**Figure 12.11**) when protected by:

- Cooper Bussmann Low-Peak KRP-C-800SP current-limiting Class L fuses
- 800-ampere low voltage power circuit breaker with STD setting of 30 cycles

To use **Table 12.6**, find the ampere rating and type of OCPD in the header of the table. Select the available short-circuit current (bolted fault current) from the left column. Select the corresponding incident energy and AFPB for the selected OCPD type and size and short-circuit current.

For the KRP-C-800SP Fuse, using Table 12.6, the calculated incident energy is 1.46 cal/cm² with an AFPB of 21 in.

Note that if *NFPA 70E* Annex D.7 and IEEE 1584 basic equations were used, the calculated incident energy (assuming ½ cycle clearing time and a short-circuit current of 22.6 kiloamperes) would be 0.47 cal/cm² with an AFPB of 10 in (see the *NFPA 70E* Annex D.7 and

IEEE 1584 Basic Equation Example for Test 3). Therefore, when the actual test data and equations for ONLY fuses are used, the result can be higher (or in some cases lower) than the calculated value per the basic equations of *NFPA 70E* Annex D.7 and IEEE 1584. Because of this, it is recommended to use the simplified fuse equations of *NFPA 70E* Annex D.7.6 and IEEE 1584 (or Cooper Bussmann Arc Flash Calculator Program or Arc Flash Incident Energy Calculator Tables) when possible to assure the most accurate result.

Select PPE using *NFPA 70E* Sections 130.7(C)(1) through (C)(16).

For the 800-ampere low-voltage power circuit breaker with STD (30 cycles). Using Table 12.6, the calculated incident energy is 46.53 cal/cm² with an AFPB of greater than 120 in. The equations used for this calculation are based on the *NFPA 70E* Annex D.7.7 and IEEE 1584 Simplified Equations for circuit breakers. These equations assume the maximum setting for STD of 30 cycles. While PPE manufacturers make arc flash suits with arc ratings over 40 cal/cm², *NFPA 70E* Section 130.7(A) FPN No. 2 recommends deenergizing before working on equipment within the **limited approach boundary** of the exposed electrical conductors or circuit parts when the incident energy exceeds 40 cal/cm² at the working distance.

If the *NFPA 70E* Annex D.7 and IEEE 1584 basic equations were used with the same short-circuit current and clearing time of 30 cycles, the incident energy would have been calculated at 30.86 cal/cm² with an AFPB of 129.59 in. Because of this, it is recommended to use the simplified equations of *NFPA 70E* Annex D.7.7 and IEEE 1584 for circuit breaker in lieu of the IEEE 1584 and *NFPA 70E* Annex D.7 basic equations.

In addition, note that if *NFPA 70E* Annex D.7 and IEEE 1584 basic equations were used, the calculated incident energy, assuming a reduced STD setting of 6 cycles and a short-circuit current of 22.6 kiloamperes, would be 5.8 cal/cm² with an AFPB of 47 in (see the *NFPA 70E* Annex D.7 and IEEE 1584 Basic Equation Example for Test 4). The decrease in the STD from 30 to 6 cycles dramatically decreases the incident energy and AFPB. Because of this, where circuit breakers are used with STD, it is best to assume the maximum permitted setting for the STD in lieu of the actual setting as done in the *NFPA 70E* Annex D.7.7 and IEEE 1584 simplified circuit breaker equations. This is because the setting can be adjusted to the maximum. If the circuit breaker is not properly maintained, extended clearing times can be experienced.

Select PPE using *NFPA 70E* Sections 130.7(C)(1) through (C)(16).

| Table 12.6 | Arc Flash Incident Energy Calculator* |

Arc Flash Incident Energy Calculator

Fuses: Bussmann® Low-Peak® KRP-C_SP (601–2000 Amp), Circuit Breakers: Low Voltage Power Circuit Breakers (w/STD)

Incident Energy (I.E.) values are expressed in cal/cm^2. Flash Protection Boundary (FPB) values are expressed in inches.

| Bolted Fault Current (kA) | 601–800 Amp | | | | 801–1200 Amp | | | |
| | Fuse | | LVPCB | | Fuse | | LVPCB | |
	I.E.	FPB	I.E.	FPB	I.E.	FPB	I.E.	FPB
1	>100	>120	>100	>120	>100	>120	>100	>120
2	>100	>120	>100	>120	>100	>120	>100	>120
3	>100	>120	>100	>120	>100	>120	>100	>120
—	—	—	—	—	—	—	—	—
—	—	—	—	—	—	—	—	—
—	—	—	—	—	—	—	—	—
20	1.70	23	38.87	>120	10.37	78	>100	>120
22	1.58	22	42.70	>120	9.98	76	>100	>120
24	1.46	21	46.53	>120	8.88	70	46.53	>120
26	1.34	19	50.35	>120	7.52	63	50.35	>120
28	1.22	18	54.18	>120	6.28	55	54.18	>120
30	1.10	17	58.01	>120	5.16	48	58.01	>120

| | 1201–1600 Amp | | | | 1601–2000 Amp | | | |
| | Fuse | | LVPCB | | Fuse | | LVPCB | |
	I.E.	FPB	I.E.	FPB	I.E.	FPB	I.E.	FPB
1	>100	>120	>100	>120	>100	>120	>100	>120
2	>100	>120	>100	>120	>100	>120	>100	>120
3	>100	>120	>100	>120	>100	>120	>100	>120
—	—	—	—	—	—	—	—	—
—	—	—	—	—	—	—	—	—
—	—	—	—	—	—	—	—	—
20	24.20	>120	>100	>120	>100	>120	>100	>120
22	23.83	>120	>100	>120	>100	>120	>100	>120
24	23.45	>120	>100	>120	29.18	>120	>100	>120
26	23.08	>120	>100	>120	28.92	>120	>100	>120
28	22.71	>120	>100	>120	28.67	>120	>100	>120
30	22.34	>120	>100	>120	28.41	>120	>100	>120

*Full table is found in *Cooper Bussmann Safety BASICs Handbook for Electrical Safety* Annex F.

HRC Method

The "Table Method" or "HRC Method" is another method permitted by *NFPA 70E* Section 130.3(B)(2) to determine the required PPE in lieu of an incident energy analysis. To use this method, three tables are consulted. The first table to be used is *NFPA 70E* Table 130.7(C)(9) to determine if the task to be performed is listed for the specific equipment and the associated HRC and requirements for use of rubber insulating gloves and tools. The second table to be used is *NFPA 70E* Table 130.7(C)(10) to determine the required PPE based on the HRC. The third table is *NFPA 70E* Table 130.7(C)(11) to determine the minimum arc rating of PPE based on the HRC.

NFPA 70E Table 130.7(C)(9) indicates the HRC for a specific work task. The table lists several types of electrical equipment such as panelboards, switchboards, MCCs, switchgear, medium voltage starters, switchgear and arc-resistant switchgear, as well as specific tasks to be performed for each type of equipment. The user first selects the type of equipment, then the specific task (if listed) that will be performed, and reads across the row to determine the HRC and if rubber insulating gloves and insulated and insulating hand tools are required. The HRCs in this table are 0, 1, 2, 2*, 3, and 4. As noted in *NFPA 70E* Section 130.7(C)(9), if the task or equipment is not listed in the table, or if the assumed short-circuit current or OCPD clearing time is more than permitted by the notes to this table, this method is not permitted. In addition, as noted in *NFPA 70E* Section 130.7(C)(9) FPN No. 4, the available short-circuit current and fault clearing time must be documented on the energized electrical work permit per *NFPA 70E* Section 130.1(B)(2)(6).

Once the HRC has been determined, *NFPA 70E* Table 130.7(C)(10) is used to select the required PPE for the task. The PPE listed in this table consists of the required PPE. *NFPA 70E* Table 130.7(C)(11) is then used to determine the required minimum arc rating of PPE. Note that the protective clothing system selected for the corresponding HRC must have an arc rating of at least the value listed in the last column of Table 130.7(C)(11). As an example, a task identified as a HRC 2 would require a protective clothing system with an arc rating of 8 cal/cm² or greater. A task identified as a HRC 4 would require a protective clothing system with an arc rating of 40 cal/cm² or greater, for example.

Example

The following is an example of how to use the HRC method to determine the required PPE for Test 4 and Test 3.

Test 4

Bolted short-circuit current = 22.6 kiloamperes, 480 volts

Clearing time = 6 cycles (0.1 seconds)

Assume work on energized parts in a combination motor controller.

First, find the correct heading in *NFPA 70E* Table 130.7(C)(9) (excerpts are printed in **Table 12.7**), which is "Other 600 V Class (277 V through 600 V, nominal) Equipment—Note 2 (except as indicated)." Notice the reference to Note 2. It is mandatory to check these notes to verify that the qualifications are met to allow the table to be used. If the qualifications are not met, the table cannot be used. Note 2 gives a maximum short-circuit current of 65 kiloampere, and a maximum fault clearing time of 0.03 seconds (2 cycles). Because the clearing time for the OCPD in Test 4 is 6 cycles, this exceeds the limit of 2 cycles in Note 2 and thus, the HRC method cannot be used.

Test 3

Bolted short-circuit current = 22.6 kiloamperes, 480 volts

Clearing time = ½ cycle (0.008 seconds)

Assume work on energized parts in a combination motor controller.

Again, find the correct heading in *NFPA 70E* Table 130.7(C)(9) (excerpts are printed in Table 12.7), which is "Other 600 V Class (277 V Through 600 V, Nominal) Equipment—Note 2 (except as indicated)." Because the clearing time for the OCPD in Test 3 is ½ cycle and the short-circuit current is not more than 65 kiloamperes, the qualifiers for Note 2 are met, and the HRC method can be used. In this case, for work on energized electrical conductors and circuit parts including voltage testing, the HRC is 2*, and rubber insulating gloves and insulated and insulating tools are required (denoted by "Y" in those columns).

Note that a 2* HRC has different PPE requirements than a 2 HRC. To determine the difference, *NFPA 70E* Table 130.7(C)(10) must be consulted. By reviewing the requirements for HRC 2 and 2*, it can be seen that difference is that a 2* HRC requires an appropriately rated balaclava if an arc flash suit hood is not selected.

However, also note the General Note (f) to Table 130.7(C)(9) applies in this case, which allows the HRC required to be reduced by one number if the equipment is protected by current limiting fuses with arcing cur-

Table 12.7	Excerpts from HRC Classifications [*NFPA 70E* Table 130.7 (C)(9)]			
Tasks Performed on Energized Equipment	**Hazard/Risk Category**	**Rubber Insulating Gloves**	**Insulated and Insulating Hand Tools**	
Other 600 V Class (277 V Through 600 V, Nominal) Equipment—Note 2 (except as indicated)				
Work on energized electrical conductors and circuit parts, including voltage testing	2*	Y	Y	

General Notes (applicable to the entire table):
(a) Rubber insulating gloves are gloves rated for the maximum line-to-line voltage upon which work will be done.
(b) Insulated and insulating hand tools are tools rated and tested for the maximum line-to-line voltage upon which work will be done, and are manufactured and tested in accordance with ASTM F 1505, *Standard Specification for Insulated and Insulating Hand Tools*.
(c) Y = yes (required), N = no (not required).
(d) For systems rated less than 1000 volts, the fault currents and upstream protective device clearing times are based on an 18 in. working distance.
(e) For systems rated 1 kV and greater, the Hazard/Risk Categories are based on a 36 in. working distance.
(f) For equipment protected by upstream current limiting fuses with arcing fault current in their current limiting range ($^1/_2$ cycle fault clearing time or less), the hazard/risk category required may be reduced by one number.
Specific Notes (as referenced in the table):
1. Maximum of 25 kA short circuit current available; maximum of 0.03 sec (2 cycle) fault clearing time.
2. Maximum of 65 kA short circuit current available; maximum of 0.03 sec (2 cycle) fault clearing time.
3. Maximum of 42 kA short circuit current available; maximum of 0.33 sec (20 cycle) fault clearing time.
4. Maximum of 35 kA short circuit current available; maximum of up to 0.5 sec (30 cycle) fault clearing time.

Reproduced with permission from *NFPA 70E*®-2009, *Electrical Safety in the Workplace*, Copyright © 2008, National Fire Protection Association. This reprinted material constitutes only a portion of the referenced table, is presented for educational purposes only, and is not the complete and official permission of the NFPA on the referenced subject, which is represented only by the standard in its entirety.

rent in their current limiting range (½ cycle fault clearing time or less). Once the HRC has been determined from Table 130.7(C)(9), Table 130.7(C)(10) is used to determine the required PPE for the task. The protective clothing selected for the corresponding HRC from Table 130.7(C)(9) must have an arc rating of at least the value listed in the last column of Table 130.7(C)(11) (Table 12.8). The required level of PPE must be marked on the equipment per *NFPA 70E* Section 130.3(C).

Arc Flash Hazard Analysis Summary

Several methods are recognized in *NFPA 70E* to determine the AFPB and PPE. For systems 600 volts and less, the AFPB can be determined by using the conditional 4 ft boundary or by calculations such as the formulas in *NFPA 70E* Annex D. AFPB calculations are always required for systems over 600 volts. Calculating the incident energy using formulas such as in *NFPA 70E* Annex D, then selecting PPE using this calculated value and the requirements of *NFPA 70E*, can be used. If the specific task and equipment are listed in the HRC tables of *NFPA 70E* and system characteristic limits are met, the PPE can be selected using the HRC tables. The incident energy or required level of PPE is then required to be field marked on the equipment.

Table 12.8	Excerpts from Protective Clothing Characteristics [*NFPA 70E* Table 130.7 (C)(11)]	
Hazard/Risk Category	**Clothing Description**	**Required Minimum Arc Rating of PPE [J/cm²(cal/cm²)]**
0	Nonmelting, flammable materials (i.e., untreated cotton, wool, rayon, or silk, or blends of these materials) with a fabric weight at least 4.5 oz/yd²	N/A
1	Arc-rated FR shirt and FR pants or FR coverall	16.74 (4)
2	Arc-rated FR shirt and FR pants or FR coverall	33.47 (8)

Reproduced with permission from *NFPA 70E*®-2009, *Electrical Safety in the Workplace*, Copyright © 2008, National Fire Protection Association. This reprinted material constitutes only a portion of the referenced table, is presented for educational purposes only, and is not the complete and official permission of the NFPA on the referenced subject, which is represented only by the standard in its entirety.

With all methods, the available 3-phase short-circuit current must be known. If the available short-circuit current is not known, it must be calculated before performing an arc flash hazard analysis. In addition, for all methods (except the use of the *NFPA 70E* Annex D.7.6 and IEEE 1584 simplified fuse equations), the arcing current must be found to determine the OCPD clearing time. Although different analysis methods may provide different results, the requirement is that arc flash hazard analysis be done in order to protect personnel from the possibility of being injured by an arc flash.

Arcing Currents in the Long Time Characteristic of Overcurrent Protective Devices

In many electrical short-circuit studies, the focus is only on the circuits with the highest or worst case short-circuit currents. However, in arc flash hazard analysis, as mentioned previously and noted in *NFPA 70E* Section 130.7(C)(9) FPN No. 2, it is important to consider both higher and lower available short-circuit currents to determine which will result in the worst-case arc flash energy. The lower short-circuit currents are important, because they could result in an arcing current value that is less than a circuit breaker's IT setting (resulting in an clearing time of up to several seconds) or could be a value less than the current-limiting range of the fuse and take several seconds to open. On lower ampere-rated circuits this is not typically a problem. However, on larger ampere-rated circuits (over 1200 amperes), this can become more of an issue. For some higher ampere rating OCPDs, the incident energy and AFPB are extremely large for some lower short-circuit currents. Examining the AFPB and incident energy for circuits with low levels of arcing current is an important consideration. Although some lower level arcing faults cannot sustain themselves, there is not much recent research in this area. In those cases where a low level arcing fault is sustained, extended clearing times can produce extremely high incident energy levels.

Additional Considerations

Other considerations should be reviewed when performing an arc flash hazard analysis. The first consideration is how often the study should be performed. *NFPA 70E* Section 130.3, requires that the arc flash hazard analysis be updated when a major modification or renovation occurs. In addition, the arc flash hazard analysis must be reviewed periodically, not to exceed 5 years, since changes to the electrical distribution systems could affect the results of the arc flash hazard analysis.

The second consideration is OCPD design, maintenance, and operation. As noted in *NFPA 70E* Section 130.3, the arc flash hazard analysis shall take into consideration the design of the OCPD and its clearing time, including its condition of maintenance. Thus, if the design or condition of maintenance could increase the clearing time (as noted by *NFPA 70E* Section 130.3 FPN No.1), the incident energy could be increased, and the arc flash hazard analysis should take this into consideration. For more information on this, see Chapter 4 on design considerations and Chapter 15, which covers maintenance considerations.

A third consideration is when an arc flash hazard analysis is not required. As noted in *NFPA 70E* Section 130.3 Exception No. 1, an arc flash hazard analysis is not required if all of the following conditions are required:

- The circuit is 240 volts or less
- The circuit is supplied by one transformer
- The transformer supplying the circuit is less than 125 kilovolt-amperes

Another consideration concerns the location of the arcing fault. Even though electrical equipment might have a main OCPD and disconnecting means, if it is possible to create a fault on the line side of the main, the clearing time and let-through characteristics of the OCPD, which feeds the main device, should be considered. For example, an industrial machine might have a main fusible disconnect switch or circuit breaker fed by a busbar plug. When the switch or circuit breaker door is open, it is possible to initiate a fault on the line terminals; therefore, the OCPD in the busbar plug must be considered in the arc flash hazard analysis.

Another consideration is the arc blast. As noted in *NFPA 70E* 130.7(A) FPN No. 1, physical trauma injuries could occur due to the explosive effect of some arc events and that the PPE requirements of Section 130.7 do not address protection against physical trauma. Little testing has been done to date on the effects and predictability of arc blast hazards, but generally, as the arc flash energy increases, the arc blast energy also increases. The more one knows about arcing faults, the more one understands that the best strategy in electrical safety is avoidance. Strive to only work on or near exposed conductors that have been placed in an electrically safe work condition.

Summary

Determining the AFPB and required PPE as part of an arc flash hazard analysis are important practices intended to provide for employee safety. This chapter has provided information on the requirements and formulas necessary to determine the AFPB. In addition, general information has been given to gain an understanding of how to select PPE by performing an incident energy analysis or using HRC tables.

Lessons *learned*

Electrician Dies in Arc Flash Incident

An electrician died as a result of burns he received when an explosion occurred while he was making a connection on a circuit breaker located in a panelboard. The victim worked for an electrical contractor that employs 130 people. The safety function is assigned to the firm's two project managers as an additional duty. The firm has no written safety policy or established safety program, and only on-the-job training is provided. The project managers meet with the foremen monthly to discuss safety issues.

On the day of the accident, the victim and his helper were to install three circuit breakers into a panelboard cabinet that supplied power to various parts of a building in an industrial park complex. The panelboard was located in a service room inside the building. Three circuit breakers had previously been installed in the panelboard and supplied power to the occupied portion of the building. The three circuit breakers to be installed the day of accident would control the heating, ventilation, and air-conditioning (HVAC) system of the unoccupied portion of the building and were located on the right side of the panelboard cabinet in a vertical fashion. Before beginning the job, the victim was instructed by the crew foreman to install the top breaker first and the bottom breaker last. The victim said he had done this type of job so many times he could "do it in his sleep." Although the victim had worked for the firm for only 4 days, he had approximately 16 years of experience as an electrician with various employers.

The victim did not follow the foreman's instructions. He secured the three breakers in the panelboard, then began to wire the bottom breaker first. The victim removed the cover from the breaker and placed it on the top of a 300-kilovolt transformer case located in the service room. The insulated wires to be connected to the breakers were fed through conduit in the bottom of the panelboard cabinet. The insulated wires then had to be fed to the upper portion of the panelboard in order to make the connections. These wires were energized. To deenergize them, the electrician would have had to cut the power to the occupied portion of the facility. The line side of the bottom breaker became energized when the connections were completed. The victim then began to make the connections on the middle breaker without replacing the cover on the bottom-breaker. This allowed the energized line side connection points to remain exposed. The victim then began to feed the wires to be used for the connections on the middle breaker through the upper portion of the panelboard cabinet. The victim had successfully fed four of the six insulated wires needed to make the connections to the upper portion of the panelboard. As the fifth wire was being fed into the upper portion of the panelboard, its uninsulated tip contacted one of the exposed energized connection points. This conductor was grounded on the other end. The contact caused an arc, and the bottom breaker exploded. The victim was standing immediately in front of the breaker when it exploded. His clothes caught fire, and he received massive burns from the explosion to the upper portion of his body.

A helper was working to the side of the panelboard cabinet. When he heard the explosion, he ran out the service room door. As he was running away, a second explosion occurred. The helper received second- and third-degree burns to his left shoulder and back and to the back of his left arm.

The crew foreman was walking toward the service room when he heard the explosions. He arrived at the service room and found the victim lying on the floor of the service room with his clothes on fire. The foreman pulled the victim outside of the service room and extinguished the victim's burning clothes with his hands, receiving second- and third-degree burns to his hands. The victim was conscious and talking at this time. The emergency rescue transported the victim to the hospital. He was transferred to a burn center, where he died the following morning. The cause of death was attributed to massive burns to 25% of the victim's upper body. The victim received burns to 80% of his body.

Recommendations/Discussion

Recommendation 1: Power distribution within a panelboard should remain consistent.

Discussion: The design of the panelboard should be reevaluated. When the middle and bottom breakers were installed in the panelboard, their line side (incoming power) was on the right and their load side (outgoing power) was on the left. The line and load sides of the other four circuit breakers in the panelboard were reversed. The inconsistency of the location of the line and load sides of the breakers might lead to confusion that would lead to an electrical accident. This might occur during installation of the breakers or at a later date when maintenance was being performed on the panelboard.

Recommendation 2: Electrical systems should be deenergized prior to any work being performed on them.

Discussion: The power on the incoming wires should have been deenergized before the installation of the circuit breakers was attempted. The incoming wires were not deenergized because it would have interfered with the power in the occupied portion of the building. This job should have been scheduled at a time (a weekend or after hours) when the incoming power could be deenergized without disrupting operations.

Recommendation 3: Safe work procedures should be followed when working in the presence of electrical energy.

Discussion: The victim did not follow the foreman's instructions to install the top breaker first and the bottom breaker last. The victim then began to make the connections on the second circuit breaker without replacing the protective cover on the bottom breaker. These two actions were both unsafe work habits. Had the victim followed the foreman's orders or replaced the protective cover, the accident might have been prevented.

Recommendation 4: The work habits and training of new employees should be closely monitored.

Discussion: Although the victim was said to have 16 years experience as an electrician, he had been working for this employer for only 4 days, and his work habits should have been more closely monitored. Employers should assure that new employees have adequate training and understand the proper procedures to perform their assigned tasks.

Adapted from: FACE 8644. Accessed 9/17/08 at http://www.cdc.gov/niosh/face/in-house/full8644.html.

Questions

1. What role did the fact that the employer did not have an electrical safety program play in this incident?
   A. The worker said that he could do the job "blindfolded," so an electrical safety program was not necessary.
   B. The monthly safety discussion with the project managers was adequate.
   C. The employer was not covered by OSHA requirements, so an electrical safety program would have no bearing on the incident.
   D. The workers were not trained as qualified persons as would be specified by an electrical safety program.

2. In the incident report, what PPE might have prevented injury to the victim and his helper?
   A. Class 4 voltage-rated gloves
   B. Adequate arc-rated FR-protective equipment
   C. Voltage-rated dielectric overshoes and rubber blankets
   D. Faraday shield clothing

3. If the victim had conducted a hazard/risk analysis, how might the outcome have been different?
   A. A hazard/risk analysis was not necessary for this simple work task.
   B. The electrician had 16 years of experience and knew what he was doing.
   C. A hazard/risk analysis would have indicated that electrocution and/or an arc flash hazard existed within the panelboard as long as the equipment remained energized.
   D. The helper had planned to remain by the electrician's side and watch carefully how the worker installed the equipment. The helper would certainly warn the electrician when he was near an energized point.

Ready for Review

- An electrical hazard analysis is required if energized electrical conductors or circuit parts at 50 volts or more are not placed in an electrically safe work condition. This must include a shock hazard analysis and an arc flash hazard analysis.
- Per *NFPA 70E* Section 130.3, an arc flash hazard analysis is required to determine the AFPB and the PPE that is required when people are within the AFPB.
- To determine the AFPB and required PPE, the available short-circuit current and the OCPD clearing time must be determined.
- To determine the OCPD clearing time, the arcing current must be determined and compared with manufacturer's time-current curves.
- An energized electrical work permit is not always required for potentially dangerous tasks, such as testing or troubleshooting. However, safe work practices and PPE are required.
- The following steps must be taken to complete an arc flash hazard analysis:
  - A short-circuit study must first be performed at the equipment to be worked on.
  - Associated arcing current must be determined and used to verify the OCPD clearing time should an arcing fault occur (typically the upstream OCPD).
  - Next, the AFPB can be determined either by using the conditional 4 ft boundary (if system limitations are met) or by calculations. For systems above 600 volts, the AFPB must be calculated.
  - The required PPE can be based on an incident energy analysis per *NFPA 70E* Section 130.3(B)(1) or HRC tables per *NFPA 70E* Section 130.3(B)(2), provided the system characteristics do not exceed the specified notes to the tables.
  - Finally, *NFPA 70E* Section 130.3(C) requires the incident energy or required level of PPE to be field marked on the equipment.
- The minimum and maximum arcing current can be determined using percentages from industry research and experience, but a more accurate calculation is found in *NFPA 70E* Annex D.7.2, from IEEE 1584.
- The arcing current value is used to determine the clearing time of the OCPD, for which the total clearing time-current curve is needed from the OCPD manufacturer. The time-current curve for an OCPD shows the minimum or average opening time and/or the total clearing time of the OCPD based on a range of current values.
- Various methods exist for determining the AFPB and PPE, ranging from a conditional value or tables to specific calculations.
- *NFPA 70E* Annex D, Table D.1 summarizes some calculation methods available to determine the AFPB and incident energy. There are advantages and disadvantages to each method.
- The Ralph Lee method was the first method developed and permitted for use by *NFPA 70E* to calculate the AFPB, but is not a preferred method when working on equipment because it only applies to open air arcs.
- The IEEE 1584 method, detailed in *NFPA 70E*, Annex D.7, is considered by many to be the preferred method for calculating the AFPB and incident energy. This is the most up-to-date method, with the widest range of system characteristics, to calculate the incident energy.
- For systems over 15 kilovolts, the *NFPA 70E* Annex D.7 and IEEE 1584 formula cannot be used, as the system limits are valid only up to 15 kilovolts.
- The Doughty/Neal/Floyd method is detailed in *NFPA 70E* Annex D.5, D.5.1, and D.5.2.
- As a precautionary measure (to ensure a 95% confidence level) when determining PPE levels per *NFPA 70E* Annex D.7 and IEEE 1584, incident energy levels should be rounded to the next highest arc rating level if the incident energy value falls between two of these levels: 1.2, 8, 25, 40 cal/cm^2.
- Fuse and circuit breaker manufacturers have developed simplified formulas in lieu of the *NFPA 70E* Annex D.7 and IEEE 1584 basic calculation

method. Simplified methods are valid only for the specific fuse manufacturer or type of circuit breaker.

- Other resources and tools for *NFPA 70E* Annex D.7 and IEEE 1584 are available from manufacturers. The Cooper Bussmann online Arc Flash Incident Energy Calculator is one example. Another example is the Arc Flash Incident Energy Calculator Table found in *Cooper Bussmann Safety BASICs Handbook for Electrical Safety*.

- The HRC method is another method permitted per *NFPA 70E* Section 130.3(B)(2) to determine the required PPE in lieu of incident energy calculations. Three tables available in *NFPA 70E* are used with this method:
  - *NFPA 70E* Table 130.7(C)(9) is used to determine if the task is listed for the specific equipment to be worked on and the associated HRC and requirements for use of rubber insulating gloves and tools.
  - *NFPA 70E* Table 130.7(C)(10) is used to determine the required PPE.
  - *NFPA 70E* Table 130.7(C)(11) is used to determine the minimum arc rating of PPE.

- With all calculation methods, the available 3-phase short-circuit current must be known. If the available short-circuit current is not known, it must be calculated before performing an arc flash hazard analysis.

- It is important to consider both higher and smaller available short-circuit currents to determine which result in the worst-case arc flash energy. Smaller short-circuit currents are important because they can result in an arcing current value that is less than a circuit breaker's IT setting or less than the current-limiting range of the fuse resulting in clearing times of several seconds, producing extremely high incident energy levels.

- Per *NFPA 70E* Section 130.3, the arc flash hazard analysis must be updated when a major modification or renovation occurs. It must also be reviewed periodically, not to exceed 5 years.

- The arc flash hazard analysis must take into consideration the design of the OCPD, its clearing time, and its condition of maintenance.

- *NFPA 70E* and IEEE 1584 have developed tools for assessing the arc flash hazard but have not addressed protection against physical trauma due to the effects of arc blast hazards.

Vocabulary

arc flash hazard analysis[*] A study investigating a worker's potential exposure to arc-flash energy, conducted for the purpose of injury prevention and the determination of safe work practices, arc flash protection boundary, and the appropriate levels of PPE.

arc flash protection boundary (AFPB)[*] When an arc flash hazard exists, an approach limit at a distance from a prospective arc source within which a person could receive a second degree burn if an electrical arc flash were to occur.

arc rating[*] The value attributed to materials that describes their performance to exposure to an electrical arc discharge. The arc rating is expressed in cal/cm^2 and is derived from the determined value of the arc thermal performance value (ATPV) or energy of breakopen threshold (E_{BT}; should a material system exhibit a breakopen response below the ATPV value) derived from the determined value of ATPV or E_{BT}.

> FPN: *Breakopen* is a material response evidenced by the formation of one or more holes in the innermost layer of flame-resistant material that would allow flame to pass through the material.

limited approach boundary[*] An approach limit at a distance from an exposed energized electrical conductor or circuit part within which a shock hazard exists.

short-circuit study An analysis of an electrical distribution system that determines the available short-circuit current at various points in the system; the purpose is to verify compliance with the *NEC* and the *NFPA 70E* and increase electrical safety.

[*]This is the *NFPA 70E* definition.

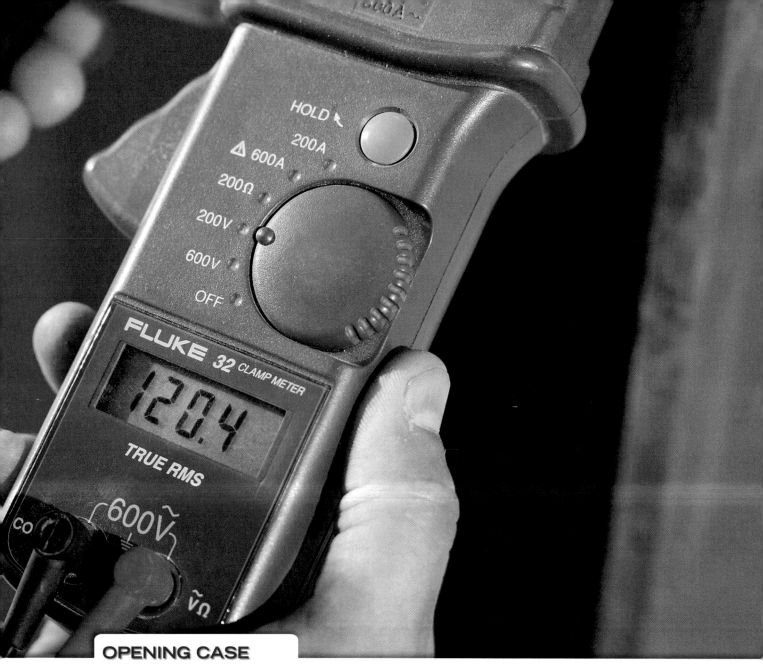

Photo courtesy of National Electrical Contractors Association (NECA).

OPENING CASE

A 40-year-old electrician died after being electrocuted while repairing a compressor unit on a freezer in a restaurant. The victim was supposed to be highly qualified and was known to be very safety conscious. According to his partner, the victim was working alone on this particular job.

Four coworkers including the manager were working in the vicinity when the incident occurred. None of the coworkers actually witnessed the incident. They all heard the victim scream, however, and came to his aid. The victim had fallen 10 ft from a ladder that he had been working from and was lying on the floor.

The victim had been working with the freezer feeder circuit when he received an electric shock (120 volts). The freezer circuit had a short to ground, which ultimately caused the electrocution. The victim had not tested this circuit for any fault in the electrical system before beginning work. It was determined that the circuit breaker boxes were not marked legibly or correctly as to the equipment they controlled. It was also found that the three-door freezer was controlled by two separate circuit breakers. One of these two circuit breakers had been deenergized, but the other one remained energized. The victim was

Testing, Overhead Lines, Other Precautions, and Other Protective Equipment

Introduction

NFPA 70E Article 130 covers a wide variety of topics addressing work involving electrical hazards. Those topics are grouped logically and covered in various chapters of this textbook. Justification for energized work, Energized Electrical Work Permit requirements, approach boundaries for shock protection, and arc flash hazard analysis requirements are covered in Chapter 10. PPE requirements from *NFPA 70E* Section 130.7 are covered in Chapter 14. This chapter introduces the *NFPA 70E* Article 130 work practice considerations not covered elsewhere, including test instrument and equipment use, work within the limited approach boundary (LAB) of uninsulated overhead lines, other precautions for worker activities, and other protective equipment.

Performing Testing Work

Electrical testing is a common task performed by electrical workers. *NFPA 70E* Section 130.4 requires workers to be qualified to perform electrical testing on electrical equipment at voltage levels of 50 volts or more if they will cross the LAB.

By definition, a qualified person performing electrical testing must have received safety training to recognize and avoid the hazards involved (**Figure 13.1**). There is no middle ground. Either a worker is qualified with respect to certain tasks or is not. Per *NFPA 70E*, qualified and unqualified persons are defined as follows:

> **Qualified Person.** One who has skills and knowledge related to the construction and operation of the electrical equipment and installations and has received safety training to recognize and avoid the hazards involved. [*NFPA 70E* Article 100 Definitions]

OPENING CASE, CONT.

not wearing any personal protective equipment (PPE) and was using an aluminum ladder when the incident occurred.

Paramedics were summoned to the scene, and the victim was taken to the hospital in full cardiac arrest. The paramedics gave the victim cardiopulmonary resuscitation (CPR) until they arrived at the hospital, where the victim was pronounced dead. The coroner's autopsy report stated the cause of death as electrocution.

Adapted from: California Fatality and Assessment and Control Evaluation (FACE) Report #92CA014, April 15, 1993. Accessed 9/22/08 at http://www.cdc.gov/niosh/face/stateface/ca/92ca014.html.

1. What additional training would have helped ensure that the victim met the *NFPA 70E* definition of a qualified person?

2. How could the worker have known that two circuits provided energy to the freezer?

CHAPTER 13

Chapter Outline

- **Introduction**
- **Performing Testing Work**
- **Uninsulated Overhead Lines**
- **Other Precautions for Personnel Activities**
- **Other Protective Equipment**
- **Summary**
- **Lessons Learned**
- **Current Knowledge**

Learning Objectives

1. Understand the requirements for work within the LAB of uninsulated overhead lines.
2. Describe other precautions for personnel activities.
3. Understand the requirements for other protective equipment.
4. Understand worker qualification requirements for testing on electrical equipment within the LAB.

References

1. *NFPA 70E*®, 2009 Edition
2. Occupational Safety and Health Administration (OSHA) 29 Code of Federal Regulations (CFR) 1910, Subpart S

OSHA Tip

1910.334(c)(1)
Only qualified persons may perform testing work on electric circuits or equipment.

Unqualified Person. A person who is not a qualified person. [*NFPA 70E* Article 100 Definitions]

Electrical testing is normally done on energized circuits, if the equipment cannot be placed in an electrically safe work condition. A hazard assessment, training, a job briefing, and selection and use of appropriate work practices and PPE are among the minimum requirements to qualify a person to perform electrical testing. *NFPA 70E* Sections 110.6(D)(1) and 110.8 were covered previously. Excerpts are offered as a brief review:

110.6(D) Employee Training in part

(1) Qualified Person. A qualified person shall be trained and knowledgeable of the construction and operation of equipment or a specific work method and be trained to recognize and avoid the electrical hazards that might be present with respect to that equipment or work method.

(a) Such persons shall also be familiar with the proper use of the special precautionary techniques, personal protective equipment, including arc flash, insulating and shielding materials, and insulated tools and test equipment. A person can be considered qualified with respect to certain equipment and methods but still be unqualified for others.

(b) Such persons permitted to work within the Limited Approach Boundary of exposed energized electrical conductors and circuit parts operating at 50 volts or

Figure 13.1 Electrical testing is energized work if the equipment is not placed in an electrically safe work condition.

more shall, at a minimum, be additionally trained in all of the following:

(1) The skills and techniques necessary to distinguish exposed energized electrical conductors and circuit parts from other parts of electrical equipment

(2) The skills and techniques necessary to determine the nominal voltage of exposed energized electrical conductors and circuit parts

(3) The approach distances specified in Table 130.2(C) and the corresponding voltages to which the qualified person will be exposed

(4) The decision-making process necessary to determine the degree and extent of the hazard and the personal protective equipment and job planning necessary to perform the task safely

(e) Employees shall be trained to select an appropriate voltage detector and shall demonstrate how to use a device to verify the absence of voltage, including interpreting indications provided by the device. The training shall include information that enables the employee to understand all limitations of each specific voltage detector that may be used.

110.8(A) General. Safety-related work practices shall be used to safeguard employees from injury while they are exposed to electrical hazards from electrical conductors or circuit parts that are or can become energized. The specific safety-related work practices shall be consistent with the nature and extent of the associated electrical hazards.

(2) Energized Electrical Conductors and Circuit Parts—Unsafe Work Condition. Only qualified persons shall be permitted to work on electrical conductors or circuit parts that have not been put into an electrically safe work condition.

Uninsulated Overhead Lines

Contact with uninsulated overhead lines is among the leading causes of electrical fatalities. *NFPA 70E* Section 130.5 addresses tasks performed within the LAB of uninsulated overhead lines (**Figure 13.2**). These requirements are, in many cases, similar to OSHA's requirements in 29 CFR 1910.333(c)(3).

Approach distances associated with this work are obtained from column 2 (exposed movable conductors) in Table 130.2(C). For example, the LAB for an energized overhead conductor operating at 5 kilovolts is 10 ft. The requirements of Section 130.5 apply once a worker gets within 10 ft of an uninsulated overhead line operating at 5000 volts for this example (**Table 13.1**).

There are five subheads in *NFPA 70E* Section 130.5 that address work done on uninsulated overhead lines:

- Uninsulated and energized
- Deenergizing or guarding
- Employer and employee responsibility
- Approach distances for unqualified persons
- Vehicular and mechanical equipment

Figure 13.2 *NFPA 70E* Section 130.5 addresses tasks performed inside the LAB of uninsulated overhead lines.

Work Within the Limited Approach Boundary

NFPA 70E Section 130.5(A) requires that precautions be taken where work is performed in locations containing uninsulated energized overhead lines that are not guarded or isolated to prevent employees from contacting such lines directly with any unguarded parts of their body or indirectly through conductive materials, tools, or equipment. The lines must be suitably guarded or deenergized and visibly grounded at the point of work when the work to be performed is such that contact with uninsulated energized overhead lines is possible.

NFPA 70E Section 130.5(B) requires that arrangements be made with the person or organization that operates or controls the lines to deenergize and visibly ground them at the point of work if the lines are to be deenergized (**Figure 13.3**).

Figure 13.3 Uninsulated overhead lines must be deenergized and visibly grounded at the point of work if the lines are to be deenergized.

Table 13.1	*NFPA 70E* Table 130.2(C), in part. Approach Boundaries to Energized Electrical Conductors or Circuit Parts for Shock Protection. (All dimensions are distance from energized electrical conductor or circuit part to employee.)	
(1)	**(2)**	**(3)**
	Limited Approach Boundary[1]	
Nominal System Voltage Range, Phase to Phase[2]	**Exposed Movable Conductor[3]**	**Exposed Fixed Circuit Part**
751 to 15 kV	3.05 m (10 ft 0 in.)	1.53 m (5 ft 0 in.)

1 See definition in Article 100 and text in 130.2(D)(2) and Annex C for elaboration.
2 For single-phase systems, select the range that is equal to the system's maximum phase-to-ground voltage multiplied by 1.732.
3 A condition in which the distance between the conductor and a person is not under the control of the person. The term is normally applied to overhead line conductors supported by poles.
Reproduced with permission from *NFPA 70E*®-2009, *Electrical Safety in the Workplace*, Copyright © 2008, National Fire Protection Association. This reprinted material constitutes only a portion of the referenced table, is presented for educational purposes only, and is not the complete and official permission of the NFPA on the referenced subject, which is represented only by the standard in its entirety.

If arrangements are made to use protective measures, such as guarding, isolating, or insulation, these precautions must prevent each employee from contacting such lines directly with any part of his or her body or indirectly through conductive materials, tools, or equipment.

NFPA 70E Section 130.5(C) details employer and employee responsibility. Both employers and employees are responsible for ensuring that guards or protective measures are satisfactory for the conditions. Employees must comply with established work practices, including the use of protective equipment.

NFPA 70E Section 130.5(D) requires that unqualified persons working on the ground or in an elevated position near overhead lines be located in such a way that the worker and the longest conductive object he or she could contact cannot come closer to any unguarded, energized overhead power line than the LAB. Objects that are not insulated for the voltage must be considered conductive.

NFPA 70E Section 130.5(E)(1) requires any vehicle or mechanical equipment structure that could be elevated near energized overhead lines to maintain the LAB distance of Table 130.2(C), column 2 (). However, under any of the following conditions, the approach distance may be reduced:

- If the vehicle is in transit with its structure lowered, the LAB to overhead lines in Table 130.2(C), column 2, may be reduced by 6 ft. If insulated barriers, rated for the voltages involved, are installed, and they are not part of an attachment to the vehicle, the clearance may be reduced to the design working dimensions of the insulating barrier.
- If the equipment is an aerial lift insulated for the voltage involved, and if the work is performed by a qualified person, the clearance (between the uninsulated portion of the aerial lift and the power line) may be reduced to the restricted approach boundary given in Table 130.2(C), column 4.

NFPA 70E Section 130.5(E)(2) details requirements for worker contact with equipment. Employees standing on the ground shall not contact the vehicle or mechanical equipment or any of its attachments, unless one of the following conditions applies:

- The employee is using protective equipment rated for the voltage.
- The equipment is located so that no uninsulated part of its structure (that portion of the structure that provides a conductive path to employees on the ground) can come closer to the line than permitted in Section 130.5(E)(1).

Figure 13.4 Insulated boom trucks are commonly used in many line construction and maintenance operations.

NFPA 70E Section 130.5(E)(3) details requirements for equipment grounding. If any vehicle or mechanical equipment capable of having parts of its structure elevated near energized overhead lines is intentionally grounded, employees working on the ground near the point of grounding must not stand at the grounding location whenever there is a possibility of overhead line contact. Additional precautions, such as the use of barricades, dielectric overshoe footwear, or insulation, shall be taken to protect employees from hazardous ground potentials (step and touch potential). Section 130.5(E)(3) contains the following Fine Print Note (FPN):

> FPN: Upon contact of the elevated structure with the energized lines, hazardous ground potentials can develop within a few feet or more outward from the grounded point.

Other Precautions for Personnel Activities

NFPA 70E Section 130.6 details numerous precautions that need to be considered for work involving electrical hazards. Many of these requirements in *NFPA 70E* are similar to what OSHA requires. The *NFPA 70E* precautions for personnel activities are broken down into the following eleven subsections:

- Alertness
- Blind reaching
- Illumination
- Conductive articles being worn
- Conductive materials, tools, and equipment being handled
- Confined or enclosed work spaces
- Housekeeping duties
- Occasional use of flammable materials
- Anticipating failure

- Routine opening and closing of circuits
- Reclosing circuits after protective device operation

Alertness

NFPA 70E Section 130.6(A) requires that employees be instructed to be alert at all times and that they be prohibited from working when their alertness is impaired due to illness, fatigue, or other reasons when they are working where electrical hazards might exist. A worker who has the flu, has been working 12 hours a day, 7 days a week, for an extended period of time, or is taking medication that could cause drowsiness are among examples of where alertness could be impaired. Employees must also be instructed to recognize any change in scope of their work that will expose them to hazards not covered in the job briefing and on the energized electrical work permit.

Blind Reaching

NFPA 70E Section 130.6(B) prohibits blind reaching. Employees must be instructed not to reach blindly into areas where an electrical hazard exists. A worker performing a task on a piece of electrical equipment that has been placed into an electrically safe work condition might reach into an adjacent piece of equipment to hang a task light and unknowingly make contact with energized equipment, for example.

Illumination

The third subsection under other precautions for personnel activities is broken down into two categories. The general requirements of *NFPA 70E* Section 130.6(C)(1) prohibit employee entrance into an area without adequate illumination so that the task can be performed safely. *NFPA 70E* Section 130.6(C)(2) prohibits employees from working where an electrical hazard exists if they cannot clearly see what they are working on.

> **OSHA Tip**
>
> 1910.333(c)(8) Conductive apparel. Conductive articles of jewelry and clothing (such a watch bands, bracelets, rings, key chains, necklaces, metalized aprons, cloth with conductive thread, or metal headgear) may not be worn if they might contact exposed energized parts. However, such articles may be worn if they are rendered nonconductive by covering, wrapping, or other insulating means.

Conductive Articles, Materials, Tools, and Equipment

NFPA 70E Section 130.6(D) directs that conductive apparel not be worn where it could come in contact with energized electrical equipment.

NFPA 70E Section 130.6(E) contains requirements for conductive materials, tools, and equipment being handled. Conductive materials, tools, and equipment that are in contact with any part of an employee's body must be handled in a manner that will prevent accidental contact with live parts. A means must be used to ensure that conductive materials do not approach exposed live parts closer than that permitted by the approach boundaries to energized electrical conductors and circuit parts requirements of *NFPA 70E* Section 130.2.

Confined or Enclosed Spaces

NFPA 70E Section 130.6(F) details confined or enclosed work space considerations.

There are responsibilities for both the employer and employee. The employer must provide, and the employee must use, protective shields, protective barriers, or insulating materials as necessary to avoid inadvertent contact with live parts and the effects of electrical haz-

> **OSHA Tip**
>
> 1910.333(c)(4)(i)
> Employees may not enter spaces containing exposed energized parts, unless illumination is provided that enables the employees to perform the work safely.
> 1910.333(c)(4)(ii)
> Where lack of illumination or an obstruction precludes observation of the work to be performed, employees may not perform tasks near exposed energized parts. Employees may not reach blindly into areas which may contain energized parts.

> **OSHA Tip**
>
> 1910.333(c)(6) Conductive materials and equipment. Conductive materials and equipment that are in contact with any part of an employee's body shall be handled in a manner that will prevent them from contacting exposed energized conductors or circuit parts. If an employee must handle long dimensional conductive objects (such as ducts and pipes) in areas with exposed live parts, the employer shall institute work practices (such as the use of insulation, guarding, and material handling techniques) which will minimize the hazard.

OSHA Tip

1910.146(b) Definitions.

Confined space means a space that:

(1) Is large enough and so configured that an employee can bodily enter and perform assigned work; and

(2) Has limited or restricted means for entry or exit (for example, tanks, vessels, silos, storage bins, hoppers, vaults, and pits are spaces that may have limited means of entry); and

(3) Is not designed for continuous employee occupancy.

OSHA Tip

1910.333(c)(5)

"Confined or enclosed work spaces." When an employee works in a confined or enclosed space (such as a manhole or vault) that contains exposed energized parts, the employer shall provide, and the employee shall use, protective shields, protective barriers, or insulating materials as necessary to avoid inadvertent contact with these parts. Doors, hinged panels, and the like shall be secured to prevent their swinging into an employee and causing the employee to contact exposed energized parts.

OSHA Tip

1910.269(x) Definitions.

Enclosed space. A working space, such as a manhole, vault, tunnel, or shaft, that has a limited means of egress or entry, that is designed for periodic employee entry under normal operating conditions, and that under normal conditions does not contain a hazardous atmosphere, but that may contain a hazardous atmosphere under abnormal conditions.

Note: Spaces that are enclosed but not designed for employee entry under normal operating conditions are not considered to be enclosed spaces for the purposes of this section. Similarly, spaces that are enclosed and that are expected to contain a hazardous atmosphere are not considered to be enclosed spaces for the purposes of this section. Such spaces meet the definition of permit spaces in 1910.146 of this Part, and entry into them must be performed in accordance with that standard.

OSHA Tip

1910.333(c)(9) Housekeeping duties.

Where live parts present an electrical contact hazard, employees may not perform housekeeping duties at such close distances to the parts that there is a possibility of contact, unless adequate safeguards (such as insulating equipment or barriers) are provided. Electrically conductive cleaning materials (including conductive solids such as steel wool, metalized cloth, and silicon carbide, as well as conductive liquid solutions) may not be used in proximity to energized parts unless procedures are followed which will prevent electrical contact.

OSHA Tip

1910.334(d)

Occasional use of flammable or ignitable materials. Where flammable materials are present only occasionally, electric equipment capable of igniting them shall not be used, unless measures are taken to prevent hazardous conditions from developing. Such materials include, but are not limited to: flammable gases, vapors, or liquids; combustible dust; and ignitable fibers or flyings.

ards where an electrical hazard exists in a confined or enclosed space. Doors, hinged panels, and the like must be secured to prevent their swinging into an employee and causing the employee to contact exposed live parts where an electrical hazard exists.

Housekeeping

NFPA 70E Section 130.6(G) prohibits employees from performing housekeeping duties inside the LAB, including the use of electrically conductive cleaning materials, unless adequate safeguards are in place to prevent electrical contact with these parts.

Flammable Materials

NFPA 70E Section 130.6(H) prohibits the use of electric equipment capable of igniting flammable materials where the flammable materials are present only occasionally unless measures are taken to prevent hazardous conditions from developing. Electrical installation requirements for locations where flammable materials are present on a regular basis are contained in the *National Electrical Code®* (*NEC®*).

Anticipating Failure

Another of the important precautions for worker activities addresses anticipating failure. *NFPA 70E* Section 130.6(I) requires electrical equipment to be deenergized when there is evidence that such equipment could fail unless the employer can demonstrate that doing so would introduce additional or increased hazards or that it is infeasible to do so. Broken or missing parts, a burning smell, an unusual noise, excessive equipment enclosure temperature, or an outdated maintenance sticker on an overcurrent device could all be cause for concern. Workers must be protected from the hazards associated with the failure of the electrical equipment when failure is imminent. Workers must not be permitted to cross approach boundaries until the equipment is placed in an electrically safe work condition by the opening of an upstream disconnecting means, for example. Workers should recognize that the PPE requirements of *NFPA 70E* do not address protection against arc blast.

> 130.7(A) FPN No. 1: The PPE requirements of 130.7 are intended to protect a person from arc flash and shock hazards. While some situations could result in burns to the skin, even with the protection selected, burn injury should be reduced and survivable. Due to the explosive effect of some arc events, physical trauma injuries could occur. The PPE requirements of 130.7 do not address protection against physical trauma other than exposure to the thermal effects of an arc flash.

Routine Opening and Closing of Circuits

Routine opening and closing of circuits is yet another activity worthy of mention under precautions for worker activities. *NFPA 70E* Section 130.6(J) cautions that load-rated devices specifically designed as a disconnecting means must be used for the opening, reversing, or closing of circuits under load. Fuses, terminal lugs, cable splice connections, and cable connectors not of the load-break type cannot be used except in an emergency.

Reclosing Circuits After Protective Device Operation

A final consideration addressed under other precautions for personnel activities deals with reclosing cir-

OSHA Tip

1910.334(b)(1)
Routine opening and closing of circuits. Load rated switches, circuit breakers, or other devices specifically designed as disconnecting means shall be used for the opening, reversing, or closing of circuits under load conditions. Cable connectors not of the load break type, fuses, terminal lugs, and cable splice connections may not be used for such purposes, except in an emergency.

OSHA Tip

1910.334(b)(2)
Reclosing circuits after protective device operation. After a circuit is deenergized by a circuit protective device, the circuit protective device, the circuit may not be manually reenergized until it has been determined that the equipment and circuit can be safely energized. The repetitive manual reclosing of circuit breakers or reenergizing circuits through replaced fuses is prohibited.

cuits after protective devices operate. *NFPA 70E* Section 130.6(K) prohibits manually reenergizing a circuit after a circuit protective device has deenergized it until it has been verified that it is safe to do so. Resetting a circuit breaker or replacing a fuse that has opened is a fairly common practice, and one that is seldom done "by the book." However, repetitively reclosing circuit breakers and replacing fuses is prohibited. Troubleshooting must ascertain the cause of the protective device operation and necessary modification, repair, and testing to assure circuit integrity needs to be completed before reenergizing the circuit unless it can be determined that the protective devices operated due to an overload condition. **Overload** and **overcurrent** are defined in *NFPA 70E* Article 100 as follows:

> **Overload.** Operation of equipment in excess of normal, full-load rating, or of a conductor in excess of rated ampacity that, when it persists for a sufficient length of time, would cause damage or dangerous overheating. A fault, such as a short circuit or ground fault, is not an overload. [*NFPA 70E* Article 100 Definitions]

> **Overcurrent.** Any current in excess of the rated current of equipment or the ampacity of a conductor. It may result from overload, short circuit, or ground fault. [70, 2008] FPN: A current in excess of rating may be accommodated by certain equipment and conductors for a given set of conditions. Therefore, the rules for overcurrent protection are specific for particular situations. [*NFPA 70E* Article 100 Definitions]

Overload relays that have tripped in a motor circuit are an example that an overload condition exists (rather than an overcurrent condition). Automatic opening of a circuit breaker is an indication that an overcurrent condition exists (rather than an overload condition) and requires an investigation of the cause, and probable modification or repair must be done before the circuit can be manually reenergized.

Other Protective Equipment

NFPA 70E Article 130 concludes by detailing PPE, other protective equipment, and alerting techniques. These requirements are located last in Article 130 and include the following categories of requirements:

- General requirements [Section 130.7(A)]
- Care of equipment [Section 130.7(B)]
- PPE [Section 130.7(C)]
- Other protective equipment [Section 130.7(D)]
- Alerting techniques [Section 130.7(E)]

NFPA 70E Section 130.7(C) is covered in Chapter 14. General requirements, care of equipment, other protective equipment, and alerting techniques are covered here.

General Requirements

Like OSHA, *NFPA 70E* Section 130.7(A) assigns responsibility to employers and employees for protective equipment use. Workers must be provided with, and they must use, protective equipment that is designed for the use and adequate for the hazard when they are performing tasks where electrical hazards are present. This is consistent with what OSHA requires in its safeguards for worker protection.

Two FPNs to *NFPA 70E* Section 130.7(A) bring important PPE limitations to light. This information drives home the important point that workers could sustain injury even while using the protective equipment determined in accordance with *NFPA 70E* Section 130.7.

> FPN No. 1: While some situations could result in burns to the skin, even with the protection selected, burn injury should be reduced and survivable. Due to the explosive effect of some arc events, physical trauma injuries could occur. The PPE requirements of 130.7 do not address protection against physical trauma other than exposure to the thermal effects of an arc flash.

> FPN No. 2: When incident energy exceeds 40 cal/cm² at the working distance, greater emphasis may be necessary with respect to deenergizing before working within the Limited Approach Boundary of the exposed electrical conductors or circuit parts.

Equipment Care

NFPA 70E Section 130.7(B) requires protective equipment to be maintained, inspected, and stored properly. It must be maintained in a safe, reliable condition, visually inspected before each use, and stored in a manner to prevent damage. Insulating equipment, such as rubber insulating gloves, must be inspected and given an air test before use and electrically tested periodically (**Figure 13.5**).

Other Protective Equipment

NFPA 70E Section 130.7(D)(1) details considerations for nine types of tools and equipment:

- Insulated tools
- Fuse or fuse holding equipment
- Ropes and handlines
- Fiberglass-reinforced plastic rods
- Portable ladders
- Protective shields
- Rubber insulating equipment
- Voltage-rated plastic guard equipment
- Physical or mechanical barriers

Workers are required to use insulated tools, handling equipment, or both, when working within the LAB (**Figure 13.6**). Like OSHA, *NFPA 70E* requires that insulated tools be protected from damage to their insulating properties.

Insulated Tools, Fuse/Fuse Holding Equipment, and Ropes and Handlines

Insulated tools must meet American Society for Testing and Materials (ASTM) F1505, *Standard Specification for Insulated and Insulating Hand Tools*, 2007, as indicated in *NFPA 70E* Table 130.7(F), and comply with *NFPA 70E* Section 130.7(D)(1)(a), which establishes the following requirements:

- Be rated for use voltage

OSHA Tip

1910.335(a)(1)(i)
Employees working in areas where there are potential electrical hazards shall be provided with, and shall use, electrical protective equipment that is appropriate for the specific parts of the body to be protected and for the work to be performed.

OSHA Tip

1910.335(a)(2)(i)
When working near exposed energized conductors or circuit parts, each employee shall use insulated tools or handling equipment if the tools or handling equipment might make contact with such conductors or parts. If the insulating capability of insulated tools or handling equipment is subject to damage, the insulating material shall be protected.

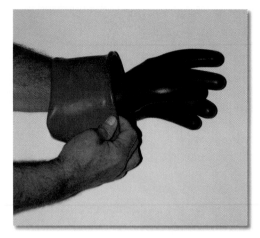
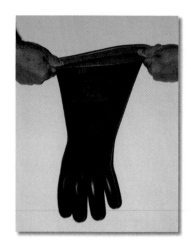

Figure 13.5 Rubber insulating gloves require inspection, including an air test, before use.

- Be designed and constructed for the environment and manner of use
- Be inspected prior to each use

Removal or replacement of fuses in electrical equipment while it is energized should be avoided where possible. *NFPA 70E* Section 130.7(D)(1)(b) requires that fuse-handling equipment that is rated for use voltage be used where fuse insertion or removal is necessary while either the fuse or its terminal is energized.

The use of a metal fish tape or conductive rope in a raceway, switchboard, panelboard, or motor control center bucket that contains exposed energized parts is not good practice. *NFPA 70E* Section 130.7(D)(1)(c) requires that nonconductive ropes and handlines be used where an electrical hazard exists.

Fiberglass-Reinforced Plastic Rods, Portable Ladders, and Protective Shields

NFPA 70E Section 130.7(D)(1)(d) indicates that fiberglass-reinforced plastic rods and tubes that are used for live line tools must meet the requirements of applicable portions of electrical codes and standards such as ASTM F711, *Standard Specification for Fiberglass-*

Figure 13.6 Workers are required to use insulated tools when working inside the LAB.

OSHA Tip

1910.335(a)(2)(ii)
Protective shields, protective barriers, or insulating materials shall be used to protect each employee from shock, burns, or other electrically related injuries while that employee is working near exposed energized parts which might be accidentally contacted or where dangerous electric heating or arcing might occur. When normally enclosed live parts are exposed for maintenance or repair, they shall be guarded to protect unqualified persons from contact with the live parts.

Reinforced Plastic (FRP) Rod and Tube Used in Live Line Tools. ASTM F711 is among the standards identified in Table 130.7(F) for other protective equipment.

NFPA 70E Section 130.7(D)(1)(e) requires portable ladders to have nonconductive side rails if they are used where an electrical hazard exists and to meet the various American National Standards Institute (ANSI) standards for ladders listed in Table 130.7(F).

NFPA 70E Section 130.7(D)(1)(f) requires that protective shields, protective barriers, or insulating materials be used to protect workers from shock, burns, or other electrically related injuries while they are working inside the LAB. Exposed energized parts must be guarded to protect unqualified persons from contact.

Rubber Insulating Equipment, Voltage Rated Plastic Guard Equipment, and Physical or Mechanical Barriers

NFPA 70E Section 130.7(D)(1)(g) requires rubber insulating equipment to be used for protection from accidental contact with energized parts to meet the requirements of the ASTM standards listed in Table 130.7(F).

NFPA 70E Section 130.7(D)(1)(h) requires that plastic guard equipment meet the requirements of the ASTM standards listed in Table 130.7(F) where it is used to protect workers from accidental contact with live parts or to protect employees or energized equipment or material from contact with ground.

NFPA 70E Section 130.7(D)(1)(i) prohibits barriers from being brought or placed closer than the restricted approach boundary distance given in Table 130.2(C). *NFPA 70E* Section 130.7(E)(2) prohibits conductive barricades from being placed closer than the LAB.

Alerting Techniques

The means used to restrict the access of unqualified persons from the work area is required to be documented on the Energized Electrical Work Permit, in accordance with *NFPA 70E* Section 130.1(B)(2)(9). The requirements of *NFPA 70E* Section 130.7(E) illustrate how restricted access of unqualified persons from the work area could be accomplished.

Safety Signs and Tags

NFPA 70E Section 130.7(E)(1) requires that safety signs, safety symbols, or accident prevention tags meet the requirements of ANSI Standard Z535 given in Table 130.7(F) and that they be used to warn employees about electrical hazards that might endanger him or her.

Barricades

NFPA 70E Section 130.7(E)(2) requires barricades to be used in conjunction with safety signs where it is necessary to prevent or limit employee access to work areas containing energized conductors or circuit parts. Conductive barricades must not be used where they might cause an electrical hazard and cannot be placed closer than the LAB.

Attendants

NFPA 70E Section 130.7(E)(3) requires an attendant:
- If signs and barricades do not provide sufficient warning and protection from electrical hazards
- To warn and protect employees
- To keep unqualified employees outside a work area where they might be exposed to electrical hazards
- To remain in the area as long as there is a potential for employees to be exposed

Reported incidents exist where a worker was killed because he or she thought successful lockout/tagout procedures were implemented on equipment when, in fact, they had locked and tagged a similar-looking piece of equipment.

Look-Alike Equipment

NFPA 70E Section 130.7(E)(4) requires the use of safety signs, safety symbols, accident prevention tags, barricades used in conjunction with safety signs, or an attendant where signs and barricades do not provide sufficient warning and protection to prevent employees from entering look-alike equipment where work performed on equipment that is deenergized and placed in an electrically safe condition exists in a work area with other energized equipment that is similar in size, shape, and construction.

Standards for Other Protective Equipment

NFPA 70E Section 130.7(F) requires that other protective equipment required in Section 130.7(D) conform to the standards provided in Table 130.7(F), which covers the following equipment:

- Ladders
- Safety signs and tags
- Blankets
- Covers
- Line hoses
- Fiberglass tools
- Fiberglass ladders
- Plastic guards
- Temporary grounding
- Insulated hand tools

Summary

NFPA 70E Article 130 covers a wide variety of topics addressing work involving electrical hazards. This chapter introduces *NFPA 70E* Article 130 precautions, protective equipment, and work practice considerations not covered in other chapters, including the following:

- Instrument and equipment use
- Work practices to be used within the LAB of uninsulated overhead lines
- Other precautions for worker activities
- Other protective equipment

Employees must be qualified to perform testing work if they are assigned to perform that task. Precautions must be taken to prevent employees from contacting unguarded parts where work is performed on unguarded or isolated uninsulated energized overhead lines. Precautions must be taken and protective equipment must be used to protect workers.

Lessons *learned*

Maintenance Electrician Dies from Burns When Electrical Panel Is Shorted

A 38-year-old employee was severely burned while attempting to test an electrical power circuit. On the day of the incident, the maintenance electrician was asked by a coworker to determine why an 85-horsepower electric motor was not functioning. The company had a full-time safety officer and a written safety program. The company had been in business for 90 years and employs 200 people. The victim had worked at the company 18 months and was not authorized to work on voltage over 440 volts. The company conducted on-the-job training, but results of the training were not measured.

The victim used a multimeter rated at 600 volts to check out the power supply source to the motor. The motor was connected to the electrical circuit inside a 14,400-volt, 600-ampere metal electrical switch box. When he attempted to check the fuse on this circuit, an electrical arc ionized the air in the cabinet, and a flash fire occurred. The injured worker was burned over 50% of his body and died 18 days later. The cause of death as determined by autopsy and listed on the death certificate was thermal burns as a consequence of an electrical flash fire.

Recommendations/Discussion

Recommendation 1: Ensure that high-voltage electrical circuits are properly labeled.

Discussion: In this incident, the voltage rating electrical switch box was not clearly marked.

Recommendation 2: Ensure that only qualified personnel are authorized to work on high-voltage circuits.

Discussion: In this case, the deceased attempted to work on a circuit that was above his level of authorization. The employer should ensure that all maintenance personnel are informed of their level of authorization and enforce this policy.

Recommendation 3: Conduct a job site survey on a regular basis to identify potential hazards, implement appropriate control measures, and provide subsequent training to employees that specifically addresses all identified hazards.

Discussion: According to 29 CFR 1926.21(b)(2), employers are required to instruct each employee in the recognition and avoidance of unsafe conditions and to control or eliminate any hazards or other exposure to illness or injury. In this and similar situations, the employer might need to provide additional training to ensure that these employees understand the hazards and how to properly use safety equipment to protect themselves.

Adapted from: Colorado FACE Investigation 92CO039. Accessed 9/1/08 at http://www.cdc.gov/niosh/face/stateface/co/92co039.html.

Questions

1. How should the electrician have known that he was unqualified to work on this motor?
   A. The nameplate of the motor would have supplied the voltage rating, and the electrician knew that his limitation was 440-volt circuits or less.
   B. The coworker should have told him.
   C. He didn't know how to tell the voltage rating of the meter.
   D. The motor supply conductors were not in conduit.

2. Why would the victim use a multimeter to check for absence of voltage on a 15 kV-class circuit?
   A. Multimeters frequently are used to check voltages up to and including 50,000 volts.
   B. When 15 kV-class leads are inserted into the multimeter, the multimeter rating changes to 15 kV-class equipment.
   C. If the multimeter had been digital, instead of analog, it would have indicated that the device was underrated.
   D. The electrician was unqualified for this task.

Ready for Review

- This chapter introduced *NFPA 70E* Article 130 work practice considerations, including test instrument and equipment use, work within the LAB of uninsulated overhead lines, other precautions for personnel activities, and other protective equipment.
- OSHA 29 CFR 1910 and *NFPA 70E* Article 130 cover many of the same points.
- *NFPA 70E* Section 130.4 requires workers to be qualified to perform electrical testing on electrical equipment at voltage levels of 50 volts or more once they cross the LAB for that energized equipment.
- Electrical testing is energized work if the equipment is not placed in an electrically safe work condition. A hazard assessment, training, a job briefing, and selection and use of appropriate work practices and PPE are among the minimum requirements to perform electrical testing as a qualified person.
- *NFPA 70E* Section 130.5 addresses tasks performed within the LAB of uninsulated overhead lines. Distances associated with work involving uninsulated overhead lines are obtained from column 2 (exposed movable conductors) in Table 130.2(C).
- When contact with an uninsulated energized overhead line is possible, *NFPA 70E* Section 130.5(A) requires that lines be suitably guarded or deenergized and visibly grounded at the point of work.
- If arrangements are made to use protective measures, such as guarding, isolating, or insulating, these precautions shall prevent each employee from contacting such lines directly with any part of his or her body or indirectly through conductive materials, tools, or equipment.
- Per *NFPA 70E* Section 130.5(C), both employers and employees are responsible for ensuring that guards or protective measures are satisfactory for the conditions.
- *NFPA 70E* Section 130.5(D) requires that unqualified persons working on the ground or in an elevated position near overhead lines be located in such a way that the employee and the longest conductive object he or she could contact do not come closer to any unguarded, energized, overhead power line than the LAB.
- *NFPA 70E* Section 130.5(E)(1) requires any vehicle or mechanical equipment structure that will be elevated near energized overhead lines to be operated so that the LAB distance of Table 130.2(C), column 2, is maintained. However, some conditions permit the clearances to be reduced.
- *NFPA 70E* Section 130.5(E)(2) details requirements for equipment contact. Employees standing on the ground must not contact the vehicle or mechanical equipment or any of its attachments except in specific circumstances.
- *NFPA 70E* Section 130.5(E)(3) details requirements for equipment grounding. If any vehicle or mechanical equipment capable of having parts of its structure elevated near energized overhead lines is intentionally grounded, employees working on the ground near the point of grounding must not stand at the grounding location whenever there is a possibility of overhead line contact. Additional precautions must also be taken.
- *NFPA 70E* Section 130.6 details precautions that should be considered for personnel activities for work involving electrical hazards.
- *NFPA 70E* Section 130.6(A) requires that employees be instructed to be alert at all times and that they be prohibited from working when their alertness is impaired due to illness, fatigue, or other reasons when they are working where electrical hazards might exist.
- *NFPA 70E* Section 130.6(B) prohibits blind reaching. Employees must be instructed not to reach blindly into areas where an electrical hazard exists.

- *NFPA 70E* Section 130.6(C)(1) prohibits employee entrance into an area without adequate illumination so that the task can be performed safely.
- *NFPA 70E* Section 130.6(C)(2) prohibits employees from working where an electrical hazard exists if they cannot clearly see what they are working on.
- *NFPA 70E* Section 130.6(D) directs that conductive apparel not be worn where it could come in contact with energized electrical equipment.
- *NFPA 70E* Section 130.6(E) contains requirements for conductive materials, tools, and equipment being handled. Conductive materials, tools, and equipment that are in contact with any part of an employee's body must be handled in a manner that will prevent accidental contact with live parts and cannot approach exposed live parts closer than that permitted by the approach boundaries to energized electrical conductors and circuit parts requirements of *NFPA 70E* Section 130.2.
- *NFPA 70E* Section 130.6(F) details confined or enclosed workspace considerations and the responsibilities for both employer and employee. Protective shields, protective barriers, or insulating materials must be provided and used. Doors, hinged panels, and the like must be secured to prevent their swinging into an employee.
- *NFPA 70E* Section 130.6(G) prohibits employees from performing housekeeping duties inside the LAB, including the use of electrically conductive cleaning materials, unless adequate safeguards are in place to prevent electrical contact with these parts.
- *NFPA 70E* Section 130.6(H) prohibits the use of electric equipment capable of igniting flammable materials where the flammable materials are present only occasionally, unless measures are taken to prevent hazardous conditions from developing.
- *NFPA 70E* Section 130.6(I) addresses anticipating failure and requires electric equipment

to be deenergized when there is evidence that such equipment could fail unless the employer can demonstrate that doing so would introduce additional or increased hazards or that is it is infeasible to do so.
- *NFPA 70E* Section 130.6(J) cautions that load-rated devices specifically designed as a disconnecting means must be used for the opening, reversing, or closing of circuits under load.
- *NFPA 70E* Section 130.6(K) prohibits manually reenergizing a circuit after a circuit-protective device has deenergized it until it has been verified that it is safe to do so.
- *NFPA 70E* Section 130.7(A) assigns responsibility to employers and employees for protective equipment use. Two FPNs highlight the limitations of PPE, specifically pointing out that PPE minimizes injury rather than completely preventing it.
- *NFPA 70E* Section 130.7(B) requires protective equipment to be maintained, inspected, and stored properly. For example, rubber insulating gloves must be inspected and given an air test before use and be electrically tested periodically.
- *NFPA 70E* Section 130.7(D)(1) details considerations for nine types of insulated tools and equipment. Workers are required to use insulated tools, handling equipment, or both, when working within the LAB, and tools must be protected from becoming damaged.
- Insulated tools must meet ASTM F1505, *Standard Specification for Insulated and Insulating Hand Tools*, 2007, and comply with *NFPA 70E* Section 130.7(D)(1)(a).
- Removal or replacement of fuses in electrical equipment while it is energized must be avoided where possible. In addition, *NFPA 70E* Section 130.7(D)(1)(c) requires that nonconductive ropes and handlines be used where an electrical hazard exists.
- *NFPA 70E* Section 130.7(D)(1)(d) indicates that fiberglass-reinforced plastic rods and tubes that are used for live line tools must meet the

requirements of applicable portions of electrical codes and standards, such as ASTM F711, *Standard Specification for Fiberglass-Reinforced Plastic (FRP) Rod and Tube Used in Live Line Tools*.

- *NFPA 70E* Section 130.7(D)(1)(e) requires portable ladders to have nonconductive side rails if they are used where an electrical hazard exists and to meet the various ANSI standards for ladders listed in Table 130.7(F).

- *NFPA 70E* Section 130.7(D)(1)(f) requires that protective shields, protective barriers, or insulating materials be used to protect workers from shock, burns, or other electrically related injuries while they are working inside the LAB.

- *NFPA 70E* Section 130.7(D)(1)(g) requires rubber insulating equipment used for protection from accidental contact with live parts to meet the requirements of the ASTM standards listed in Table 130.7(F).

- *NFPA 70E* Section 130.7(D)(1)(h) requires that plastic guard equipment meet the requirements of the ASTM standards listed in Table 130.7(F) where it is used to protect workers from accidental contact with live parts or to protect employees or energized equipment or material from contact with ground.

- *NFPA 70E* Section 130.7(D)(1)(i) prohibits barriers from being brought or placed closer than the restricted approach boundary distance given in Table 130.2(C). *NFPA 70E* Section 130.7(E)(2) prohibits conductive barricades from being placed closer than the LAB.

- The Energized Electrical Work Permit must document how unqualified persons were restricted from the work area. *NFPA 70E* Section 130.7(E) provides an example of how restricted access of unqualified persons from the work area could be accomplished.

- *NFPA 70E* Section 130.7(E)(1) requires that safety signs, safety symbols, or accident prevention tags be used to warn employees about electrical hazards.

- *NFPA 70E* Section 130.7(E)(2) requires barricades to be used in conjunction with safety signs where it is necessary to prevent or limit employee access to work areas containing energized conductors or circuit parts.

- *NFPA 70E* Section 130.7(E)(3) requires an attendant in certain circumstances.

- *NFPA 70E* Section 130.7(E)(4) requires the use of safety signs, safety symbols, accident prevention tags, barricades used in conjunction with safety signs, or an attendant where signs and barricades do not provide sufficient warning and protection to prevent employees from entering look-alike equipment where work performed on equipment that is deenergized and placed in an electrically safe condition exists in a work area with other energized equipment that is similar in size, shape, and construction.

- *NFPA 70E* Section 130.7(F) requires that other protective equipment required in Section 130.7(D) conform to the standards provided in Table 130.7(F).

Vocabulary

confined space§ A space that is large enough and so configured that an employee can bodily enter and perform assigned work; has limited or restricted means for entry or exit (for example, tanks, vessels, silos, storage bins, hoppers, vaults, and pits are spaces that may have limited means of entry); and is not designed for continuous employee occupancy.

enclosed space§ A working space, such as a manhole, vault, tunnel, or shaft, that has a limited means of egress or entry, that is designed for periodic employee entry under normal operating conditions, and that under normal conditions does not contain a hazardous atmosphere, but that may contain a hazardous atmosphere under abnormal conditions.

overcurrent* Any current in excess of the rated current of equipment or the ampacity of a conductor. It may result from overload, short circuit, or ground fault.

FPN: A current in excess of rating may be accommodated by certain equipment and conductors for a given set of conditions. Therefore, the rules for overcurrent protection are specific for particular situations.

overload* Operation of equipment in excess of normal, full-load rating, or of a conductor in excess of rated ampacity that, when it persists for a sufficient length of time, would cause damage or dangerous overheating. A fault, such as a short circuit or ground fault, is not an overload.

*This is the *NFPA 70E* definition.
§This is the OSHA definition.

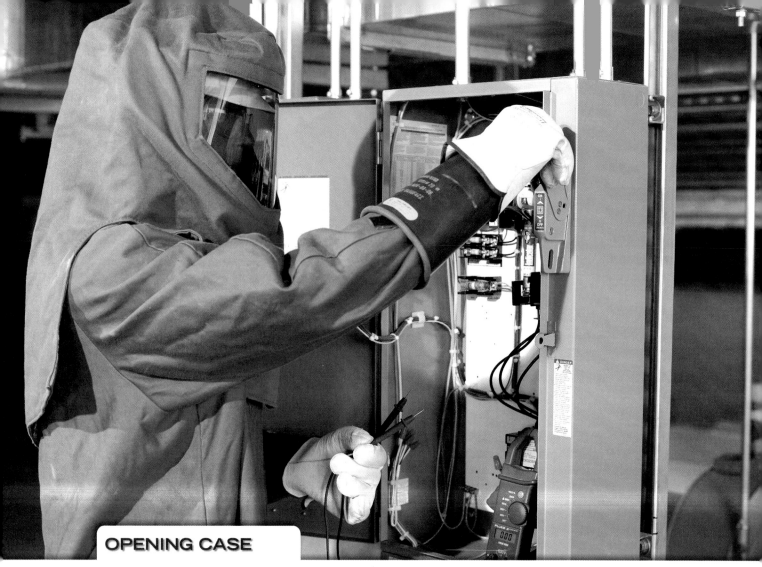

Photo courtesy of National Electrical Contractors Association (NECA).

OPENING CASE

A 32-year-old electrician was electrocuted while connecting a hydraulic press brake. The victim was a self-employed, licensed, Journeyman electrician in business for 2 years and 6 months at the time of his death. In order to obtain his license, the victim had completed classroom training and a 4000-hour on-the-job apprenticeship. He had also passed the state-mandated licensing examination.

The victim was contracted to connect new equipment and install new lights in the plant where he was working. The electrical system in the plant consisted of a 480-volt, 3-phase, 3-wire underground Delta transformer. A 200-ampere, 3-phase main disconnect switch supplied a 3-phase distribution panel that, in turn, fed a bus duct within the facility. The bus duct powered plant machinery, two dry-type step-down transformers, a welding machine, and temporary wiring that the victim used the day of the incident.

Before leaving the plant on Friday, June 18, the victim and coworker disconnected the temporary wiring on a hydraulic press brake and made a permanent connection to the brake with another cable. They then taped and rolled up the end of the temporary lead and left it in the building rafters for use the next Monday.

On Monday morning, the plant president and an assistant routinely reenergized the building electrical equipment. Plant policy called for deactivating most equipment nights, weekends, and holidays. Arriving at the plant at approximately 9:30 a.m., the victim and coworker resumed the work they had left on Friday. Making sure to deactivate the circuit breaker that powered all the electrical lines they were to work on, the victim and coworker spent approximately 2 hours wiring light switches and a leaf-break lockout. The two then took a half-hour break.

Personal Protective Equipment

CHAPTER 14

Introduction

NFPA 70E Article 130 covers a wide variety of topics addressing work involving electrical hazards. Those topics are grouped by topic similarity and covered in various chapters throughout this textbook. Justification for energized work, energized electrical work permit requirements, approach boundaries for shock protection, and arc flash hazard analysis requirements are covered in Chapter 10.

NFPA 70E classifies personal protective equipment (PPE) into two categories: PPE, such as hard hats, safety glasses, and flame-resistant (FR) clothing, and other protective equipment, such as insulated tools, protective barriers, and rubber insulating equipment. Other protective equipment requirements from *NFPA 70E* Section 130.7(D) are covered in Chapter 13.

This chapter details the PPE requirements in *NFPA 70E* Section 130.7(C). There are two different methods to determine the protective equipment, incident energy analysis, and hazardous/risk category (HRC).

Chapter Outline

- **Introduction**
- **Personal Protective Equipment**
- **HRC Classifications to Select PPE**
- **Summary**
- **Lessons Learned**
- **Current Knowledge**

Learning Objectives

1. Understand the types of PPE needed for shock and arc flash.
2. Understand the requirements that apply for the selection of PPE based on incident energy analysis.
3. Understand the requirements that apply for the selection of PPE based on HRC.

References

1. *NFPA 70E®*, 2009 Edition
2. Occupational Safety and Health Administration (OSHA) 29 Code of Federal Regulations (CFR) 1910

OPENING CASE, CONT.

Returning from their break around noon, the workers began preparing the previously taped and coiled temporary lead for connection to a new hydraulic press for a test run. The coworker recalled asking the victim if he thought the breaker was still deactivated. The victim reassured him that it was, and began to cut into the taped end of the lead. Shortly afterwards, the victim touched an energized conductor and yelled to turn off the breaker, which was only 20 ft away. When this was done, the victim collapsed to the floor. Apparently, someone had energized the bus duct to use the welding machine while the two were on break.

Hearing a plea for help, two plant workers responded from their lunchroom to see what had happened. Finding the victim lying on the floor, open-armed and on his back, one of the plant workers detected no pulse and called 911 for emergency response. When no one answered, he directly called some emergency responders who arrived on-site in minutes. The victim was transported to the local hospital, where he was officially pronounced dead approximately 45 minutes following the incident. The medical examiner listed the cause of death as electrocution.

Adapted from: Massachusetts Fatality and Assessment and Control Evaluation (FACE) MA-93-09. Accessed 9/7/08 at http://www.cdc.gov/niosh/face/stateface/ma/93ma009.html.

1. Why didn't the electrician install a lock and tag on the temporary service?

2. What was the role of the facility owner in preventing this fatality?

Personal Protective Equipment

NFPA 70E Article 130 concludes by detailing PPE, other protective equipment, and alerting techniques. *NFPA 70E* Sections 130.7(A), 130.7(B), 130.7(D), and 130.7(E) are discussed in Chapter 13. Portions of *NFPA 70E* Section 130.7(A) and 130.7(B) are included here as a review. However, the primary focus of this chapter is the PPE requirements of Section 130.7(C).

General Requirements

NFPA 70E Section 130.7(A) requires that workers be provided with and use protective equipment that is designed for the use and adequate for the hazard when performing tasks where electrical hazards are present (**Figure 14.1**).

Two fine print notes (FPNs) to *NFPA 70E* Section 130.7(A) point out that workers could sustain injury even while using the protective equipment determined in accordance with *NFPA 70E* Section 130.7.

> FPN No. 1: While some situations could result in burns to the skin, even with the protection selected, burn injury should be reduced and survivable. Due to the explosive effect of some arc events, physical trauma injuries could occur. The PPE requirements of 130.7 do not address protection against physical trauma other than exposure to the thermal effects of an arc flash.

> FPN No. 2: When incident energy exceeds 40 cal/cm² at the working distance, greater emphasis may be necessary with respect to deenergizing before working within the Limited Approach Boundary of the exposed electrical conductors or circuit parts.

Equipment Care

NFPA 70E Section 130.7(B) requires protective equipment to be maintained, inspected, and stored properly. It must be maintained in a safe, reliable condition, be visually inspected before each use, and stored in a manner to prevent damage.

PPE Selection

There are generally two methods to determine the PPE that workers need to wear once they cross the arc flash boundary requiring such protection; namely incident energy analysis and HRCs. Appropriate PPE is selected by the requirements of *NFPA 70E* Section 130.7(C) once the incident energy is determined per Section 130.3(B)(1) or the HRC is determined.

NFPA 70E Section 130.7(C)(1) requires all parts of the body inside the AFPB to be protected. PPE must be selected based on the hazards associated with the work task. Shock protective equipment is selected based on the circuit voltage. Two methods may be used to determine the PPE that workers must wear once they cross the AFPB. For arc flash protection, one method is calculating incident energy, and the second method is determining the HRC from the tables included in Article 130. Appropriate PPE may be selected by determining the incident energy, according to Section 130.3(B)(1), or by determining the HRC as permitted by Section 130.3(B)(2).

Arc-Flash Protective Equipment (FR Clothing)

Arc-rated **flame-resistant (FR)** clothing and other PPE is selected based on the incident energy exposure in accordance with the requirements of *NFPA 70E* Sections 130.7(C)(1) though (8) and 130.7(C)(12) though (16), unless the HRC method is used, as permitted by Section 130.3(B)(2).

The HRC method is used for the selection of personal and other protective equipment in conjunction with the requirements of *NFPA 70E* Sections and Tables 130.7(C)(9), 130.7(C)(10), and 130.7(C)(11). PPE must conform to the standards provided in Table 130.7(C)(8).

General PPE Requirements

NFPA 70E Section 130.7(C)(1) requires all parts of the body inside the AFPB to be protected in accordance with *NFPA 70E* Section 130.3. *NFPA 70E* Section 130.3 requires three things:
- The AFPB per Section 130.3(A)
- The incident energy or HRC for the selection of personal and other protective equipment per Section 130.3(B)
- Equipment labeling containing the available incident energy or required level of PPE

The AFPB, incident energy or HRC, and protective equipment must be documented on the Energized Electrical Work Permit.

Figure 14.1 Workers must be provided with and use protective equipment that is designed for the use and adequate for the hazard.

130.1(B) Energized Electrical Work Permit, in part.

(2) Elements of Work Permit. The energized electrical work permit shall include, but not be limited to, the following items:

(6) Results of the arc flash hazard analysis

(7) The arc flash protection boundary

(8) The necessary personal protective equipment to safely perform the assigned task

Movement and Visibility

NFPA 70E Section 130.7(C)(2) requires FR clothing to cover all ignitable clothing and allow for movement and visibility. *NFPA 70E* Section 130.7(C)(12) provides factors to be considered for the selection of protective clothing, and 130.7(C)(14) details clothing material characteristics.

Figure 14.2   Nonconductive head protection is required to protect against electric shock or burns caused by contact with live parts or from flying objects.

70E Highlights

NFPA 70E Section 130.7(C)(12)(d), in part
Clothing shall cover potentially exposed areas as completely as possible. Shirt sleeves shall be fastened at the wrists, and shirts and jackets shall be closed at the neck.
NFPA 70E Section 130.7(C)(12)(e), in part
Tight-fitting clothing shall be avoided. Loose fitting clothing provides additional thermal insulation because of air spaces. FR apparel shall fit properly such that it does not interfere with the work task.
NFPA 70E Section 130.7(C)(12)(f), in part
The garment selected shall result in the least interference with the task but still provide the necessary protection. The work method, location, and task could influence the protective equipment selected.
NFPA 70E Section 130.7(C)(14) FPN No. 2
Non-FR cotton, polyester-cotton blends, nylon, nylon-cotton blends, silk, rayon, and wool fabrics are flammable. These fabrics could ignite and continue to burn on the body, resulting in serious burn injuries.

or flashes or from flying objects resulting from electrical explosion (**Figure 14.3**).

• Hairnets and/or beard nets must be nonmelting and FR if they are worn.

Eye Protection

NFPA 70E Section 130.7(C)(4) requires employees to wear protective equipment for the eyes whenever there is danger of injury from electric arcs, flashes, or from flying objects resulting from electrical explosion (**Figure 14.4**). Eye protection is required in conjunction with face protection.

Body Protection

NFPA 70E Section 130.7(C)(5), body protection, requires employees to wear FR clothing (**Figure 14.5**) wherever there is possible exposure to an electric arc flash

Head, Face, Neck, and Chin Protection

NFPA 70E Section 130.7(C)(3) details head, face, neck, and chin protection:

• Nonconductive head protection must be worn by workers wherever there is a danger of head injury from electric shock or burns caused by contact with live parts or from flying objects caused by an electrical explosion (**Figure 14.2**).

• Nonconductive protective equipment for the face, neck, and chin is required whenever there is a danger of injury from exposure to electric arcs

Figure 14.3   Nonconductive protective equipment for the face, neck, and chin is required to protect against electric arcs or flashes or from flying objects.

Figure 14.4 Protective equipment for the eyes is required whenever there is danger of injury from electric arcs, flashes, or from flying objects.

Figure 14.5 FR clothing is required wherever there is possible exposure to an electric arc flash above 1.2 cal/cm².

70E Highlights

NFPA 70E Section 130.7(C)(15), in part
Clothing and other apparel (such as hard hat liners and hair nets) made from materials that do not meet the requirements of 130.7(C)(14) regarding melting, or made from materials that do not meet the flammability requirements shall not be permitted to be worn.
FPN: Some flame-resistant fabrics, such as non-FR modacrylic and nondurable flame-retardant treatments of cotton, are not recommended for industrial or utility applications.

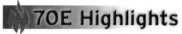

70E Highlights

NFPA 70E Section 130.7(C)(13)(b), in part
Face shields shall have an arc rating suitable for the arc flash exposure. Face shields without an arc rating shall not be used. Eye protection (safety glasses or goggles) shall always be worn under face shields or hoods.

above 1.2 cal/cm², which is the threshold incident energy level for a second-degree burn.

Two methods are available to determine the PPE needed to cross the AFPB. The worker must use one of the following methods to select the PPE:

- *An incident energy analysis per Section 130.3(B) (1)*. The worker must use arc-rated FR clothing that is based on the incident energy exposure associated with the specific task.
- *An HRC permitted per Section 130.3(B)(2)*. The requirements of Sections 130.7(C)(9), 130.7(C)

(10), and 130.7(C)(11) may be used to select personal and other protective equipment.

A FPN to *NFPA 70E* Section 130.7(C)(5) provides examples of FR body protection that may be used to meet this requirement:

- An arc flash suit jacket and arc flash suit pants
- Shirts and pants
- Coveralls
- A combination of jacket and pants
- Overalls with jacket and pants

Generally, the higher degree of protection is provided by heavier weight fabrics and/or by combinations of one or more layers of FR clothing.

Hand and Arm Protection

NFPA 70E Section 130.7(C)(6) divides hand and arm protection requirements into several categories:

- Hand injury from electric shock
- Hand and arm injury from electric shock
- Hand and arm protection arc flash protection
- Electrical protective equipment use
- Electrical protective equipment maintenance

Shock Protection for Hands and Arms

NFPA 70E Section 130.7(C)(6)(a) requires rubber insulating gloves to be rated for the voltage to which the gloves will be exposed (**Figure 14.6**). Employees are required to wear rubber insulating gloves with leather protectors where there is a danger of hand injury from electric shock due to contact with live parts.

Employees must generally wear rubber insulating gloves with leather protectors and rubber insulating sleeves (**Figure 14.7**) where there is a danger of hand and

ASTM Labeling Chart
SALISBURY's Natural Rubber and SALCOR® Rubber Electrical Protective Equipment

Class Color	Proof Test Voltage AC / DC	Max. Use Voltage AC / DC	Rubber Molded Products Label	Insulating Rubber Glove Label	Insulating Rubber Dipped Sleeve Label	ASTM Specification Reference	
00 Beige	2,500 / 10,000	500 / 750		10 SALISBURY ANSI/ASTM MADE IN D120 CLASS 00 U.S.A TYPE I MAX USE VOLT 500V AC		D120	Rubber Insulating Gloves
0 Red	5,000 / 20,000	1,000 / 1,500	SALISBURY MAX USE VOLTAGE: 1,000 V AC CLASS 0 TYPE I	10 SALISBURY ANSI/ASTM D120 CLASS 0 TYPE I MAX USE VOLT 1000V AC	SALISBURY ANSI/ASTM D1051 CLASS 0 TYPE I MAX USE VOLT 1000V AC	D178	Rubber Insulating Matting
1 White	10,000 / 40,000	7,500 / 11,250	SALISBURY MAX USE VOLTAGE: 7,500 V AC CLASS 1 TYPE I	10 SALISBURY ANSI/ASTM MADE IN D120 CLASS 1 U.S.A TYPE I MAX USE VOLT 7500V AC	SALISBURY ANSI/ASTM D1051 CLASS 1 U.S.A TYPE I MAX USE VOLT 7500V AC	D1048	Rubber Insulating Blankets
2 Yellow	20,000 / 50,000	17,000 / 25,500	SALISBURY MAX USE VOLTAGE: 17,000 V AC CLASS 2 TYPE I	10 SALISBURY ANSI/ASTM D120 CLASS 2 TYPE I MAX USE VOLT 17000V AC	SALISBURY ANSI/ASTM D1051 CLASS 2 TYPE I MAX USE VOLT 17000V AC	D1049	Rubber Insulating Covers
3 Green	30,000 / 60,000	26,500 / 39,750	SALISBURY MAX USE VOLTAGE: 26,500 V AC CLASS 3 TYPE I	10 SALISBURY ANSI/ASTM MADE IN D120 CLASS 3 TYPE I MAX USE VOLT 26500V AC	SALISBURY ANSI/ASTM MADE IN D1051 CLASS 3 U.S.A TYPE I MAX USE VOLT 26500V AC	D1050	Rubber Insulating Line Hose
4 Orange	40,000 / 70,000	36,000 / 54,000	SALISBURY MAX USE VOLTAGE: 36,000 V AC CLASS 4 TYPE II	10 SALISBURY ANSI/ASTM MADE IN D120 CLASS 4 TYPE I MAX USE VOLT 36000V AC	SALISBURY ANSI/ASTM MADE IN D1051 CLASS 4 U.S.A TYPE I MAX USE VOLT 36000V AC	D1051	Rubber Insulating Sleeves

ASTM Specification Reference	
D120	Rubber Insulating Gloves
D178	Rubber Insulating Matting
D1048	Rubber Insulating Blankets
D1049	Rubber Insulating Covers
D1050	Rubber Insulating Line Hose
D1051	Rubber Insulating Sleeves
F478	In-Service Care of Line Hose & Covers
F479	In-Service Care of Insulating Blankets
F496	In-Service Care of Gloves & Sleeves
F696	Leather Protectors for Insulating Gloves and Mittens
F1236	Inspection Guide for Rubber Products
F1742	PVC Insulating Sheeting
F2320	Rubber Insulating Sheeting
F2677	Electrically Insulating Aprons
Type I-	Designates natural rubber.
Type II-	Designates SALCOR® UV and ozone resistant rubber.

SALISBURY by Honeywell
7520 N. Long Ave. Skokie IL 60077
toll free ph. 877.406.4501
toll free fax 866.824.4922
whsalisbury.com

Insulating Gloves and Sleeves must have a color coded label to meet appropriate ASTM Specifications.

REQUEST THE BEST WITH SALISBURY

Part #ASTMCHART
updated 110608

Figure 14.6 Rubber insulating gloves must be rated for the voltage to which the gloves will be exposed.

Figure 14.7 Rubber insulating gloves with leather protectors and rubber insulating sleeves are required where there is a danger of hand and arm injury from electric shock due to contact.

OSHA Tip

1910.137(b)(2)(vii) Protector gloves shall be worn over insulating gloves, except as follows:

1910.137(b)(2)(vii)(A) Protector gloves need not be used with Class 0 gloves, under limited-use conditions, where small equipment and parts manipulation necessitate unusually high finger dexterity.

Note: Extra care is needed in the visual examination of the glove and in the avoidance of handling sharp objects.

1910.137(b)(2)(vii)(B)

Any other class of glove may be used for similar work without protector gloves if the employer can demonstrate that the possibility of physical damage to the gloves is small and if the class of glove is one class higher than that required for the voltage involved. Insulating gloves that have been used without protector gloves may not be used at a higher voltage until they have been tested under the provisions of paragraphs (b)(2)(viii) and (b)(2)(ix) of this section.

70E Highlights

NFPA 70E Section 130.7(C)(13)(c) Hand Protection.
(1) Leather or FR gloves shall be worn where required for arc flash protection.
(2) Where insulating rubber gloves are used for shock protection, leather protectors shall be worn over the rubber gloves.

arm injury from electric shock due to contact with energized electrical conductors or circuit parts.

Flash Protection for Hands and Arms

NFPA 70E Section 130.7(C)(6)(b) requires that hand and arm protection be worn where there is possible exposure to arc flash burn. Section 130.7(C)(13)(c) describes the requirements necessary to protect the hands from burns.

Arm protection is addressed by the body protection requirements in Section 130.7(C)(5). The arm, like all parts of the body, must be protected by FR clothing where the possibility of an exposure to an electric arc flash above 1.2 cal/cm^2 exists.

PPE Maintenance and Use Requirements

NFPA 70E Section 130.7(C)(6)(c) details the maintenance and use requirements for PPE. Electrical protective equipment must be maintained in a safe, reliable condition and requires periodic electrical tests (**Figure 14.8**). Test voltages and the maximum intervals between tests shall be in accordance with Table 130.7(C)(6)(c).

Figure 14.9 Insulating gloves must be given an air test prior to use.

Insulating equipment is required to be inspected for damage daily before it is used and immediately following incidents that can reasonably be suspected of having caused damage. Insulating gloves must be given an air test as part of the daily inspection (**Figure 14.9**).

Foot Protection

NFPA 70E Section 130.7(C)(7) details general requirements for foot protection. Insulated soles must not be used as primary electrical protection. Dielectric overshoes are required where insulated footwear is used as protection against step and touch potential (**Figure 14.10**). *NFPA 70E* Section 130.7(C)(13)(d) addresses foot protection where arc flash is a concern. Heavy-duty leather work shoes are required for tasks identified as HRC 2 and higher in Table 130.7(C)(9) and for incident energy exposures greater than 4 cal/cm^2 as determined by the incident energy analysis per Section 130.3(B)(1).

Standards for PPE

Do not confuse the PPE standards in Table 130.7(C)(8) with the other protective equipment standards of Table 130.7(C)(F), such as rubber insulating blankets,

Figure 14.8 Electrical protective equipment requires periodic electrical tests.

Figure 14.10 Dielectric overshoes are required where insulated footwear is used as protection against step and touch potential.

OSHA Tip

1910.137(b)(2)(viii) Electrical protective equipment shall be subjected to periodic electrical tests. The maximum intervals between tests shall be in accordance with Table I-6.

Table I-6   Rubber Insulating Equipment Test Intervals

Type of Equipment	When to Test
Rubber insulating line hose	Upon indication that insulating value is suspect.
Rubber insulating covers	Upon indication that insulating value is suspect.
Rubber insulating blankets	Before first issue and every 12 months thereafter(1).
Rubber insulating gloves	Before first issue and every 6 months thereafter(1).
Rubber insulating sleeves	Before first issue and every 12 months thereafter(1).

Footnote(1) If the insulating equipment has been electrically tested but not issued for service, it may not be placed into service unless it has been electrically tested within the previous 12 months.

line hose, and insulated hand tools, covered in Chapter 13. *NFPA 70E* Section 130.7(C)(8) requires use of PPE that meets ANSI standards, identified in Table 130.7(C)(8):

- Head protection
- Eye and face protection
- Rubber insulating gloves
- Rubber insulating sleeves
- In-service care of rubber insulating gloves and sleeves
- Leather protectors for rubber insulating gloves and mittens
- Footwear
- Visual inspection of electrical protective rubber products
- Apparel for momentary electric arc and related thermal hazards
- Arc-resistant and FR rainwear
- Arc rating and standard specification for face protective products

Non-FR and other flammable fabrics are not covered by a standard in Table 130.7(C)(8). Guidance for those fabrics can be found in *NFPA 70E* Sections 130.7(C)(14)(a), 130.7(C)(14)(b), and 130.7(C)(15).

Factors in Selection of Protective Clothing

NFPA 70E Section 130.7(C)(12) details factors the employer should consider when selecting protective clothing and equipment. Clothing and equipment must be used to provide protection for the degree of shock and arc flash hazard exposure. The required protection can be worn alone or integrated with flammable, nonmelting apparel. PPE items are normally used in combination as a system to provide the appropriate level of protection.

If FR clothing is required, it must cover associated parts of the body and all flammable apparel. It must allow for movement and visibility and be maintained in a sanitary and functionally effective condition. A FPN to this section offers additional insight.

> FPN: Protective clothing includes shirts, pants, coveralls, jackets, and parkas worn routinely by workers who, under normal working conditions, are exposed to momentary electric arc and related thermal hazards. Flame-resistant rainwear worn in inclement weather is included in this category of clothing.

Layering

The *NFPA 70E* Section 130.7(C)(12)(a) requirements for layering indicate that nonmelting, flammable fiber garments are permitted to be used as underlayers in conjunction with FR garments in a layered system for added protection. However, if nonmelting, flammable fiber garments are used as underlayers, the system arc rating must be sufficient to prevent breakopen of the innermost FR layer at the expected arc exposure incident energy level to prevent ignition of flammable underlayers.

> Breakopen is a material response evidenced by the formation of one or more holes in the innermost layer of flame-resistant material that would allow flame to pass through the material [70E-2009, FPN to Article 100 definition of Arc Rating].

The FPN to Section 130.7(C)(12)(a) informs that a typical layering system might include cotton underwear, a cotton shirt and trousers, and an FR coverall, and that specific tasks might call for additional FR layers to achieve the required protection level.

Outer and Under Layers

NFPA 70E Section 130.7(C)(12)(b) requires that garments worn as outer layers over FR clothing must also be made from FR material (**Figure 14.11**).

NFPA 70E Section 130.7(C)(12)(c) prohibits meltable fibers such as acetate, nylon, polyester, polypropylene, and spandex in underwear next to the skin except for an incidental amount of elastic used with nonmelting fabric underwear or socks. Two FPNs to this section offer additional insight into these requirements.

Figure 14.11 Garments worn as outer layers over FR clothing must also be made from FR material.

FPN No. 1: FR garments (e.g., shirts, trousers, and coveralls) worn as underlayers that neither ignite nor melt and drip in the course of an exposure to electric arc and related thermal hazards generally provide a higher system arc rating than nonmelting, flammable fiber underlayers.

FPN No. 2: FR underwear or undergarments used as underlayers generally provide a higher system arc rating than nonmelting, flammable fiber underwear or undergarments used as underlayers.

Coverage, Fit, and Interference

NFPA 70E Sections 130.7(C)(12)(d),(e), and (f) address coverage, fit, and interference, respectively. Clothing is required to cover potentially exposed areas as completely as possible. To accomplish this coverage, shirtsleeves must be fastened at the wrists, and shirts and jackets must be closed at the neck.

FR apparel must fit properly so that it does not interfere with the task to be performed. The garment selected must result in the least interference with the task but still provide the necessary protection. Loose-fitting clothing provides additional thermal insulation because of air spaces; therefore, tight-fitting clothing must be avoided. The work method, location, and task could influence the protective equipment selected.

Arc Flash Protective Equipment

NFPA 70E Section 130.7(C)(13) covers the requirements for four types of arc flash protective equipment: Arc flash suits (**Figure 14.12**), face protection, hand protection, and foot protection.

Arc Flash Suits

NFPA 70E Section 130.7(C)(13)(a) requires arc flash suits to be easily and rapidly removed and have an arc rating that is suitable for the arc flash exposure.

> **Arc Flash Suit.** A complete FR clothing and equipment system that covers the entire body, except for the hands and feet. This includes pants, jacket, and beekeeper-type hood fitted with a face shield.

If exterior air is supplied into a flash suit hood, the air hoses and pump housing must be covered by FR materials or be constructed of nonmelting and nonflammable materials.

Face Protection

When face shields are worn within an AFPB, *NFPA 70E* Section 130.7(C)(13)(b) requires them to have an arc rating that is adequate for the arc flash exposure. Eye protection is always required to be worn under face shields, including face shields in flash suit hoods (**Figure 14.13**).

FPN information warns that face shields made with energy-absorbing formulation providing protection

Figure 14.12 Arc flash protective equipment includes arc flash suits.

from the radiant energy of an arc flash are tinted and can reduce visual acuity and color perception and that additional illumination might be necessary when these types of arc-protective face shields are used.

Hand Protection

NFPA 70E Section 130.7(C)(13)(c) details considerations for hand protection to protect against arc flash. Either leather or FR gloves must be worn within the AFPB.

Boundary, Arc Flash Protection. When an arc flash hazard exists, an approach limit at a distance from a prospective arc source within which a person could receive a second degree burn if an electrical arc flash were to occur.

130.7(C) (1) General. When an employee is working within the Arc flash protection boundary he/she shall wear protective clothing and other personal protective equipment in accordance with 130.3. All parts of the body inside the Arc Flash Protection Boundary shall be protected.

130.7(C)(5) Body Protection. Employees shall wear FR clothing wherever there is possible exposure to an electric arc flash above the threshold incident-energy level for a second-degree burn [5 J/cm^2 (1.2 cal/cm^2)].

Figure 14.13 Eye protection is always required to be worn under face shields.

Gloves made from layers of FR material provide hand protection against arc flash (**Figure 14.14**). FR gloves must be worn for protection against arc flash where rubber insulating gloves are not required to be worn for shock protection.

Rubber insulating gloves provide hand protection against arc flash. Leather protectors worn over insulating rubber gloves provide additional arc flash protection for the hands. Leather protectors must be worn in conjunction with rubber insulating gloves for protection against arc flash, whereas insulating rubber gloves are used for shock protection. However, leather can shrink and cause a decrease in protection during high arc flash exposures.

Figure 14.14 Gloves made from layers of FR material provide hand protection against arc flash.

Foot Protection

NFPA 70E Section 130.7(C)(13)(d) defines the foot protection required to protect the worker from arc flash. This is for arc flash protection which differs from the Section 130.7(C)(7) requirements for foot protection used as protection against step and touch potential. Heavy-duty leather work shoes provide some arc flash protection to the feet and must be used where the incident energy analysis of Section 130.3(B)(1) determines exposures greater than 4 cal/cm².

Clothing Material Characteristics

FR clothing must meet the requirements described in Section 130.7(C)(14) and (15). *NFPA 70E* Section 130.7(C)(14) begins with three FPNs offering nonmandatory clothing material characteristic information.

> FPN No. 1: FR materials, such as flame-retardant treated cotton, meta-aramid, para-aramid, and poly-benzimidazole (PBI) fibers, provide thermal protection. These materials can ignite but will not continue to burn after the ignition source is removed. FR fabrics can reduce burn injuries during an arc flash exposure by providing a thermal barrier between the arc flash and the wearer.

> FPN No. 2: Non-FR cotton, polyester-cotton blends, nylon, nylon-cotton blends, silk, rayon, and wool fabrics are flammable. These fabrics could ignite and continue to burn on the body, resulting in serious burn injuries.

> FPN No. 3: Rayon is a cellulose-based (wood pulp) synthetic fiber that is a flammable but nonmelting material.

NFPA 70E Section 130.7(C)(14) generally prohibits clothing made from flammable synthetic materials (either alone or in blends) that melt at temperatures below 600°F (316°C). Examples include acetate, acrylic, nylon, polyester, polyethylene, polypropylene, and spandex.

FPN information explains that these materials melt as a result of arc flash exposure, form intimate contact with the skin, and aggravate burn injury.

An exception to Section 130.7(C)(14) allows fiber blends that contain materials that melt, such as acetate, acrylic, nylon, polyester, polyethylene, polypropylene, and spandex if the requirements of ASTM F1506 are met and the fiber blends do not exhibit evidence of a melting and sticking hazard during arc testing according to ASTM F1959.

Prohibited Apparel

NFPA 70E Section 130.7(C)(15) addresses prohibited apparel. Clothing and other apparel, such as hard hat liners and hair nets, made from materials that do not meet the flammability requirements in the ASTM standards or the requirements of *NFPA 70E* Section 130.7(C)(14) regarding melting, are prohibited. An exception allows nonmelting, flammable non-FR materials, as described in Section 130.7(C)(14), as underlayers to FR clothing and for HRC 0 as described in Table 130.7(C)(10).

A second exception permits non-FR PPE under special permission by the authority having jurisdiction (AHJ) work performed inside the AFPB that exposes the worker to multiple hazards, such as airborne contaminants, where it can be shown that the protection is adequate to address the arc flash hazard.

> ## 70E Highlights
>
> *NFPA 70E* defines special permission in Article 100 as follows:
> **Special Permission.** The written consent of the authority having jurisdiction.

The FPN to Section 130.7 (C)(15) advises that some FR fabrics are not recommended for industrial electrical or utility applications. Examples include non-FR modacrylic and nondurable flame-retardant treatments of cotton.

Care and Maintenance of FR Clothing and FR Arc Flash Suits

NFPA 70E Section 130.7(C)(16) details inspection, care, maintenance, storage, and cleaning requirements for FR clothing and arc flash suits.

- FR apparel must be inspected before each use.
- FR apparel must be stored in a manner that prevents damage to it from deteriorating agents and from contamination from flammable or combustible materials.

Continued on page 268

BACKGROUND

Protective Clothing

The following information was contributed by Westex, Inc.

The Need for FR Clothing

Workers in the electrical maintenance, utility, oil, gas, petrochemical, and steel industries work in environments that can expose them to hazards that could cause severe or fatal burn injuries. In the event of a momentary electric arc, flash fire, or molten metal splash exposure, non-FR work clothes can ignite and continue to burn even after the source of ignition has been removed. Untreated natural fabrics continue to burn until the fabric is totally consumed, and non-FR synthetic fabrics will burn with melting and dripping causing severe contact burns to the skin. Government reports note that the majority of severe and fatal burn injuries are due to the individual's clothing igniting and continuing to burn, not to the exposure itself. The use of FR, arc-resistant clothing provides thermal protection at the exposure area. The level of protection typically rests in the fabric weight and composition. After the source of the ignition is removed, FR garments self-extinguish, limiting the body burn percentage.

Flame Resistance Defined

NFPA 70E Article 100 defines FR as follows:

> The property of a material whereby combustion is prevented, terminated, or inhibited following the application of a flaming or non-flaming source of ignition, with or without subsequent removal of the ignition source.

> FPN: Flame resistance can be an inherent property of a material, or it can be imparted by a specific treatment applied to the material.

Flame resistance is the characteristic of a fabric that causes it to self-extinguish when the source of ignition is removed. The most commonly used test method is ASTM D6413, *Standard Test Method for Flame Resistance of Textiles* (vertical flame test). The vertical flame test is a test method with no pass/fail requirements. Industry established standards range from 4-inch to 6-inch maximum char lengths. It is very important for FR fabrics to self-extinguish. Fabrics that self-extinguish after the source of ignition is removed can dramatically reduce body burn percentage and increase the chance for survival. However, char length measurements by themselves have no correlation to the protection afforded by a FR fabric.

Protection from thermal events is better measured by testing the thermal resistance of fabrics against exposures to simulated hazards, such as the flash fire manikin test or the arc-thermal performance test.

Key to Evaluating and Comparing FR Fabrics

The first step is for the worker to search out and evaluate information that was generated using the following three criteria. By doing this, a worker can evaluate different types of FR fabrics on a level playing field and compare "apples to apples."

1. *Identify the potential hazard.* Exposures such as electric arc flash and flash fire are unique hazards with vastly different characteristics, and the test results do not directly correlate to one another. The results from flash fire testing should not be substituted for electric arc flash testing when evaluating products.
2. *Identify industry consensus standards for the exposure.* Industry standards have been developed for electric arc flash and flash fire testing. For electric arc flash, ASTM has developed F1959, which produces an arc-thermal performance value (ATPV). *NFPA 2112* covers potential fuel-based flash-fire hazards.
3. *Make sure the testing is conducted at independent laboratories.* This helps ensure that unbiased and scientifically valid data is being produced. While it is often helpful and interesting to witness testing conducted by a company that has a vested interest in the FR business, there is no substitute for information generated at an independent laboratory.

Electric Arc Flash Protection

The intense energy and very short duration of an electric arc flash represents a very unique exposure. *NFPA 70E* Annex K.3 says that the temperature of an electric arc flash can reach 35,000°F:

> Exposure to these extreme temperatures both burns the skin directly and causes ignition of clothing, which adds

(continued)

to the burn injury. The majority of hospital admissions due to electrical accidents are from arc-flash burns, not from shocks. Each year more then 2,000 people are admitted to burn centers with severe arc-flash burns. Arc-flashes can and do kill at distances of 3 m (10 ft).

The thermal energy released in an electric arc flash is expressed in calories per centimeter squared (cal/cm^2), and potential exposures exceed 40 cal/cm^2. *Everyday work clothes made from regular cotton or poly/cotton fabrics can be readily ignited at exposure levels as low as 4 to 5 cal/cm^2 and, once ignited, can continue to burn, adding to the extent of injury sustained from the arc alone.* Many people consider non-FR 100% cotton as an acceptable option for protection from an electric arc flash because it does not contain a synthetic component that can melt, drip, and adhere to the skin. However non-FR 100% cotton can ignite just as easily as poly/cotton fabric in an electric arc flash. While 100% cotton will not melt and drip, it burns hotter than poly/cotton fabrics and typically is heavier, providing more fuel, which would allow it to burn longer and be harder to extinguish.

ASTM 1506

ASTM F1506, *Standard Performance Specification for Textile Materials for Wearing Apparel for Use by Electrical Workers Exposed to Momentary Arc and Related Thermal Hazards*, was developed to provide minimum performance specifications for protective clothing. The major requirement of this specification is that the fabric used in garments is FR and has been tested to ASTM F1959 to receive an arc rating or ATPV.

ASTM F1959

ASTM F1959, *Standard Test Method for Determining the Arc Rating of Materials for Clothing*, describes testing that exposes panels of FR fabrics to electric arc flashes of varying energies. The tests measure both the heat transmission through the fabric and the energy released by the electric arc. The data is evaluated against the Stoll Curve (or second-degree burn curve) through logistic regression techniques to determine the probability of burn injury. This data enables the researcher to determine the arc rating of the fabric or fabric system.

An Option to Simplify *NFPA 70E* Compliance

Many companies have decided to simplify *NFPA 70E* compliance by implementing everyday uniform programs using garments that meet the requirements of *NFPA 70E* HRC 0, 1, and 2 as a single-layer. To supplement everyday uniforms, arc flash suits and hoods in double-layer combinations are available for higher energy HRC 3 and 4 level tasks. *NFPA 70E* Annex H Simplified; Two Category, Flame Resistant (FR) Clothing System provides information on this.

Continued from page 266

- Protective items may not be used if they are contaminated with grease, oil, flammable liquids, combustible materials, or damaged to the extent their protective qualities are reduced.
- The FR apparel manufacturer's instructions for cleaning, care, and maintenance must be followed to retain protective qualities.
- The same FR materials that were used to manufacture FR clothing must be used to make repairs to it.
- ASTM F1506, *Standard Performance Specification for Textile Material for Wearing Apparel for Use by Electrical Workers Exposed to Momentary Electric Arc and Related Thermal Hazards*, must be followed for affixing trim, name tags, and/or logos to FR clothing.

HRC Classifications to Select PPE

Appropriate PPE is selected by the requirements of *NFPA 70E* Section 130.7(C) once the incident energy is determined per Section 130.3(B)(1) or the HRC is determined as permitted per Section 130.3(B)(2). The HRC method is used for the selection of personal and other protective equipment in conjunction with the requirements of *NFPA 70E* Sections and Tables 130.7(C)(9), 130.7(C)(10), and 130.7(C)(11).

HRC Classifications and Use of Rubber Insulating Gloves and Insulated and Insulating Hand Tools

NFPA 70E Section 130.3(B)(2) permits personal and other protective equipment to be selected based on the requirements of Sections 130.7(C)(9), 130.7(C)(10), and 130.7(C)(11).

NFPA 70E Sections 130.7(C)(9), 130.7(C)(10), and 130.7(C)(11) addresses the selection of personal and other protective equipment when an incident energy analysis is not performed. Table 130.7(C)(9) is used to determine the HRC and the need for other protective equipment including rubber insulating gloves and insulated and insulating hand tools for a task performed for a particular category of energized equipment in lieu of the Incident Energy Analysis of Section 130.3(B)(1). Table 130.7(C)(10) is used to determine the PPE for the task based on the HRC determined from Table 130.7(C)(9).

The available short-circuit current and maximum fault clearing time of the overcurrent protective device (OCPD) must be known in order to use Table 130.7(C)(9) and the arc flash hazard analysis must take into consideration the design of the OCPD and its opening time, including its condition of maintenance. The assumed maximum short-circuit current capacities and maximum fault clearing times for various tasks identified in Table 130.7(C)(9) are listed in the notes to that table.

Table 130.7(C)(9) cannot be used if all of the following conditions are not met:

70E Highlights

NFPA 70E Section 130.3, in part
The arc flash hazard analysis shall take into consideration the design of the overcurrent protective device and its opening time, including its condition of maintenance.

- The task must be listed for the particular category of equipment in Table 130.7(C)(9).
- The maximum short-circuit current capacity identified in the notes to Table 130.7(C)(9) must not be exceeded.
- The maximum fault clearing times identified in the notes to Table 130.7(C)(9) must not be exceeded.

Table 14.1 shows a portion of *NFPA 70E* Table 130.7(C)(9). The title of Table 130.7(C)(9) is "Hazard/Risk Category Classifications and Use of Rubber Insulating Gloves and Insulated and Insulating Hand Tools." That table identifies a number of tasks performed on energized equipment where the Hazard/Risk Category Classifications (HRC) and the need for

Table 14.1	*NFPA 70E* Table 130.7(C)(9), in part. Hazard/Risk Category Classifications and Use of Rubber Insulating Gloves and Insulated and Insulating Hand Tools

Tasks Performed on Energized Equipment	Hazard/Risk Category	Rubber Insulating Gloves	Insulated and Insulating Hand Tools
Panelboards or Other Equipment Rated 240 V and Below—Note 1			
Perform infrared thermography and other non-contact inspections outside the restricted approach boundary	0	N	N

General Notes (applicable to the entire table):
(a) Rubber Insulating Gloves are gloves rated for the maximum line-to-line voltage upon which work will be done.
(b) Insulated and Insulating Hand Tools are tools rated and tested for the maximum line-to-line voltage upon which work will be done, and are manufactured and tested in accordance with the ASTM F1505, *Standard Specification for Insulated and Insulating Hand Tools*.
(c) Y = yes (required), N = no (not required).
(d) For systems rated less than 1000 volts, the fault currents and upstream protective device clearing times are based on an 18 in. working distance.
(e) For systems rated 1 kV and greater, the Hazard/Risk Categories are based on a 36 in. working distance.
(f) For equipment protected by upstream current limiting fuses with arcing fault current in their current limiting range ($^1/_2$ cycle fault clearing time or less), the hazard/risk category required may be reduced by one number.
Specific Notes (as referenced in the table):
1. Maximum of 25 kA short circuit current available, and maximum of 0.03 second (2 cycle) fault clearing time.
2. Maximum of 65 kA short circuit current available, and maximum of 0.03 second (2 cycle) fault clearing time.
3. Maximum of 42 kA short circuit current available, and maximum of 0.33 second (20 cycle) fault clearing time.
4. Maximum of 35 kA short circuit current available, and maximum of up to 0.5 second (30 cycle) fault clearing time.

Reproduced with permission from *NFPA 70E*®-2009, *Electrical Safety in the Workplace*, Copyright © 2008, National Fire Protection Association. This reprinted material constitutes only a portion of the referenced table, is presented for educational purposes only, and is not the complete and official permission of the NFPA on the referenced subject, which is represented only by the standard in its entirety.

rubber insulating gloves and insulated and insulating hand tools can be determined within the maximum short-circuit current and the maximum fault clearing time outlined in the "specific notes" to the table.

Let's review an example. Can Table 130.7(C)(9) be used to determine the HRC and the need for rubber insulating gloves and insulated and insulating hand tools to perform infrared thermography outside the restricted approach boundary in an energized 240-volt panelboard where there is 18,000 amperes of short-circuit current available at the panelboard and the properly maintained upstream OCPD has a clearing time of one cycle? First, review the requirements for using Table 130.7(C)(9). In accordance with the provisions detailed in Section 130.7(C)(9):

- The task must be listed for the particular category of equipment in Table 130.7(C)(9).
- The maximum short-circuit current capacity identified in the notes to Table 130.7(C)(9) must not be exceeded.
- The maximum fault clearing times identified in the notes to Table 130.7(C)(9) must not be exceeded.

The category of equipment in this example is "Panelboards or Other Equipment Rated 240 Volts and Below." This category of equipment has "Note 1" associated with it as the short-circuit and fault clearing time parameters in the specific notes to the table. Note 1 restricts the use of Table 130.7(C)(9) to a maximum of 25 kiloamperes short-circuit current available and maximum two-cycle fault clearing time.

- Performing infrared thermography outside the restricted approach boundary in an energized 240-volt panelboard is a task listed in Table 130.7(C)(9).
- The available short-circuit current available is 18,000 amperes. The maximum short-circuit current capacity of 25,000 amperes identified in the Note 1 to Table 130.7(C)(9) is not exceeded.
- The upstream OCPD has a clearing time of one cycle. The maximum fault clearing time of two cycles identified in Note 1 to Table 130.7(C)(9) is not exceeded.

The answer to this example is yes, Table 130.7(C)(9) can be used in this case to determine the HRC and the need for rubber insulating gloves and insulated and insulating hand tools to perform infrared thermography outside the restricted approach boundary in an energized 240-volt panelboard.

Per Table 14.1, the HRC is 0. An "N" in the columns for "Rubber Insulating Gloves" and "Insulated and Insulating Hand Tools" indicate that rubber insulating gloves and insulated and insulating hand tools are not required to perform this task [per Note (c) to Table 130.7(C)(9)]. Table 130.7(C)(10) is used to determine the PPE for HRC 0.

The required level of PPE (HRC) must be marked on a label on the equipment to meet the requirements of *NFPA 70E* Section 130.3(C), and the HRC and protective equipment must be documented on the Energized Electrical Work Permit to meet *NFPA 70E* Section 130.1(B)(2).

> 130.1(B) Energized Electrical Work Permit.
>
> (2) Elements of Work Permit. The energized electrical work permit shall include, but not be limited to, the following items:
>
> (6) Results of the arc flash hazard analysis:

Three FPNs to Table 130.7(C)(9) offer important information to help understand the origin and protection of Table 130.7(C)(9). The first FPN discusses the origin of the tasks and protection contained in Table 130.7(C)(9). The protection indicated is generally based on determination of estimated exposure levels and, in most cases, working with the door to the equipment closed does not eliminate the need for PPE if a worker is interacting with the equipment. The second FPN cautions that both larger and smaller available short-circuit currents could result in higher available arc flash energies. This is covered in detail in Chapter 12. The third FPN addresses circuits operating at less than 50 volts and cautions that these circuits may need to be deenergized to reduce or eliminate exposure to electrical burns or to explosion from an electric arc.

> FPN No. 1 The work tasks and protective equipment identified in Table 130.7(C)(9) were identified by a task group and the protective clothing and equipment selected was based on the collective experience of the task group. The protective clothing and equipment is generally based on determination of estimated exposure levels.
>
> In several cases where the risk of an arc flash incident is considered low, very low, or extremely low by the task group, the hazard/risk category number has been reduced by 1, 2 or 3 numbers, respectively. The collective experience of the task group is that in most cases closed doors do not provide enough protection to eliminate the need for PPE for instances where the state of the equipment is known to readily change (e.g., doors open or closed, rack in or rack out). The premise used by the Task Group is considered to be reasonable, based on the consensus judgment of the full NFPA 70E Technical Committee.
>
> FPN No. 2: Both larger and smaller available short-circuit currents could result in higher available arc-flash energies.

If the available short-circuit current increases without a decrease in the opening time of the overcurrent protective device, the arc-flash energy will increase. If the available short-circuit current decreases, resulting in a longer opening time for the overcurrent protective device, arc-flash energies could also increase.

FPN No. 3: Energized electrical conductors or circuit parts that operate at less than 50 volts may need to be de-energized to satisfy an "electrically safe work condition." Consideration should be given to the capacity of the source, any overcurrent protection between the energy source and the worker, and whether the work task related to the source operating at less than 50 volts increases exposure to electrical burns or to explosion from an electric arc.

Protective Clothing and PPE Selection

NFPA 70E Section 130.7(C)(10) requires that Table 130.7(C)(10) be used to determine the PPE for working inside the AFPB for the HRC once the HRC and the need for rubber insulating gloves and insulated and insulating hand tools has been identified from Table 130.7(C)(9).

Table 130.7(C)(10) lists the protective clothing and other protective equipment required based on the HRC determined from Table 130.7(C)(9).

Using Table 130.7(C)(10), let's work through the following example. Determine the protective equipment required to perform infrared thermography outside the restricted approach boundary in an energized 240-volt panelboard where there is 18,000 amperes of short-circuit current available at the panelboard and the properly maintained upstream OCPD has a clearing time of one cycle.

Table 130.7(C)(9) indicates that the HRC is 0 and that rubber insulating gloves and insulated and insulating hand tools are not required to perform this task. Table 130.7(C)(10) is used to determine the PPE for HRC 0. The PPE is required to be documented on the Energized Electrical Work Permit to meet *NFPA 70E* Section 130.1(B)(2)(8).

130.1(B) Energized Electrical Work Permit.

(2) Elements of Work Permit. The energized electrical work permit shall include, but not be limited to, the following items:

(8) The necessary personal protective equipment to safely perform the assigned task

Per Table 130.7(C)(10), the necessary PPE to safely perform the assigned task is:

Protective Clothing, Nonmelting (according to ASTM F 1506-00) or Untreated Natural Fiber:

- Long sleeve shirt
- Long pants

FR Protective Equipment:

- Safety glasses or safety goggles
- Hearing protection (ear canal inserts)
- Leather gloves (AN) (Note 2)

AN = As needed (optional)

Note 2. If rubber insulating gloves with leather protectors are required by Table 130.7(C)(9), additional leather or arc-rated gloves are not required. The combination of rubber insulating gloves with leather protectors satisfies the arc flash protection requirement.

NFPA 70E Section 130.7(C)(10) FPN No. 2 warns that the PPE requirements of Section 130.7(C)(10) do not address protection against physical trauma other than exposure to the thermal effects of an arc flash and are intended to protect a person from arc flash and shock hazards only. Some incidents could result in burns to the skin, even with the protection indicated in Table 130.7(C)(10). Physical trauma injuries could occur due to the explosive effect of some arc events. However, burn injury should be reduced and survivable.

Protective Clothing Characteristics

NFPA 70E Section 130.7(C)(11) requires that the FR clothing selected from Table130.7(C)(10) have an **arc rating** of at least the value listed in Table 130.7(C)(11). The arc rating for a particular garment is the arc rating assigned to that garment while the arc rating for a particular clothing system is obtained from the FR clothing manufacturer. Arc rating is in cal/cm^2 by definition.

> **Arc Rating.** The value attributed to materials that describes their performance to exposure to an electrical arc discharge. The arc rating is expressed in cal/cm^2 and is derived from the determined value of the thermal performance value (ATPV) or energy of breakopen threshold (E_{BT}) (should a material system exhibit a breakopen response below the ATPV value) derived from the determined value of ATPV or E_{BT}. FPN: *Breakopen* is a material response evidenced by the formation of one or more holes in the innermost layer of flame-resistant material that would allow flame to pass through the material. [*NFPA 70E*-2009, Article 100]

The note to Table 130.7(C)(11) indicates that arc rating can be either ATPV or E_{BT}. Arc rating is reported as either ATPV or E_{BT}, whichever is the lower value. Both are defined terms in ASTM F1959, *Standard Test Method for Determining the Arc Thermal Performance Value of Materials for Clothing.*

> ATPV is the incident energy on a material or a multilayer system of materials that results in a 50% probability that sufficient heat transfer through the tested specimen is predicted to cause the onset of a second degree skin burn injury based on the Stoll curve, cal/cm^2.

Table 14.2   *NFPA 70E* Table 130.7(C)(11), in part. Protective Clothing Characteristics

Hazard/Risk Category	Clothing Description	Required Minimum Arc Rating of PPE [J/cm²(cal/cm²)]
0	Nonmelting, flammable materials (i.e., untreated cotton, wool, rayon, or silk, or blends of these materials) with a fabric weight at least 4.5 oz/yd²	N/A

E_{BT} is the incident energy on a material or material system that results in a 50% probability of breakopen.

NFPA 70E Table 130.7(C)(11) identifies the HRC, examples of protective clothing and clothing systems, and the minimum arc rating required for a particular HRC in both J/cm² and cal/cm².

Using , determine the arc rating associated with HRC 0 from Table 130.7(C)(11).

- There is no arc rating associated with HRC 0.

Summary

There are generally two methods to determine the PPE that workers need to wear once they cross the arc flash boundary; namely an incident energy analysis and the HRCs.

Arc-rated FR clothing and other PPE is selected based on the incident energy exposure in accordance with the requirements of *NFPA 70E* Sections 130.7(C)(1) through (8) and 130.7(C)(12) through (16) unless the HRC method is used as permitted by Section 130.3(B)(2).

The HRC method is used for the selection of personal and other protective equipment in conjunction with the requirements of *NFPA 70E* Sections and Tables 130.7(C)(9), 130.7(C)(10), and 130.7(C)(11). PPE must conform to the standards given in Table 130.7(C)(8).

Lessons *learned*

Electrician's Helper Electrocuted

A 21-year-old electrician's helper was electrocuted after contacting the exposed 480-volt bus that supplies power to a movable overhead crane. The incident occurred while the worker was running cables for surveillance cameras at a factory.

The employer is a small electrical contractor who has been in business since 1970. The company employed four people at the time of the incident, including the owner, two electrical helpers, and an office worker. The victim was a 21-year-old male who had been employed by the company for 8 months. He had completed his apprenticeship and had previous experience as an electrician's helper.

The factory had hired the electrical contractor to install video surveillance cameras in different areas of the plant. On the day of the incident, the first day of the contracted work at the factory, the owner of the contracting company walked with his two helpers (the victim and his coworker) through the plant to explain the job. The owner states that he does this at each job in order to point out the locations of important equipment (such as breaker boxes) and to identify any safety hazards. During the walk, he showed his helpers the bus for the overhead cranes, explaining that the wire voltage was 480 volts and that they should be careful with them. After walking through the job, he left them to do the work.

The job required the victim and his coworker to wire coaxial cable over and around the ceiling supports of the factory. To reach the ceiling, the factory provided a scissor lift. The workers would raise the platform, pull the cables over the supports, and move the lift forward as needed. The job also required installing the cables over three overhead cranes. These cranes are designed to move approximately 15 ft overhead along rails, drawing power from the exposed 3-phase 480-volt bus. As the crane moves forward, brushes on the crane make contact with the exposed bus conductors, supplying power to the crane. The bus conductors are mounted to the side of a large steel I-beam that also serves to support the crane.

Throughout the day, the two electrical helpers labored to install the cable over the supports and overhead cranes. When they reached the first two cranes, they ran the cables over the energized bus without incident. At about 4:30 p.m., they reached the third crane and positioned the scissor lift directly under the I-beam supporting the bus. After raising the lift up to the I-beam, the helpers stood at opposite ends of the lift platform to install the cable over the beam. As they ran the cables, the victim warned his coworker about the 480 volts they were working near.

Due to the differing heights of the roof on each side of the beam, the two were unable to see each other as they worked. The coworker stated that he was attempting to pass the cables over the I-beam to the victim when he heard a bang. The bang was apparently caused by the victim contacting the bus conductors. The coworker then called to the victim two or three times before he saw him fall flat on his back onto the lift platform. At this point the lift was lowered and the emergency medical service (EMS) was notified.

The police arrived a few minutes later and attended to the victim, with the assistance of factory first-aid personnel. The victim, who was breathing and had a weak pulse, went into cardiac arrest at the scene. The police immediately began cardiopulmonary resuscitation (CPR) on the victim until the EMS arrived. The EMS continued CPR and transported the victim to the local hospital emergency room where he was declared dead.

The victim apparently climbed onto the safety railing of the platform in order to reach over the I-beam, due to the differing heights of the roof. Burn marks on his chest and elbow indicate that he might have leaned or fallen onto the bus conductors while reaching for the coaxial cables.

The cause of death was attributed to electrocution. The medical examiner's report stated that there were electrical burns on the chest and right elbow of the victim's body.

Recommendations/Discussion

Recommendation 1: Employers must ensure that employees deenergize electrical systems before performing any work near them. Employers should also ensure that employees implement lockout/tagout procedures and test the system to verify that it has been deenergized before beginning work.

Discussion: In this incident, the helper was electrocuted after taking the unnecessary risk of working near energized wires. This is a violation of the federal OSHA standard 29 CFR 1926.416(a)(l), which prohibits employees from working in the proximity of energized power circuits unless the circuit is deenergized or guarded. It is imperative that employers identify all potential electrical hazards and, if possible, deenergize circuits before working on or near them. After deenergizing, a lockout/tagout procedure should be used by the workers to ensure that electrical systems are not inadvertently reenergized while they are working on it. Finally, all circuits should be tested to verify that they are deenergized.

Recommendation 2: Employers should provide and enforce the use of PPE to protect employees from electrical hazards. Guarding and shielding should also be used to prevent contact with energized conductors.

Discussion: In this case, the helpers were not issued or used any type of electrical PPE. In situations where workers might come in contact with energized conductors, the employer should require the use of PPE such as insulating gloves, aprons, and sleeves. Guarding and shielding equipment (such as insulating blankets and line hoses) might also prevent inadvertent contact with energized circuits.

Recommendation 3: Employers should develop, implement, and enforce a comprehensive safety program that includes worker training in avoiding electrical and other safety hazards.

Discussion: Although the buses feeding the cranes had been identified as a hazard, it appears that the helpers become complacent after installing cables over the first two cranes, leading them to become careless with the third. In addition, it appears that the victim misused the lift by climbing up onto the safety railings in order to reach over the beam. The employer should institute a comprehensive safety training program to reinforce appropriate work practices. This program should also include training for the appropriate use of special equipment, such as the scissor lift that was provided by the site owner.

Recommendation 4: Employers of electrical workers should ensure that all workers are trained in basic CPR.

Discussion: One of the most dangerous effects of electric shock is disruption of the natural heart rhythms that can lead to cardiac arrest and death. It is generally recommended that the employers of electrical workers should train their employees in CPR. The timely use of CPR is the only effective first-aid treatment for cardiac and respiratory arrest pending the arrival of advanced life-support personnel.

Adapted from: New Jersey Case Report: 91NJ009 (formerly NJ9106). Accessed 9/7/08 at http://www.cdc.gov/niosh/face/stateface/nj/91nj009.html.

Questions

1. When the contractor owner walked around with his two employees explaining the job and the hazards, what did he not do that would have avoided the electrocution?
   A. He could have provided written authorization to do the job on energized circuits.
   B. He could have parked his pickup truck closer to the cafeteria.
   C. He should have told the workers to lock out and tag out the circuit supplying the cranes.
   D. He could have joined the workers in doing the work.

2. What could the victim have done to avoid contacting the exposed energized conductor?
   A. He could have installed rubber blankets to cover the exposed bus.
   B. He could have moved the crane to a position in which it supported his body weight while he pulled the coaxial cable.
   C. He could have installed rubber blankets under the wheels of the scissor lift.
   D. He could have worn an arc flash suit to do the work.

3. Why did the electrician fail to measure the voltage on the bus supplying the cranes?
   A. He did not have a voltmeter.
   B. The owner of the facility did not permit contractor employees to bring their own voltmeters onto the site.
   C. Because his supervisor had told him that the cranes were energized, he had no need to measure the voltage.
   D. The electrician thought his coworker had checked the voltage.

Ready for Review

- *NFPA 70E* Section 130(C) covers PPE requirements and is the main topic of this chapter.
- Highlights of subsections of *NFPA 70E* Section 130.7 include:
  - Section 130.7(A) requires workers to be provided with and use protective equipment that is designed for the use and adequate for the hazard when performing tasks where electrical hazards are present.
  - Section 130.7(B) requires PPE to be maintained, inspected, and stored properly.
- It is important to note that workers can still sustain injury even when using protective equipment; when used appropriately, PPE can reduce the extent of injury but does not eliminate the possibility of injury. It does not protect against physical trauma other than exposure to the thermal effects of an arc flash and is intended to protect a person from arc flash and shock hazards only.
- Two methods may be used to determine the appropriate PPE: determining incident energy and determining the HRCs. Appropriate PPE is selected by the requirements of *NFPA 70E* Section 130.7(C).
- *NFPA 70E* Section 130.7(C)(2) requires that FR clothing cover all ignitable clothing and allow for movement and visibility.
- *NFPA 70E* Section 130.7(C)(3) details head, face, neck, and chin protection and when it must be worn.
- *NFPA 70E* Section 130.7(C)(4) requires employees to wear protective equipment for the eyes whenever there is danger of injury from electric arcs, flashes, or from flying objects resulting from electrical explosion. Eye protection is required in conjunction with face protection.
- *NFPA 70E* Section 130.7(C)(5), body protection, requires workers to wear FR clothing wherever there is possible exposure to an incident energy level above 1.2 cal/cm^2.
- *NFPA 70E* Section 130.7(C)(6) covers hand and arm protection requirements.
  - Employees must wear rubber insulating gloves with leather protectors when there is a danger of hand and/or arm injury from electric shock. Rubber insulating gloves must be rated for the voltage to which the gloves will be exposed.
  - Leather or FR gloves must be worn to protect hands/arms when there is possible exposure to arc flash.
  - Arms must be protected by FR clothing where there is possible exposure to an electric arc flash above 1.2 cal/cm^2.
- *NFPA 70E* Section 130.7(C)(6)(c) details the maintenance and use requirements for PPE, and requires periodic electrical tests. Test voltages and the maximum intervals between tests shall be in accordance with Table 130.7(C)(6)(c).
- *NFPA 70E* Section 130.7(C)(7) details general requirements for foot protection. Dielectric overshoes are required where insulated footwear is used as protection against step and touch potential.
- PPE must conform to the standards provided in Table 130.7(C)(8).
- Note that *NFPA 70E* handles PPE and other protective equipment separately. Do not confuse the PPE standards in Table 130.7(C)(8) with the other protective equipment standards of Table 130.7(F), such as rubber insulating blankets, line hose, and insulated hand tools.
- *NFPA 70E* Section 130.7(C)(12) details factors to consider when selecting protective clothing and equipment. Clothing and equipment must be used to provide protection for the degree of shock and arc flash hazard exposure.
- If FR clothing is required, it must cover associated parts of the body and all flammable apparel. It must allow for movement and visibility and be maintained in a sanitary and functionally effective condition.
- Subsections (a), (b), and (c) of *NFPA 70E* Section 130.7(C)(12) cover requirements for layering (underlayers and outer layers) and prohibited fibers/fabrics.
- Subsections (d), (e), and (f) of *NFPA 70E* Section 130.7(C)(12) discuss coverage, fit, and interfer-

ence. Clothing must cover potentially exposed areas as completely as possible.

- *NFPA 70E* Section 130.7(C)(13) details the requirements for four types of arc flash protective equipment: Arc flash suits, face protection, hand protection, and foot protection.
- Arc flash suits must be able to be easily and rapidly removed. Arc flash suits and face shields must both have an arc rating that is suitable for the arc flash exposure.
- Eye protection must always be worn under face shields.
- Either leather gloves or FR gloves must be worn where the incident energy analysis of Section 130.3(B)(1) determines an incident energy exposure greater than 1.2 cal/cm^2.
- *NFPA 70E* Section 130.7(C)(13)(d) details the foot protection required to protect against arc flash. Heavy-duty leather work shoes are required for tasks identified as HRC 2 and higher in Table 130.7(C)(9) and for incident energy exposures greater than 4 cal/cm^2 as determined by the incident energy analysis per Section 130.3(B)(1).
- *NFPA 70E* Section 130.7(C)(14) details clothing material characteristics.
- Flame resistance is the characteristic of a fabric that causes it to self-extinguish when the source of ignition is removed. The majority of severe and fatal burn injuries are due to the individual's clothing igniting and continuing to burn, not to the exposure itself. Thus the use of FR clothing is critical. The level of protection typically rests in the fabric weight and composition.
- Everyday work clothes made from regular cotton or poly/cotton fabrics can be readily ignited at exposure levels as low as 4 to 5 cal/cm^2, and once ignited, continue to burn, adding to the extent of injury sustained from the arc alone. Non-FR 100% cotton will not melt and drip, but burns hotter than poly/cotton fabrics and can burn longer.
- ASTM standards establish performance specifications for electrical worker apparel and testing methods for determining the arc rating of materials for clothing.

- *NFPA 70E* requires employees to wear FR clothing that meets the requirements of ASTM F1506 and is appropriate to the potential energy of the hazard where exposed to an arc flash hazard.
- FR clothing must meet the requirements described in Section 130.7(C)(14) and (15). Section 130.7(C)(14) generally prohibits clothing made from flammable synthetic materials (either alone or in blends) that melt at temperatures below 600°F (316°C).
- Many companies have decided to simplify *NFPA 70E* compliance by implementing everyday uniform programs using garments that meet the requirements of *NFPA 70E* HRC 0, 1, and 2. To supplement everyday uniforms, arc flash suits and hoods in double-layer combinations are available for higher energy HRC 3 and 4 level tasks. *NFPA 70E* Annex H, Simplified; Two Category, Flame Resistant (FR) Clothing System, provides information on this.

Vocabulary

arc rating* The value attributed to materials that describes their performance to exposure to an electrical arc discharge. The arc rating is expressed in cal/cm^2 and is derived from the determined value of the thermal performance value (ATPV) or energy of breakopen threshold (E_{BT}) (should a material system exhibit a breakopen response below the ATPV value) derived from the determined value of ATPV or E_{BT}.

> FPN: *Breakopen* is a material response evidenced by the formation of one or more holes in the innermost layer of flame-resistant material that would allow flame to pass through the material.

flame-resistant (FR)* The property of a material whereby combustion is prevented, terminated, or inhibited following the application of a flaming or non-flaming source of ignition, with or without subsequent removal of the ignition source.

> FPN: Flame resistance can be an inherent property of a material, or it can be imparted by a specific treatment applied to the material.

*This is the *NFPA 70E* definition.

A.
B.

Figure C15.1   **A.** Cubicle and racking handle after the incident. **B.** Circuit breaker and racking handle after the incident.

A.
B.

Figure C15.2   Bus connection/breaker primary disconnects after the incident. **A.** The inside of the cubicle. **B.** The breaker's primary disconnects.

OPENING CASE

On Sunday October 22, 2006, at 7:44 a.m., a utility experienced an arc flash in a 4.16 kilovolt air/retrofit/vacuum circuit breaker. The circuit involved fed Forced Draft Fan 3, and operations personnel had tried to rack the breaker in, unsuccessfully. Operations called an electrician, who removed the breaker and connected it to a Test Station, where it was determined that the circuit breaker had a mechanical problem and needed to be replaced. The spare breaker was connected to the Test Station and cycled several times to ensure it was working properly.

The spare breaker was then inserted into the cubicle and racked in. As the racking was taking place, the electrician heard a loud "clunk." The electrician assumed this noise was the floor trippers discharging the springs, which would be a normal event. The electrician did not verify the status of the circuit breaker. The breaker had actually closed instead of discharging the springs. Later discussions with the manufacturer revealed this situation could occur once in every 10,000 racking operations. As the electrician racked the now-closed breaker onto the bus, the circuit breaker tried to carry load current. However, since it was only partially racked in, the breaker did not have adequate contact to carry the current. The resulting arc flash blew an arc plasma ball and molten copper out of the cubicle.

This utility had received a new 40 cal/cm^2 arc flash suit the week before. The suit it replaced was an obsolete, unrated arc flash suit. Fortunately, the electrician had elected to wear the new arc flash suit to

Existing Electrical Equipment: Maintenance and Safety

CHAPTER 15

Introduction

This chapter covers several important topics. For existing electrical systems, it might be possible to improve electrical safety conditions for workers by upgrading components. The company should provide safe work practices regarding overcurrent protective devices (OCPDs) and train workers in their use. The *National Electrical Code® (NEC®)* and *NFPA 70E®* have arc flash labeling requirements, and some owners have system-wide labeling of equipment. *NFPA 70E* has requirements for maintenance of electrical equipment that are important for worker safety.

Work Practices and System Upgrades

Specific safe work practices should be used when working with existing electrical equipment. This section mentions a few related to OCPDs. Other safe work practices are covered in other chapters.

Resetting Circuit Breakers or Replacing Fuses

Do not reset a circuit breaker or replace fuses until the cause is known and rectified.

Both Occupational Safety and Health Administration (OSHA) 1910.334(b)(2) and *NFPA 70E*, 2009 Edition Section 130.6(K) state the following regulation:

OPENING CASE, CONT.

perform this task. **Figure C15.1A** and **Figure C15.1B** show the condition of the breaker and cubicle as found after the accident. Note that the door is open, even though it could have been closed during the racking operation. This could have provided some shielding for the electrician, even though it probably would have been forced open by the pressure wave (blast) of the arc. The arc flash was estimated at 33 cal/cm^2.

Figure C15.2A shows the inside of the cubicle where the breaker's primary disconnects touched the bus connections. Note the vaporized metal. **Figure C15.2B** shows the breaker's primary disconnects after contact with the energized bus.

The electrician received minor burns across the back of his neck, due to the hood being worn incorrectly. The electrician sustained no other injuries.

Source: This case study was contributed by Jim White of Shermco Industries.

1. What should have been done to prevent this incident?

2. Did the proper rated PPE save the worker's life or prevent severe injuries?

Chapter Outline

- **Introduction**
- **Work Practices and System Upgrades**
- **Maintenance: How It Relates to Electrical Safety**
- **Lessons Learned**
- **Current Knowledge**

Learning Objectives

1. Understand work practices with regard to OCPDs that can enhance worker safety.
2. Understand the arc flash hazard labeling requirements that are important for worker safety.
3. Understand that upgrades and adjustments can be made to existing equipment to improve electrical safety.
4. Understand that electrical safety is directly tied to the condition of electrical power systems equipment, especially OCPDs, that maintenance of equipment directly affects electrical safety, and the *NFPA 70E* Chapter 2 maintenance requirements.

References

1. *NFPA 70E*, 2009 Edition
2. *NFPA 70B, Recommended Practice for Electrical Equipment Maintenance*
3. American National Standards Institute/InterNational Electrical Testing Association (ANSI/NETA) MTS-07, *Maintenance Testing Specifications*

Reclosing circuits after protective device operation. After a circuit is deenergized by a circuit protective device, the circuit **may not be manually reenergized** until it has been determined that the equipment and circuit can be safely energized. The repetitive manual reclosing of circuit breakers or reenergizing circuits through replaced fuses shall be prohibited. When it can be determined from the design of the circuit and the overcurrent devices involved that the automatic operation of a device was caused by an overload rather than a fault condition, no examination of the circuit or connected equipment shall not be required before the circuit is reenergized.

This is an important safety practice. If an OCPD opened under fault conditions, damage at the point of the fault could result. If the fault is not located and rectified, reclosing into the fault again might result in an even more severe fault than the first fault. In addition, if the protective device is a circuit breaker, it could have been damaged or degraded on the initial interruption. Reclosing a degraded circuit breaker into a fault might cause the circuit breaker to fail in an unsafe manner (explode). Therefore, following proper procedures after an OCPD has interrupted a fault is important.

Circuit Breaker Evaluation

After fault interruption, circuit breakers must be evaluated for suitability before being placed back into service by a person qualified for circuit breaker evaluations. This evaluation requires visual inspection and electrical testing to specifications according to the manufacturer's procedures or industry standards.

70E Highlights

NFPA 70E requires the following with regard to circuit breakers after interrupting a fault:

225.3 Circuit Breaker Testing.

Circuit breakers that interrupt faults approaching their interrupting ratings shall be inspected and tested in accordance with the manufacturer's instructions.

After a circuit breaker interrupts a fault, it might not be suitable for further service. A fault can possibly erode the circuit breaker's contacts and the arc chutes or weaken the circuit breaker's case. If the short-circuit (fault) current is high, circuit breaker manufacturers recommend that a circuit breaker receive a thorough inspection with replacement, if necessary. Some difficulties in the evaluation process are not knowing a circuit breaker's service history, what level of short-circuit current a circuit breaker interrupted, or what degradation occurred on the inside of the circuit breaker. That is one reason why periodic maintenance by a qualified maintenance person is necessary.

The following is an insightful quote by Vince A. Baclawski, Technical Director, Power Distribution Products, National Electrical Manufacturers Association (NEMA); published in *EC&M magazine*, pp. 10, January 1995:

"After a high level fault has occurred in equipment that is properly rated and installed, it is not always clear to investigating electricians what damage has occurred inside encased equipment. The circuit breaker may well appear virtually clean while its internal condition is unknown. For such situations, the NEMA AB4 "Guidelines for Inspection and Preventive Maintenance of MCCBs Used in Commercial and Industrial Applications" may be of help. Circuit breakers unsuitable for continued service may be identified by simple inspection under these guidelines. Testing outlined in the document is another and more definite step that will help to identify circuit breakers that are not suitable for continued service."

"After the occurrence of a short circuit, it is important that the cause be investigated and repaired and that the condition of the installed equipment be investigated. A circuit breaker may require replacement just as any other switching device, wiring or electrical equipment in the circuit that has been exposed to a short circuit. Questionable circuit breakers must be replaced for continued, dependable circuit protection."

(For more information, see the discussion on circuit breaker maintenance later in this chapter.)

Fuses

There are several work practice considerations that should be evaluated when an electrical installation includes fuses. A discussion of these follows.

Testing Fuses

When a person suspects that a fuse has opened, he or she must first follow their written procedures for deenergizing the fuse from all sources of power, including lockout/tagout procedures and then remove both indicating and non-indicating fuses from the circuit and check them for continuity. Taking resistance measurements on deenergized fuses is every bit as reliable as taking voltage measurements on energized fuses, but without the potential of accidentally coming in contact with a live part or starting an arc-flash incident.

Testing Knife-Blade Fuses

It is important to properly test **knife-blade fuses**. A common mistake that workers make when testing knife-blade fuses (**Figure 15.1**) is to touch the end caps

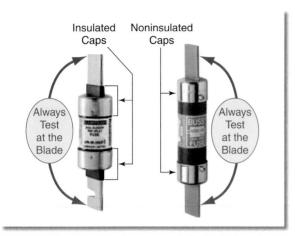

Figure 15.1 A continuity test across any knife-blade fuse should be taken only along the fuse blades. Do not test a knife-blade fuse with meter probes to the fuse caps.

of the fuse with their probes. Contrary to popular belief, fuse manufacturers do not generally design their knife-blade fuses to have electrically energized fuse caps during normal fuse operation. Electrical inclusion of the caps into the circuit occurs as a result of the coincidental mechanical contact between the fuse cap and terminal extending through it. In most brands of knife-blade fuses, this mechanical contact is not guaranteed; therefore, electrical contact is not guaranteed. Thus, a resistance reading taken across the fuse caps is not indicative of whether or not the fuse is open.

In a continuing effort to promote safer work environments, Cooper Bussmann® has introduced different versions of knife-blade Fusetron® Fuses (Class RK5) and knife-blade Low-Peak® Fuses (Class RK1) for some of the ampere ratings. The improvement is that the end caps are insulated to reduce the possibility of accidental contact with a live part. With these improved fuses, the informed worker knows that the end caps are isolated. With older style non-insulated end caps, the worker doesn't really know if the fuse is energized or not by simply taking readings at the end caps. A portion of all testing-related injuries could be avoided by proper testing procedures.

Replacing Fuses

Fuses that interrupt a circuit should be replaced with the proper fuse type and ampere rating. Modern current-limiting fuses are always recommended. When using modern current-limiting fuses, new factory calibrated fuses are installed in the circuit, and the original level of overcurrent protection is maintained for the life of the circuit. In most newer building systems and utilization equipment, the fuse mountings only accept current-

limiting fuses of a specific Underwriters Laboratories Inc. (UL) Class fuse. This is a unique fuse electrical safety system feature that ensures high interrupting rating, the proper voltage rating, and depending upon the available short-circuit current, a specific level of current limitation for arc flash hazard reduction. For older systems, where the fuse clips can accept older style fuses (Class H), it is recommended to only store and use modern current-limiting style fuses (Class RK1) that also can be used in those clips. See the section in this chapter on upgrading existing fuses.

Moving People Outside the Arc Flash Protection Boundary

Numerous injuries and deaths occur to workers who rack in/out low-voltage power circuit breakers or medium-voltage vacuum circuit breakers. For instance, if the working distance from the potential arc flash is 36 in. rather than 18 in., the potential incident energy at 36 in. is typically about one-fourth that at 18 in. The following means could move the worker farther from the potential arc source or outside the flash protection boundary for hazardous operations:

1. Remote-controlled motorized devices that rack in and out low- and medium-voltage circuit breakers
2. Extended length, hand-operated racking tools (**Figure 15.2**)

Marking Equipment with the Arc Flash Hazard

Marking or labeling electrical equipment with the arc flash hazard warning is required by *NFPA 70E* and the *NEC*. These markings are intended to reduce the occurrence of serious injury or death due to arcing faults. The warning label reminds a qualified worker that a serious hazard exists when they are exposed to an arc flash hazard if electrical equipment is not in an electrically safe work condition. Qualified workers must follow appropriate work practices and wear appropriate PPE

Figure 15.2 Racking tools of varying lengths are available to move the worker further from the potential arc source.

for the specific hazard (a nonqualified worker must not open or be near open energized equipment). The marking (labeling) systems range from only a simple warning label to more informative labels with data on the level of hazards and other pertinent information that helps ensure safe work practices.

The subject of arc flash hazard marking or labeling has several aspects, as follows:

1. *NEC* Section 110.16, requires field marking with arc flash warning label on new equipment installations.

2. *NFPA 70E* Section 130.3(C), requires marking equipment with incident energy or level of PPE when an arc flash hazard analysis is performed.

3. For new installations or existing installations, many owners mark their equipment in a facility with the shock protection boundaries, arc flash protection boundary (AFPB), and incident energy or required level of PPE.

4. Updating the arc flash hazard information if system changes occur, as required in *NFPA 70E*, Section 130.3.

These labeling aspects are discussed further in the next sections.

Arc Flash Hazard Warning Labels on New Installations

The *NEC* covers new installations and has an arc flash hazard warning labeling requirement:

> *NEC* **110.16 Flash Protection.** Electrical equipment, such as switchboards, panelboards, industrial control panels, meter socket enclosures, and motor control centers, that are in other than one- and two-family dwelling occupancies, and are likely to require examination, adjustment, servicing, or maintenance while energized shall be field marked to warn qualified persons of potential electric arc flash hazards. The marking shall be located so as to be clearly visible to qualified persons before examination, adjustment, servicing, or maintenance of the equipment.

This requirement, which was new in the *NEC* 2002, only requires that this marking be a warning label of an arc flash hazard as shown in **Figure 15.3**. It is not required to provide the shock or arc flash hazard detail information. Note, any electrical equipment as described in *NEC* Section 110.16 that is likely to be serviced or worked on while not in an electrically safe work condition must be field labeled if this equipment has been installed according to the 2002 *NEC* or later editions. The intent of this requirement is to be field marked, although many manufacturers add this basic arc flash hazard warning label on equipment prior to shipping the electrical equipment.

Figure 15.3 Arc flash warning label complying with *NEC* Section 110.16.

Marking New or Existing Equipment with the Arc Flash Hazard

The 2009 edition of *NFPA 70E* added a new labeling requirement, as follows:

> **130.3 (C) Equipment Labeling.** Equipment shall be field marked with a label containing the available incident energy or required level of PPE.

If workers are involved in work where they are exposed to an arc flash hazard, an arc flash hazard analysis is required to be conducted, according to Section 130.3, which includes the following:

1. Determination of the arc flash protection boundary, according to Section 130.3(A)

2. Determination of the incident energy or hazard risk category and associated personal and other protective equipment required, according to Section 130.3(B)

3. Field marking the equipment with the available incident energy or required level of PPE, according to Section 130.3(C).

Chapter 3 indicates that 29 Code of Federal Regulations (CFR) 1910.132(d) *Hazard Assessment and Equipment Selection*, requires that if a hazard is present, the employer must document by written certification. This OSHA regulation requires employers to assess for hazards that might require PPE, to select and have affected employees use the appropriate PPE for the hazards, and to communicate the PPE requirement to each affected employee.

> **§1910.132(d)(2)**
>
> **The employer shall verify that the required workplace hazard assessment has been performed through a written certification** that identifies the workplace evaluated; the person certifying that the evaluation has been performed; the date(s) of the hazard assessment; and, which identifies the document as a certification of hazard assessment.

The *NFPA 70E* Section 130.3(C) labeling requirement, along with the elements required in the *Energized Elec-*

trical Work Permit in *NFPA 70E* Section 130.1(B), are a means to comply with a major portion of the OSHA written certification requirement for the hazard assessment. A benefit is that a qualified worker can readily know the hazard, which helps foster safe electrical work practices. Figure 15.4 illustrates an example of such a label. Electrical equipment is not required to be labeled with the detailed arc flash hazard information until an arc flash hazard analysis is conducted. At that time, the incident energy or level of PPE shall be marked on the label and installed on the equipment subject to the arc flash hazard analysis.

Note: Not all electrical equipment is likely to be worked on while energized or poses an arc flash hazard. Therefore, in those cases, an arc flash hazard analysis and a detailed arc flash hazard marking are not necessary. Section 130.3, Exception No. 1, does not require an arc flash hazard analysis; therefore, a detailed arc flash label for 240-volt or less equipment that is on circuits fed by a transformer rated less than 125 kilovolt-amperes is not necessary. For instance, it is not likely that an arc flash hazard exists at a 15- or 20-ampere, 120-volt receptacle. In addition, *NEC* Section 110.16 and Institute of Electrical and Electronics Engineers (IEEE) 1584, *Guide for Arc-Flash Hazard Analysis*, provides some guidance on the types of equipment and system parameters where flash hazards should be calculated.

The party responsible for the marking should also include the shock hazard and other vital information on the label to enhance safe work practices. In this way, the qualified worker and his/her management can more readily assess the risk and better ensure appropriate safe work practices, PPE, and tools. The example label in Figure 15.4 includes more of the vital information that fosters safe work practices.

Specific additional information that could be added to the label includes the following:
- Voltage level
- Limited approach boundary
- Restricted approach boundary
- Prohibited approach boundary
- AFPB
- Incident energy expressed in cal/cm^2 at a working distance in inches from arc source and/or level of PPE (Hazard/Risk Category)
- PPE required
- Analysis date
- Clearing time of the upstream OCPD
- Available three-phase, short-circuit current

Arc Flash Hazard and Shock Hazard Labeling for Entire Premise

Many owners are conducting arc flash hazard analysis and shock hazard analysis for their entire premises and labeling the electrical equipment. The labels are similar to those in Figure 15.4. Premise-wide implementation of labeling can facilitate a more efficient operation, especially when qualified workers are doing routine tasks, such as checking for the absence of voltage where an energized work permit is not required. The typical process is to update or create a complete and accurate single-line diagram of the electrical facilities, calculate the short-circuit currents throughout the premise, calculate the arc flash incident energies throughout the premises, and affix labels on the electrical equipment. Chapter 12 covers arc flash hazard analysis in greater depth.

Updating Arc Flash Hazard Information

Labels or other documentation records that include the specific arc flash hazard level data such as available incident energy or level of PPE must be updated when changes occur to the electrical system. This practice is required in *NFPA 70E*:

130.3 Arc Flash Hazard Analysis, in part.

The arc flash hazard analysis shall be updated when a major modification or renovation takes place. It shall be reviewed periodically, not to exceed five years to account for changes in the electrical distribution system that could affect the results of the arc flash hazard analysis.

Many changes can alter the available short-circuit current (as shown in Chapter 11) and/or alter the time

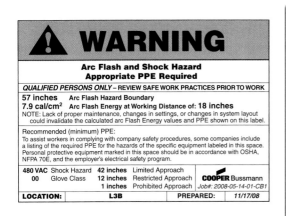

Figure 15.4 An example of an arc flash label that provides vital information on the level of arc flash hazard as well as other information important for safe work practices. The selection of PPE based on *NFPA 70E* Section 130.7(C) can vary based on the facility safety program. In addition, for specific types of work, other protective equipment needed must be in accordance with *NFPA 70E* Section 130.7(D). See Chapter 16 for an in-depth discussion on PPE requirements.

duration for fuses or circuit breakers to interrupt the arcing current. These changes may affect the available arc flash incident energy and required level of PPE for a specific piece of electrical equipment. Some changes that can affect the arc flash hazard level:

1. If a transformer upstream in the electrical system is replaced with a different kilovolt-ampere rating or impedance, the available short-circuit current could substantially change.

2. If a fuse or circuit breaker type, ampere rating, or overcurrent protection setting is changed, the duration of the arcing current under arc fault conditions could be substantially different.

3. If a circuit's conductor length is changed to a substantially longer or shorter distance, the available short-circuit current will be altered.

4. If a circuit's conductor size is changed, such as due to voltage drop considerations, the available short-circuit current will be altered.

5. If generators are added to supply part of the system load, the available short-circuit current might change.

6. If primary or secondary distribution systems are modified or "beefed up," the available short-circuit current will likely change.

Therefore, it is recommended that employers include the date of the analysis on the label and retain all the pertinent data for the analysis in a file. If a software program was used for the analysis, all computer data, files, and libraries for the calculated system should be retained. In this way, if an alteration does occur, determining the new available short-circuit currents, available incident energies, and level of PPE will be easier.

Evaluating Overcurrent Protective Devices in Existing Facilities for Proper Interrupting Rating

Fuses, circuit breakers, and other OCPDs must be selected based on the comparison of the interrupting rating with available short-circuit current.

Just as changes in an electrical system mentioned in the previous section can affect the arc flash hazard, changes can result in higher available short-circuit currents that can exceed the interrupting rating of existing OCPDs. For example, when a service transformer is increased in kilovolt-amperes size or the transformer is replaced with a lower impedance transformer, available short-circuit current might increase. In addition to *NEC* Section 110.9 requiring fuses and circuit breakers to have adequate interrupting rating, *NFPA 70E* Section

210.5 and OSHA require that OCPDs have adequate interrupting ratings, irrespective of the installation age of the system.

§OSHA 29 CFR 1910.303(b)(4):

Interrupting rating. Equipment intended to interrupt current at fault levels shall have an interrupting rating sufficient for the nominal circuit voltage and the current that is available at the line terminals of the equipment. Equipment intended to interrupt current at other than fault levels shall have an interrupting rating at nominal circuit voltage sufficient for the current that must be interrupted.

Whenever system changes occur in a premise or by the utility that might increase the available short-circuit currents, the existing OCPDs must be reevaluated as to whether they have sufficient interrupting rating. If a short-circuit analysis and/or an arc flash analysis is performed for an existing facility, the employer must evaluate whether all the fuses and circuit breakers have sufficient interrupting rating for the available short-circuit current at their line terminals. Replace fuses or circuit breakers that do not have an adequate interrupting rating with fuses or circuit breakers that have an adequate interrupting rating.

Interrupting rating is the maximum current that a fuse or circuit breaker can safely interrupt under standard test conditions. An OCPD that attempts to interrupt a short-circuit current beyond its interrupting rating can violently rupture. This in itself can be an arc flash and arc blast hazard, plus the violent rupturing can cause an arcing fault in other parts of the equipment. (Chapter 11 covers how to calculate the maximum available short-circuit current.)

Modern current-limiting fuses have interrupting ratings of 200,000 and 300,000 amperes, which virtually eliminates this hazard contributor. However, renewable and Class H fuses only have a 10,000-ampere interrupting rating, some Class K fuses have a 50,000-ampere interrupting rating, and Class G fuses have 100,000-ampere interrupting rating.

Circuit breakers have varying interrupting ratings, so they need to be assessed accordingly.

Figure 15.5 A–F and **Figure 15.6 A–F** show laboratory testing illustrating when fuses and circuit breakers attempt to interrupt short-circuit currents beyond their interrupting rating. This is a violation of *NEC* Section 110.9 and OSHA §1910.303 (b)(4).

Figure 15.7 A and B show a laboratory test where the fuses have an interrupting rating greater than the available short-circuit current and therefore safely interrupt the current. Note there is no violence or emitted byproducts.

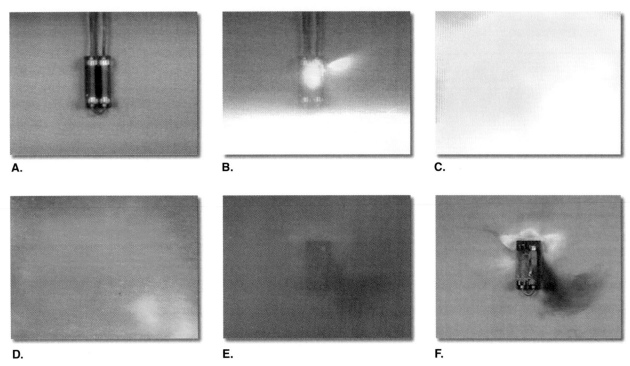

Figure 15.5 This figure illustrates what happens when Class H Fuses, which have an interrupting rating of only 10,000 amperes, are subjected to a 50,000-ampere fault. Obviously, this is a misapplication, but this emphasizes how important proper interrupting rating is for arc flash protection and proper application of OCPDs. In a fraction of a second the fuses can violently rupture.

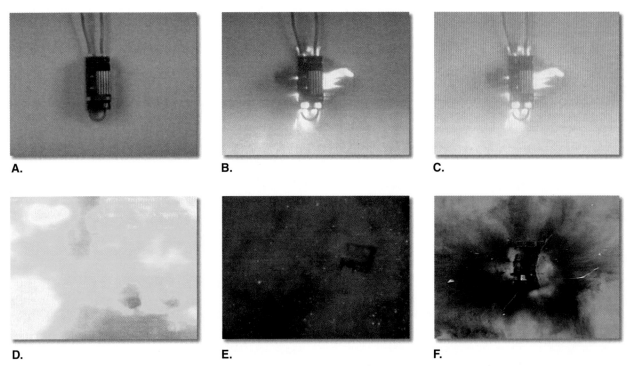

Figure 15.6 This figure illustrates what happens when a circuit breaker with an interrupting rating of 14,000 amperes is subjected to a 50,000-ampere fault. This is a misapplication, but illustrates the sudden violence that occurs. In a fraction of a second the circuit breaker violently ruptured.

 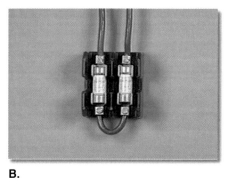

A. **B.**

Figure 15.7 This figure illustrates Class J, Low-Peak LPJ fuses, safely interrupting a 50,000-ampere available short-circuit current. The LPJ fuses have an interrupting rating of 300,000 amperes. **A.** The fuses before the test. **B.** The fuses during and after the test. This is a safe interruption of a short-circuit current.

Reduce Arc Flash Hazard for Existing Fusible Systems

If the electrical system is an existing fusible system, consider replacing or upgrading the existing fuses with fuses that are more current limiting. This can reduce the arc flash hazard.

Owners of existing fusible systems should consider upgrading Class H, K5, K9, and RK5 fuses to Class RK1 fuses and verify that Class J and Class L fuses are the most current limiting available. An assessment of many facilities can uncover that the installed fuse types are not the most current limiting, or that fuses were installed decades ago and new fuses with better current limitation are now available.

Figure 15.8 pictorially illustrates this concept for replacing all existing Class H, Class K, and Class RK5 fuses with Class RK1 fuses.

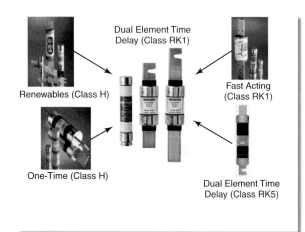

Dual Element Time
Delay (Class RK1)

Renewables (Class H)

One-Time (Class H)

Fast Acting
(Class RK1)

Dual Element Time
Delay (Class RK5)

Figure 15.8 Upgrade to Class RK1 fuses.

Adjusting the Instantaneous Trip Setting on Existing Circuit Breakers

Some circuit breakers have adjustable instantaneous trip settings, allowing adjusting the setting at which it will cause the circuit breaker to operate in its instantaneous mode for fault conditions. This adjustment is used to improve coordination. However, if the instantaneous trip is set too high, it might not operate in its instantaneous mode during an arcing fault. The result is a longer opening time, and therefore a much higher level of energy release during an arcing fault is possible. An engineer should assess whether the instantaneous trip settings should be adjusted to be as low as possible without incurring nuisance tripping. One consideration regarding whether to lower a circuit breaker's instantaneous trip setting is the effect this will have on selective coordination of the OCPDs. Selective coordination is required for some circuits supplying life-safety loads and desired for some circuits supplying mission-critical loads and general purpose loads.

Sizing Under-Utilized Circuits with Lower Ampere Rated Fuses or Circuit Breakers

In cases in which the circuit ampacity is significantly larger than actually necessary, the actual current under the maximum load conditions should be measured, and the most current-limiting fuses should be sized for the load. For instance, if an 800-ampere feeder to a motor control center draws only 320 amperes, 400-ampere current-limiting fuses could be considered. For circuit breakers that incorporate ampere-rating plugs, some benefit might be achieved by lowering the plug rating. A lower circuit breaker plug rating can lower the arc flash hazard for lower level arcing fault currents. Downsiz-

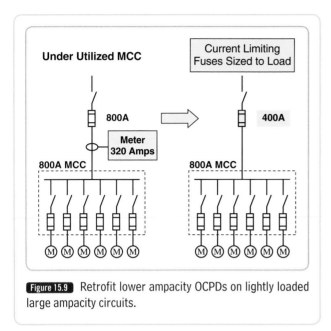

Figure 15.9 Retrofit lower ampacity OCPDs on lightly loaded large ampacity circuits.

ing fuses or circuit breakers is a strategy that can be utilized on some new systems or retrofitting existing systems (**Figure 15.9**).

Relaying Schemes

Protective relays schemes can be used on new installations or existing systems to mitigate the arc flash hazard. When an arc flash event occurs that causes the relay to call for the circuit to be interrupted, the relay output signals a properly rated disconnecting means to interrupt the circuit. The disconnecting means can be a vacuum interrupter, circuit breaker, or disconnect switch with a shunt trip. The following are two examples.

1. **Current relay in combination with sensors monitoring other physical parameters**

   A current relay with current transformers (CTs) in combination with one or more sensors in an enclosure that are monitoring sound, pressure, and/or light can detect an arcing fault in an enclosure. If an arc flash event occurs within the enclosure, the sound, pressure, and light all rapidly escalate to high levels. When the current escalates as well with one or more of these other monitors reaching a trip point, the relay signals the disconnecting means to interrupt. This can be an effective method on equipment where the normal OCPDs are too large in ampacity to quickly respond to the arcing current and mitigate the arc flash hazard to an acceptable level. For instance, on a 480-volt unit substation with large ampacity main OCPD, this technology could be used to signal the medium voltage disconnecting means to interrupt.

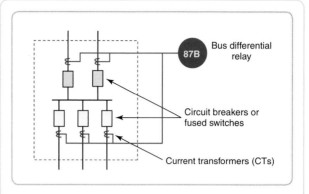

Figure 15.10 The CTs and relay are configured to sum the currents entering and subtract the sum of the currents exiting via the normal conducting paths. If a fault occurs on the bus with a current magnitude greater than the relay set point, the relay signals the disconnects to interrupt.

2. **Bus differential relays**

   A bus differential relay and CTs can be utilized to detect and signal the disconnecting means to interrupt when the current coming in via the normal paths is not equal to current leaving via the normal paths. When there are multiple large power sources supplying a bus, this approach can reduce the arc flash hazard substantially compared to other means. This can also be used for a single power source with several feeders when the normal OCPDs are too large in ampacity to quickly respond to the arcing current and mitigate the arc flash hazard to an acceptable level (**Figure 15.10**).

The following section, *Maintainenance: How it Relates to Electrical Safety*, was contributed by Jim White, Shermco Industries Training Director. Jim is a certified Level IV NETA Technician and is a technical committee member on *NFPA 70E* and *NFPA 70B* in addition to several maintenance related industry standard committees. Shermco Industries provides training, testing, repair, maintenance, and analysis of power distribution systems and related equipment. NETA (InterNational Electrical Testing Association) is an independent, third-party electrical testing association that establishes world standards in electrical maintenance and acceptance testing.

Maintenance: How It Relates to Electrical Safety

How does maintaining electrical power system equipment and devices become a worker safety issue? Many mid- and upper-level managers (especially those from a management or accounting background) would consider maintenance to be an overhead cost unrelated to worker safety. This is far from accurate and is why the 2009 edition of *NFPA 70E* in Chapter 2, *Safety-Related Maintenance Requirements,* specifically requires electrical protective devices to be maintained properly. Section 205.3 states,

> **General Maintenance Requirements.** Overcurrent protective devices shall be maintained in accordance with the manufacturers' instructions or industry consensus standards.

In addition, *NFPA 70E* Section 200.1(3) states,

> (3) For the purpose of Chapter 2, maintenance shall be defined as preserving or restoring the condition of electrical equipment and installations, or parts of either, for the safety of employees who work on, near, or with such equipment. Repair or replacement of individual portions or parts of equipment shall be permitted without requiring modification or replacement of other portions or parts that are in a safe condition.
>
> Fine Print Note (FPN): Refer to NFPA 70B, *Recommended Practice for Electrical Equipment Maintenance,* and ANSI/NETA MTS-2007, *Standard for Maintenance Testing Specification,* for guidance on maintenance frequency, methods, and tests.

Note that the above specifically calls out that maintenance is preserving or restoring the condition of electrical power systems equipment or devices *for the safety of employees who work on, near, or with such equipment.* Electrical safety is directly tied to the condition of electrical power systems equipment, especially protective devices, such a circuit breakers and protective relays. One topic of discussion that repeatedly came up during the *NFPA 70E* committee meetings was that power systems are safe if they are:

1. Properly designed and engineered
2. Properly installed (using all applicable national and local codes and standards)
3. Properly maintained

The consensus of the *NFPA 70E* committee was that if all three of these elements were not in place, no procedure or method to determine how to work safely on that system exists. If equipment and protective devices are not properly maintained, engineering analysis, such as short-circuit current studies and coordination studies, no longer have any relevance. This also includes the calculations performed as part of an arc flash hazard analysis. IEEE-1584, *NFPA 70E,* and other similar methods of determining the arc flash hazard and choosing PPE

presume properly maintained and operating electrical systems. When we encounter systems that are not, bad things can happen.

How a Lack of Maintenance Can Increase Hazards

When we or our employees operate electrical power equipment, we assume that when the handle is operated everything will work as advertised. The National Institutes of Occupational Safety and Health (NIOSH) presented a paper at the 11th Annual IEEE/IAS Electrical Safety Workshop: *Non-Contact Electric Arc-Induced Injuries in the Mining Industry: A Multi-Disciplinary Approach,* by Kathleen Kowalski-Trakofler, PhD. **Figure 15.11** and **Figure 15.12** are slides from this presentation, which illustrate the problem.

Figure 15.11 shows that 34% of electrically-related accidents are caused by component failure and 19% of those accidents are caused by components that failed during normal operation.

Figure 15.12 shows that among electrical workers, 24% of accidents are the result of troubleshooting-type activities. This makes sense, especially when the compactness of newer equipment is considered, and the fact that the conductors often have to be pushed out of the way to get test probes or tools into position. 18% of accidents were the result of failure during repair or repair-related activities. In this study, five types of electrical devices were implicated in two thirds of the cases studied:

1. Circuit breakers   17%
2. Conductors   16%
3. Non-powered hand tools   13%
4. Electrical test meters and leads   12%
5. Connectors and plugs   11%

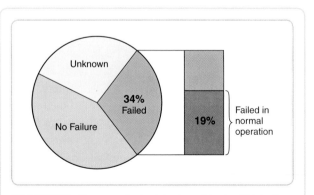

Figure 15.11 Causes of electrical accidents. Thirty-four percent resulted from component failure. Nineteen percent involved an electrical component that failed in routine operation.

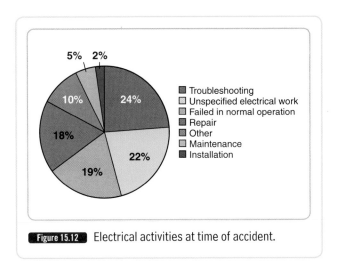

Figure 15.12 Electrical activities at time of accident.

We can be fairly certain that most of the devices that failed in this study did so due to lack of maintenance.

Lack of Lubrication

Electrical power system protective devices require regular maintenance, and when they do not receive that maintenance, dangerous situations are created for electrical workers. Lack of maintenance can take on many forms. The circuit breaker repair shop of Shermco Industries sees an astonishing number of circuit breakers and switches of all voltage ratings that suffer from a lack of lubrication. One such example is shown in **Figure 15.13**. Note that this operating mechanism has severely rusted components and spider webs hanging off it. Obviously it has been neglected for years. Many equipment owners don't realize that the conductive (current-carrying) paths of circuit breakers and many switches are lubricated when they are built and that

the lubrication dries out over time, even if the breaker (or switch) has not had to operate. They think that the device should be like brand new, because it rarely operates.

An example of this is shown in **Figure 15.14**. These are components of a circuit breaker moving contact assembly that are in the current path. These parts are normally lubricated at the factory and require periodic cleaning and relubrication in order to function properly. As the lubricant on these parts ages, it dries and becomes gummy, finally drying to the point where it flakes off the components. The rollers in Figure 15.14 show deep gouging caused by lack of lubrication, and the arcing contact assembly also shows significant wear at the point where it contacts the rollers. This breaker operated very slowly and would occasionally seize in mid-operation.

Consider the effect on operating time if this occurs. According to *NFPA 70E* Annex D, incident energy is proportional to operating time. If a circuit breaker's clearing time doubles, due to lack of maintenance, the resultant arc flash incident energy also doubles. As an example, a worker is about to service a device in a panel protected by a molded-case circuit breaker (approxi-

Figure 15.13 Would you bet your life on this working?

Figure 15.14 Lack of lubrication on contact components caused damage.

mately two-cycle clearing time) and has an incident energy of 8 cal/cm² [at 18 in.]. If the worker wears arc-rated PPE for a maximum 10 cal/cm², he or she should be protected. If that device delays opening, his/her protection might not be adequate. The operating time of OCPDs is critical to worker safety.

Maintenance Can Affect Safety

NFPA 70E Section 130.3, *Arc Flash Hazard Analysis*, requires that the AFPB and PPE be determined for people who will be within the AFPB. Section 130.3 also contains requirements for considering the condition of maintenance for OCPDs when conducting arc flash hazard analysis. The reliability of OCPDs can directly impact arc flash hazards. One portion of Section 130.3 reads as follows:

> The arc flash hazard analysis shall take into consideration the design of the overcurrent protective device and its opening time, including its condition of maintenance.

In addition, Section 130.3 refers to two FPNs concerning the importance of overcurrent protective device maintenance:

> FPN No. 1: Improper or inadequate maintenance can result in increased opening time of the overcurrent protective device, thus increasing the incident energy.

> FPN No. 2: For additional direction for performing maintenance on overcurrent protective devices see Chapter 2, Safety-Related Maintenance Requirements.

Poorly maintained OCPDs result in higher arc flash hazards. **Figure 15.15** illustrates the dangerous arc flash hazard consequences due to poorly maintained OCPDs. In this example, the panel is protected by an upstream 800-ampere circuit breaker that has a short-time delay setting of six cycles. This is an intentional delay feature of the circuit breaker in order to have this circuit breaker selectively coordinate with the circuit breakers in the panel it is supplying. Figure 15.15A shows the calculated arc flash hazard using the six-cycle opening time for the circuit breaker short-time delay. However, if the circuit breaker has not been maintained, it may take longer to interrupt an arcing fault. Figure 15.15B shows that the arc flash hazard is much higher if the circuit breaker took 30 cycles to open due to poor maintenance. Obviously, if the circuit breaker took even longer to open, the actual arc flash hazard would be even worse. This illustrates why maintenance of OCPDs is so important.

What if the ability of an OCPD to function properly is questioned? Often times, as part of the hazard/risk analysis, assuming that the OCPD will not function

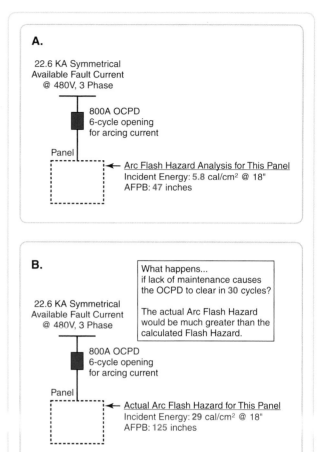

Figure 15.15 Calculated arc flash hazard versus actual hazard, and the corresponding AFPBs. **A.** Assuming the circuit breaker has been maintained and operates as specified by manufacturer's performance data, the incident energy would be 5.8 cal/cm² at 18 in. The AFPB is therefore 47 in. (calculation per IEEE 1584). **B.** If lack of maintenance causes the circuit breaker to clear in 30 cycles rather than in six cycles, the actual arc flash hazard would be an incident energy of 29 cal/cm² at 18 in., and the AFPB would be 125 in. (calculation per IEEE 1584).

properly is safer. In these cases, it is best to deenergize. The time spent in a planned outage is far less than what would be required to repair and replace damaged equipment, not to mention the possible effects on the worker.

However, if the decision is made that the circuit or equipment cannot be placed in an electrically safe work condition, the next device upstream that is deemed reliable has to be considered as the protective device that will operate. For example, since the 800-ampere OCPD in Figure 15.15 is not maintained, the next device up-

A.

B.

Figure 15.16 Damage to a switch caused by misapplied renewable fuses.

stream that is maintained properly would be the one used in determining the arc flash hazard. It is probable that, due to the increase in operating time, the incident energy will be substantially higher. If the next device upstream is not reliable, the device upstream from it would have to be used to assess the arc flash hazard.

Misapplied Fuse Application

Figure 15.16 shows a disconnect that had renewable type (replaceable element) fuses installed. The fuses clearly state they have a maximum of 200-ampere full-load current rating, but the equivalent of 400-ampere element (double-linked) was used. The loading on the switch exceeded the switch rating for an extended time and resulted in severe overheating of the switch. To remove the switch from service, the technician had to pry the mechanism open with his screwdriver!

OSHA regulation 29 CFR 1910.334(b)(3) states:

"Overcurrent protection modification." Overcurrent protection of circuits and conductors may not be modified, even on a temporary basis, beyond that allowed by 1910.304(e).

Renewable fuses are an older type in which the body of the fuse can be disassembled and renewable fuse links inserted. These types fuses are not recommended for use and should be replaced with modern Class RK1 current-limiting fuses.

The 2008 edition of the *NEC* also weighs in on replaceable element fuses in Article 240.60(D),

Renewable Fuses. Class H cartridge fuses of the renewable type shall be permitted to be used only for replacement in existing installations where there is no evidence of overfusing or tampering.

General Maintenance Requirements

As stated before, OCPD interrupting times are one of the key factors affecting the arc flash incident energy. Circuit breakers are mechanical devices and require periodic maintenance to ensure proper operation. Preventive maintenance for circuit breakers should include exercising the mechanism by opening and closing circuit breakers every 6 to 12 months, periodic visual and mechanical inspections, and periodic calibration tests. The trip latch mechanism is important to exercise also, since it can seize due to lack of use. The trip latch mechanism can be exercised by primary injection testing or if a circuit breaker is equipped with a push-to-trip or similar button (usually red in color), which directly operates the trip latch. This is preferable to just opening and closing the breaker. In addition, periodically check conductor terminations for signs of overheating, poor connections and/or insufficient conductor ampacity. Infrared thermographic scans are one method that can be used to monitor these conditions. Records on the maintenance tests and conditions should be retained to establish a trend.

The internal parts of current-limiting fuses do not require maintenance. However, periodically checking fuse bodies, fuse mountings, and adjacent conductor terminations for signs of overheating, poor connections, or insufficient conductor ampacity is important. Fuses are typically used in conjunction with disconnects. Disconnects also require periodic inspection and maintenance. If a disconnect has lubricated mechanisms, then is necessary to maintain the lubrication properly.

In applications where disconnects are equipped with a ground-fault protection relay, disconnects should be periodically inspected and maintained, and the ground-fault relay calibrated. Also, after a disconnect interrupts a ground fault, the disconnect should be inspected and maintained.

Often, engineering studies are performed analyzing protective device coordination and arc flash hazard potential for an electrical system. As part of the engineering analysis, the type of OCPDs and breaker or relay settings are determined. OCPDs must be maintained and their settings not altered from those determined from these studies. If a circuit breaker or a fuse operates, it is a clear indication that there is a problem in the electrical system, not that there is a problem with the circuit breaker or fuse. Unfortunately, many production people do not understand that replacing correctly-rated fuses with larger ampacity fuses, automatically reclosing circuit breakers after they open or changing a breaker's settings is dangerous to the equipment being protected and to the people working on or near it.

Maintenance Requirements for OCPDs in *NFPA 70E*

Important requirements for OCPD maintenance are contained in Chapter 2 of *NFPA 70E*:

205.3 General Maintenance Requirements. Overcurrent protective devices shall be maintained in accordance with the manufacturers' instructions or industry consensus standards.

210.5 Protective Devices. Protective devices shall be maintained to adequately withstand or interrupt available fault current.

FPN: Failure to properly maintain protective devices can have an adverse effect on the arc flash hazard analysis incident energy values.

225.1 Fuses.

Fuses shall be maintained free of breaks or cracks in fuse cases, ferrules, and insulators. Fuse clips shall be maintained to provide adequate contact with fuses. Fuseholders for current-limiting fuses shall not be modified to allow the insertion of fuses that are not current-limiting.

225.2 Molded-Case Circuit Breakers.

Molded-case circuit breakers shall be maintained free of cracks in cases and cracked or broken operating handles.

225.3 Circuit Breaker Testing.

Circuit breakers that interrupt faults approaching their interrupting ratings shall be inspected and tested in accordance with the manufacturer's instructions.

NFPA 70E Chapter 2 is dedicated to maintenance requirements of electrical equipment for worker safety. In addition to maintenance requirements for OCPDs, *NFPA 70E* also contains requirements for other parts of the electrical power system. The following are Articles of *NFPA 70E* Chapter 2:

200 Introduction
205 General Maintenance Requirements
210 Substations, Switchgear, Switchboards, Panelboards, Motor Control Center (MCC) and Disconnect Switches
215 Premise Wiring
220 Controller Equipment
225 Fuses and Circuit Breakers
230 Rotating Equipment
235 Hazardous (Classified) Locations
240 Batteries and Battery Rooms
245 Portable Electric Tools and Equipment
250 Personal Safety and Protective Equipment

Maintenance Programs and Frequency of Maintenance

The scope of *NFPA 70E* Chapter 2 clarifies that these maintenance requirements are only those associated with employee safety, and these requirements are not prescriptive. The employer must choose the maintenance and methods that satisfy the requirements. Sources for guidance are provided in setting up maintenance programs, determining the frequency of maintenance, and providing prescriptive procedures. Equipment manufacturer's maintenance manuals should be one of the first sources to use.

NFPA 70B, Recommended Practice for Electrical Equipment Maintenance, provides some frequency of maintenance guidelines as well as guidelines for setting up an Electrical Preventative Maintenance (EPM) program, including sample forms and requirements for electrical system maintenance. Appendix D of this book is an example of the maintenance and tests recommended for molded case circuit breakers from MTS-07. ANSI/NETA MTS-07, *Maintenance Testing Specifica-*

tions, is another standard that is prescriptive about what maintenance and testing is required for electrical power systems devices and equipment. Mechanical, visual, and electrical maintenance and tests are specified by equipment type, as well as what results are acceptable. ANSI/NETA MTS-07, *Maintenance Testing Specifications* includes guidelines for the frequency of maintenance required for electrical system power equipment in their Appendix B, *Frequency of Maintenance Tests*. This is a very useful tool for determining how often testing and maintenance are required, based on the condition of the equipment and its criticality. This Maintenance Frequency Matrix is reproduced as Appendix C in this book, courtesy of NETA.

Maintenance and the Hazard/Risk Analysis

NFPA 70E Section 110.8(B)(1) directs a worker to perform a hazard/risk analysis prior to work beginning when the electrical equipment is energized. There are two parts to this requirement—the hazard and the risk. The hazard would be the voltage level (shock) and the expected incident energy exposure (arc flash). When blast hazard calculations become available, they should be included as well. This is where many stop their assessment. However, the risk must also be evaluated.

As an example, in approaching a lineup of switchgear, several factors must be evaluated:

1. The environment in which the equipment is operating. Is it exposed to the elements? Is it in a positive-pressure, air-conditioned area that is generally clean? Equipment in clean, dry environments is much less likely to be deteriorated than equipment that is exposed to the elements or contaminants.

2. How heavily loaded is the equipment? Lightly loaded equipment is under much less thermal stress than equipment that is carrying its maximum rated load (or more).

3. What is the overall condition of the equipment? Clean, well-maintained equipment is more likely to operate in accordance to the manufacturer's specifications than equipment in distress.

4. What is the operating history of the equipment? If certain types or brands of equipment are known to be troublesome, take that history into account.

5. When was the equipment last maintained and tested? If the equipment is out of calibration or if there is no calibration sticker present, the chances of a problem increase greatly.

6. What is the configuration of the equipment and enclosure? Equipment with vents, such as those in **Figure 15.17**, poses an arc flash hazard to operators, even though the door is closed. These vents cannot be closed, as the breakers and bus current ratings depend on the airflow through such vents. Closing them would cause rapid deterioration of the insulation or nuisance tripping in this type of switchgear. Figure 15.17A illustrates how the arc plasma ball exited through the louvered vents. In Figure 15.17B, the expanded metal air vents exposes arcing components directly to the operator. In either case, an arc flash would expose the operator to the arc-plasma ball, causing severe burns if not properly protected. For the purposes of a hazard/risk analysis, these doors are closed for shock hazards, but opened for the arc flash hazard.

Typically, workers will find that more arc flash PPE is needed after a hazard/risk evaluation, not less. *NFPA 70E* is intended to make injuries survivable, not to prevent injury. If a worker is exposed to the maximum incident energy for which their PPE is rated, there is a 50 percent probability of a second-degree burn on bare skin under the arc flash personal protective clothing. If the incident energy is greater than the arc rating of the PPE, the chances of injury increase.

The best course of action is to deenergize the equipment if there is serious doubt about the condition and functionality of the equipment to be worked on. This is not always possible, especially if troubleshooting electrical equipment and devices. If this is the case, the following items should be considered:

1. Distance is your friend. The incident energy generally decreases by the inverse square of the distance. Simply put, if the distance from the arc source is doubled, the incident energy exposure to the worker is decreased by approximately one-fourth. If the distance is tripled, the incident energy exposure is decreased by approximately one-ninth. Consider using longer tools or test probes, wearing leather gloves (even for 120-volt panelboards), and keeping your arms stretched to a comfortable distance.

2. If the equipment fails, it does not matter what task is being performed. Chances are the worker will be exposed to the full arc flash and its effects. Downsizing arc-rated PPE and equipment usually is not a safe course of action, unless the PPE itself poses a risk or increases the risk.

A.   **B.**

Figure 15.17   Vented switchgear. **A.** Louvered venting showing the arc-plasma ball damage done during a failure at the primary bus connections. **B.** Front-vented low-voltage switchgear.

3. If troubleshooting a problem, the chances of failure are probably higher than normal. That's why you're troubleshooting.

4. The chances of injury increase with the number of workers present. Only the people directly involved with the task should be exposed to the hazards. People who are monitoring or auditing the task should be outside either the limited approach boundary or the AFPB, whichever distance is greater.

5. Each layer of nonmelting clothing under arc-rated PPE reduces the arc flash heat to the body. The air that is trapped between layers of clothing acts in the same way that feathers act in a down jacket. However, consider the *NFPA 70E* Section 130.7(C)(12)(a) requirements described in Chapter 14.

> *NFPA 70E* Section 130.7(C)(12)(a), in part. If nonmelting, flammable fiber garments are used as underlayers, the system arc rating must be sufficient to prevent breakopen of the innermost FR layer at the expected arc exposure incident energy level to prevent ignition of flammable underlayers.

6. Incident energy is proportional to the time of exposure. If the exposure time is doubled, the incident energy received by the worker is doubled.

Qualifications for Performing Maintenance on Electrical Equipment

Several factors should be considered when deciding whether or not to perform electrical power systems maintenance and testing in-house or to use an outside company. Depending on the circumstances of your particular company, these may or may not be an issue:

1. The cost of test equipment. If the equipment is already purchased, there is no issue. If it has to be purchased, it also has to be maintained, which would include periodic calibration and repair for field damage.

2. Training of the test technician. Tests must be performed in a standardized manner that does not affect the results. Test technicians must be well trained and familiar with the specific test equipment used and the equipment being tested. Factors that affect test results must also be part of the knowledge base. Experience goes a long way with this item.

3. Having the correct test equipment for the tests being performed. Trying to "make do" with certain test equipment does not work well when some-

thing more sophisticated or with a larger capacity is needed. Test equipment is like any other tool; it must fit the task.

4. Documentation. Customers require accurate documentation of both the as-found and as-left test results. Hand-written notes or forms are not acceptable for most companies. Several companies offer software for this purpose.

5. Elevated pay levels. Test technicians must be highly trained, motivated, and accurate. This may require more training and precision than what is required for an average electrician.

6. Frequency of use. The old saying, "Use it or lose it" is certainly applicable in testing. To be good at testing, technicians must perform work frequently. People tend to lose some expertise when performing tasks infrequently, especially when performing more complex types of testing, such as insulation power factor.

7. Financial exposure. If things go bad during testing, the testing company can expect to be sued by the client. The first item that is suspected when there is a major failure is the testing and maintenance that may have been done in the past. Protective device calibration is always a potential issue.

When deciding to outsource electrical maintenance needs, several items should be considered when using an outside company:

1. Reputation. Does this company have a good reputation in the industry? Contact other customers in similar businesses where they have performed maintenance services. Electrical service companies sometimes tend to specialize in certain industries. They can shine in one type of industry but have no practical experience in others.

2. Are they independent of manufacturers? Companies that are closely associated with manufacturers can be a two-edged sword. On the one hand they can provide technical expertise that might not be generally available to the industry. On the other hand they might try to steer work to other parts of their company that might not have the necessary level of expertise. Manufacturers also tend to ask for a higher hourly rate, because they consider themselves "experts." Judge companies by the quality of the technicians that are sent to the site, not their uniforms.

3. When was their equipment calibrated? What are their requirements for calibration? For example,

NETA-member companies are required to calibrate analog test equipment every 6 months and digital test equipment every 12 months by a National Institute of Standards and Technology (NIST) traceable lab.

4. What training are the company's technicians required to have? Are they qualified under OSHA's 29 CFR 1910.399 definition of a qualified electrical worker, 29 CFR 1910.332, and by *NFPA 70E* Article 110.6? OSHA's Multi-Employer Worksite Policy (CPL-2-0.124 Rev. 15.00) ensures that under most circumstances, the premises (host) owner has a shared responsibility with the company that is actually performing the work. If a fatality occurs or OSHA chooses to investigate an accident, using contractors does not absolve the host employer of responsibility or liability. To be qualified, the worker must have the technical skills and expertise to perform the work and the safety training to understand the hazards and avoid them. In addition, *NFPA 70E*, Section 205.1, *Qualified Persons*, states,

> Employees who perform maintenance on electrical equipment and installations shall be qualified persons as required in Chapter 1 and shall be trained in, and familiar with, the specific maintenance procedures and tests required.

5. Is the electrical service company financially stable? Does it have adequate insurance if something goes wrong? There are a myriad of horror stories of small testing companies performing inadequate work, the equipment failing and the service company closing its doors, only to reopen under a new name.

6. Does the company have the technical expertise to evaluate problems that might arise? NETA-member companies are required to have at least one staff electrical engineer who is also a registered engineer, for example.

Legal Repercussions

Since safety is related to electrical power systems maintenance, it seems reasonable to assume there could be legal issues if maintenance is not performed. OSHA has not yet taken the stand that not performing maintenance as required by the manufacturer, *NFPA 70B* or ANSI/NETA MTS-07 constitutes a willful violation. However, the 2009 edition of *NFPA 70E* requires this maintenance, and OSHA has stated on their website that *NFPA 70E* is "a guide for meeting the requirements of the OSHA electrical regulations." In addition, Federal courts have found that *NFPA 70E* is "standard industry practice." Once a company receives and accepts a willful citation, especially if received as the result of an accident investigation, their Workman's Compensation protection no longer shields them. One definition given by a trial attorney for a willful citation was that it is equal to negligent behavior. OSHA defines a willful citation as one in which:

> the employer knowingly commits with plain indifference to the law. The employer either knows that what he or she is doing constitutes a violation, or is aware that a hazardous condition existed and made no reasonable effort to eliminate it.

Safe Versus Unsafe Example

Industrial insurers also require maintenance of electrical power systems and their protective devices. Often the risk an insurer assumes will not only cover a company for loss and repair of equipment, but also their financial losses. An incident can cost the insurer several million dollars. At a midwest refinery that flooded in 2007, the insurer paid for loss of profitability at the rate of $4,000,000 per day. The insurer paid that company for several weeks while they made repairs to their electrical systems. It is not unusual for companies to request expedited services to meet an insurer's maintenance requirements.

An excellent example of the importance of maintenance is one customer's experience by the service team of Shermco Industries. On a yearly basis, Shermco Industries calibrated a customer's protective relays on a 13,800-volt system. Each year its field service technicians would discuss the maintenance needed on an air-magnetic circuit breakers, and each year the customer would decline, saying, "They never operate and should be like brand new." Then one year, several months after servicing their protective relays, the customer had a failure on one of their 13.8-kilovolt underground feeder cables. The fault cascaded through six levels of circuit breakers before it finally cleared. The arc lasted only 2 to 3 seconds, but it destroyed not only the switchgear, but a 20 cal/cm^2 arc-rated arc flash suit that was hanging on the wall some 15 ft away. **Figure 15.18A** through **Figure 15.18D** show the damage caused. Because this company could not show it had performed the maintenance required by the insurer, its claim for financial loss was denied.

Summary

This chapter has explored key areas that can enhance worker safety: safe work practices related to OCPDs, arc flash hazard labeling requirements, existing system upgrades and adjustments, and the importance of maintenance of electrical equipment. Various work practices can improve the level of safety when working on electrical equipment; these include properly evaluating circuit breakers prior to resetting after a fault interruption, using proper fuse types, moving people outside the arc flash protection boundary, and ensuring that fuses and circuit breakers have interrupting ratings equal to or greater than the available short-circuit current. Finally, it is necessary to mark equipment with the arc flash hazard markings according to the *NEC* and *NFPA 70E*, and update this information any time a change occurs.

This chapter additionally discussed how electrical equipment maintenance directly affects worker safety. Poorly maintained equipment can create hazards in electrical equipment. *NFPA 70E* contains requirements for maintaining electrical equipment. Preventive maintenance of electrical equipment is essential, especially for OCPDs since the arc flash hazard can increase if the lack of maintenance delays an OCPD's opening time. If the condition of maintenance is questionable, extra precautions must be taken. In these cases, deenergization should occur prior to working, whenever possible. If that is not possible, additional PPE and more cautious work practices must be considered.

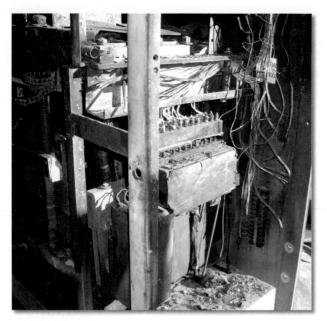

A.

B.

C.

D.

Figure 15.18 Damage from failure to clear a fault. **A.** 13.8kV breaker and cubicle damaged by fault that did not clear. **B.** Ceramic portions of arc chutes were damaged by extreme heat. **C.** All metal inside the breaker was melted or vaporized. **D.** Arc flash suit hanging on wall was reduced to ashes.

Lessons *learned*

Switchgear Failure Due to Lack of Maintenance

A major customer had a new extension built onto its electrical system to supply an additional line. A very well-known manufacturer won the project bid, and the work included the installation of the equipment and all start-up and acceptance testing. The project proceeded smoothly, and the work was completed. Several weeks after the extension was placed in service, a failure occurred in a 13.8-kilovolt underground feeder cable. Medium voltage circuit breakers failed to open, and approximately $5.2 million dollars was lost between replacing equipment and lost production. What went wrong?

Shermco Industries forensic investigation determined that a simple $100, two-pole, 100-ampere molded-case circuit breaker would not carry rated current. This breaker tripped in 70 to 90 seconds when 45 amperes was injected into it, which is a characteristic that would be expected from a 15-ampere circuit breaker. This breaker fed the battery charger that supplied direct current (dc) control power, including tripping power, for the 13.8-kilovolt circuit breakers. **Figure L15.1** shows the affected breaker and battery bank.

The client's maintenance personnel did not check the batteries, because they were "brand new." The battery bank slowly lost dc voltage, and when the cable failure occurred, there was no power to trip the breaker. The result was **Figure L15.2**, which is a power transformer and a reactor (used to limit short-circuit current). Note that the transformer windings are literally blown away from the core, which is characteristic of large through-faults.

So, is the manufacturer's testing company responsible for this incident? Not really. It turns out that the purchasing agent for the client company did not specify that these

Figure L15.1 The culprit—the little breaker that couldn't.

Figure L15.2 Transformer and reactor damaged by through-fault current.

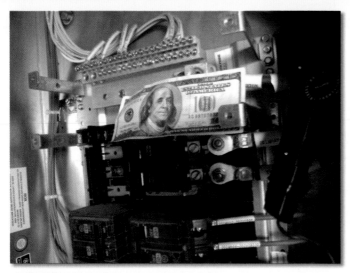

Figure L15.3 The winner and still champion!

small breakers be tested in the Request for Quote (RFQ). After all, why spend $150 to test a $100 circuit breaker? When the breaker feeds a critical load, is why a $100 circuit breaker should be tested at a cost of $150. When we assume that devices are going to function properly or that someone else did their job or that the circuit we're about to test is deenergized without testing it, we are set up for an incident. **Figure L15.3** shows the real winner!

Just because electrical equipment, especially an OCPD, is brand-new does not mean it is good. "Brand-new" might mean that the equipment has not been tested and verified. Acceptance testing is used to ensure that electrical devices function according to the manufacturer's specifications on all settings and all functions. It is a thorough series of tests that detects many problems such as the one illustrated. How was this issue resolved? It is probably still in court.

Source: This case study was contributed by Jim White of Shermco Industries.

Questions

1. In the incident report, why was the fault in the cable not opened by the immediate upstream circuit breaker and required a circuit breaker further upstream to interrupt the fault, resulting in significantly longer fault clearing time?
   A. The magnetic circuit breaker had been tested and shown to be satisfactory.
   B. The magnetic field in the two-pole breaker interfered with the radio transmission from the security detail.
   C. The operating handle on the magnetic circuit breaker failed to move the contacts.
   D. The circuit breaker feeding the battery charger would not carry rated full-load current.

2. What should have been done to avoid this catastrophe?
  A. Periodically check to see if the circuit breaker has tripped.
  B. For critical circuits, new equipment must be field-tested and operational characteristics verified and then periodically inspected, operated, and maintained.
  C. Nothing. This is an unavoidable occurrence.
  D. Change out the molded-case circuit breaker every six months.

3. What happens to the arc flash hazard when an OCPD that is suppose to operate in a specified time takes much longer to interrupt or does not operate (larger upstream OCPD must interrupt the arcing fault)?
  A. The arc flash hazard is likely to be significantly greater.
  B. The arc flash hazard is the same.
  C. The opening time of an OCPD has nothing to do with the arc flash hazard.
  D. The arc flash hazard is lower.

Ready for Review

- Workers must be trained and qualified in safe work practices including those related to OCPDs.

- After interrupting a fault current, OSHA and *NFPA 70E* require that a circuit may not be manually reenergized by resetting a circuit breaker or replacing fuses until it has been determined that the equipment and circuit can be safely energized.

- Fuses that interrupt a circuit should be replaced with the proper fuse type and ampere rating. Modern current-limiting fuses are recommended, where possible, to reduce the arc flash hazard.

- After fault interruption, circuit breakers should be evaluated for suitability before being placed back into service by a person qualified for circuit breaker evaluations. After interrupting a fault, a circuit breaker might not be suitable for further service. It must be inspected according to manufacturer instructions or industry standards.

- It is important that OCPDs not be modified, even temporarily, beyond what is allowed by OSHA 29 CFR 1910.304(e).

- It is important to move workers further from the potential arc source or outside the AFPB for hazardous operations. Methods to achieve this include remote-controlled motorized racking devices and extended length hand-operated racking tools.

- Marking or labeling electrical equipment with an arc flash hazard warning for arcing faults can reduce the occurrence of serious injury or death.
  - *NEC* Section 110.16 requires field marking with arc flash warning label on new equipment installations.
  - *NFPA 70E* Section 130.3(C) requires marking equipment with incident energy or level of PPE when an arc flash hazard analysis is performed. Installations can also be marked with the shock protection boundaries, the AFPB, and the shock hazard.
  - *NFPA 70E* Section 130.3 requires that arc flash hazard information be updated if system changes occur.
  - Many premise owners are conducting arc flash hazard analysis and shock hazard analysis for their entire premises and labeling the electrical equipment. Premise-wide implementation of arc flash hazard labeling can facilitate a safer operation.

- Electrical system changes can alter the available short-circuit current and/or alter the time duration for fuses or circuit breakers to interrupt the arcing current. These changes can affect the available arc flash incident energy and required level of PPE for a specific piece of electrical equipment. It is recommended that employers include the date of the analysis on the label and retain all the pertinent data for the analysis in a file.

- Electrical system changes can result in higher available short-circuit currents, which then might exceed the interrupting rating of existing OCPDs. Existing OCPDs must be reevaluated as to whether they have a sufficient interrupting rating.

- It is suggested to upgrade to fuses with the highest degree of current limitation to reduce the arc flash hazard. If the electrical system is an existing fusible system, owners should consider replacing the older fuses or less current-limiting fuses with newer, more current-limiting fuses where possible.

- A direct relationship exists between maintaining electrical equipment and worker safety. *NFPA 70E* recommends maintaining electrical equipment to keep it in a safe condition. If equipment and protective devices are not properly maintained, engineering analysis, such as short-circuit current studies and coordination studies, no longer have any relevance.

- CDC and NIOSH studies revealed the following:
  - Among electrical activities, 24 percent of accidents result during troubleshooting.

- ○ Eighteen percent of accidents were the result of failure during repair or repair-related activities.
- ○ Thirty-four percent of all general industry electrically related accidents are caused by component failure and 19 percent of those accidents are caused by components that failed during normal operation. Most of the devices that failed in this study did so due to lack of maintenance.
- A lack of lubrication can lead to equipment not functioning properly. For example, a circuit breaker's clearing time could increase due to lack of lubrication, leading to a greater arc flash incident energy and a greater hazard. Equipment that is lubricated when initially manufactured will dry out even if the breaker or switch has not had to operate. Periodic cleaning and re-lubrication are required to ensure proper functioning.
- *NFPA 70E* Section 130.3 contains requirements for considering the condition of maintenance for OCPDs when conducting arc flash hazard analysis. The reliability of OCPDs can directly impact arc flash hazards.
- If the ability of an OCPD to function properly is questionable, deenergizing the equipment is the best practice. If the decision is made that the circuit or equipment cannot be placed in an electrically safe work condition, the next device upstream that is deemed reliable must be considered as the protective device that would operate. This could mean, due to the increase in operating time, the incident energy would be substantially higher.
- Preventive maintenance for circuit breakers should include exercising the mechanism by opening and closing circuit breakers every 6 to 12 months, periodic visual and mechanical inspections, and periodic calibration tests. The trip latch mechanism is important to exercise also, since it can seize, due to lack of use. Conductor terminations must be checked periodically for signs of overheating, poor connections, and/or insufficient conductor ampacity. Records of the maintenance tests must be kept.

- Fuse bodies must be checked for integrity (not cracked or damaged). Fuse mountings and adjacent conductor terminations must be checked periodically for signs of overheating, poor connections, or insufficient conductor ampacity. Disconnects also require periodic inspection and maintenance.
- *NFPA 70E* Chapter 2 is dedicated to maintenance requirements of electrical equipment for safety. In addition to the requirement for maintenance of OCPDs, *NFPA 70E* also contains requirements for other parts of the electrical power system.
- Employers must choose the maintenance and methods that satisfy the requirements of *NFPA 70E*, determine the frequency of maintenance, and provide procedures. Equipment manufacturer's maintenance manuals should be used, but if not available, *NFPA 70B* and ANSI/NETA MTS-07 are industry standards that can be used.
- *NFPA 70E* Section 110.8(B)(1) directs a worker to perform a hazard/risk analysis before beginning work when the electrical equipment is energized (not in an electrically safe work condition). Assessing both the hazard and the risk are critical.
- If deenergizing the equipment is not possible and there is some question on the reliability of the circuit's protective devices, the following should be considered:
  - ○ Use longer tools or test probes, wear leather gloves (even for 120-volt panelboards), and keep your arms stretched to a comfortable distance.
  - ○ The incident energy will be higher than normal for that circuit and will probably require the use of arc-rated PPE and equipment that has a higher arc rating.
  - ○ Be aware that when troubleshooting, the chances of failure are probably higher than normal.
  - ○ Only include the people directly involved with the task; the chances of injury increase with the number of workers present. Others should be outside either the limited approach boundary or the AFPB, whichever is the greater distance.

- Wear underlayers of clothing; each layer of non-melting clothing under arc-rated PPE reduces the heat to the body. If nonmelting, flammable fiber garments are used as underlayers, the system arc rating must be sufficient to prevent breakopen of the innermost FR layer at the expected arc exposure incident energy level to prevent ignition of flammable underlayers.
  - Note that if the exposure time is doubled, the incident energy received by the worker is approximately doubled.
- When considering maintaining electrical power systems with in-house staff, the following should be considered:
  - Cost of purchasing and maintaining the correct test equipment for the task being performed
  - Training and compensation of the test technician
  - Maintaining in-house employee skills for tests that are performed infrequently
  - Documentation of the testing
  - Financial liability in the event of something going wrong
- When considering hiring an outside company to maintain electrical power systems, the following should be considered:

- Reputation
- Independence from manufacturers
- Requirements for calibration of test equipment
- Training and qualifications of the company's technicians
- Financial stability/insurance
- Technical expertise

Vocabulary

interrupting rating (IR)* The highest current at rated voltage that a device is intended to interrupt under standard test conditions.

> FPN: Equipment intended to interrupt current at other than fault levels may have its interrupting rating implied in other ratings, such as horsepower or locked rotor current.

knife-blade fuses A type of fuse that contains blades that are inserted into fuse clips; it is important not to rely on test measurements by touching the end caps of the fuse with probes. Touch the probes to the fuse blades.

*This is the *NFPA 70E* definition.

⚠ WARNING

Arc Flash and Shock Hazard
Appropriate PPE Required

_____ Flash Hazard Boundary
_____ cal/cm2 Flash Hazard at 18 inches
_____ PPE Level, _____
_____
_____ Shock Hazard when Cover is_____
_____ Limited Approach
_____ Restricted Approach- _____
_____ Prohibited Approach- _____

Equipment Name:_____ Review Date:_____

OPENING CASE

A 23-year-old male apprentice electrician was electrocuted while making a connection for a light fixture in a junction box. The victim had previously worked for the company for 2 years. He left the company for about 2 years and returned to work at the beginning of 1989. While working for another employer the victim received electrical burns on his hands. The victim had recently taken the examination to be classed as a Journeyman electrician.

The employer, an electrical contractor with 40 employees, has been in business for 15 years. The company has a designated part-time safety officer but has no formal safety program. The company does a wide range of electrical work. Safety is handled through on-the-job training, observation of work practices by supervisors, and printed handouts of safety topics that management deems appropriate.

The employer had a contract to install wiring and fixtures in an office complex that was located behind a new shopping mall. The third floor of the office complex was being hurriedly prepared for a tenant. The off-site designer made daily changes (e.g., fixture locations) to electrical system blueprints and transmitted the revised drawings to the job site via fax machine.

The lighting system is a 3-phase, 4-wire, 277/480-volt system. The wires for two sets of conductors (three hot wires and one neutral wire per circuit) were run in one conduit down a central hall with junction boxes installed for branch circuits to individual lights. The lights are mounted in a metal gridwork, flush with ceiling tiles. System grounding is achieved through the metal conduit. Work had been completed on one side of the hall, and the victim was installing conduit and fixtures on the other side.

At the time of the incident, the victim was wiring in a light fixture in a junction box and was in contact with the gridwork. During the work, he received a shock, came down from the fiberglass ladder, said, "Cut the juice," and collapsed. Two coworkers immediately started CPR while another worker called the emergency service number. The phone call was made from an elevator that had the same number as other elevators in the mall-office complex. The person at the emergency service answering location asked if the call was coming from the place where the victim was located. (The number and location were available to the emergency service person.) The caller replied that it was, neither one realizing that the mall was the listed address for the phone number.

Review and Implementation

Introduction

The first 15 chapters have discussed a wide variety of topics and requirements such as electrical safety culture, electrical hazards, electrical system design and maintenance, the electrical safety program and training, lockout/tagout and achieving an electrically safe work condition, shock hazard analysis, other protective equipment, and a flash hazard analysis including **arc flash protection boundary (AFPB)**, personal protective equipment (PPE), and labeling. The first part of this chapter briefly revisits many of those requirements as a review of them. The second part of this chapter describes how a shock hazard analysis and flash hazard analysis are integrated into an Energized Electrical Work Permit.

OPENING CASE, CONT.

The emergency rescue team went to the mall searching for the victim, while the coworkers waited with the victim for assistance in the office complex. Approximately 20 minutes elapsed before the rescue team reached the victim. The victim was then transported to a local hospital where he was pronounced dead-on-arrival. The medical examiner ruled that the cause of death was electrocution.

Work was stopped for that day, and the area secured until the company could make an investigation. The next day, no energized lines could be detected in the junction box where the victim was working when he received the fatal shock. The panel box was in a closet, but it was not locked. After work in the area was finished, further investigation revealed that the two neutral conductors in the same conduit run had been cross-wired at a junction box. That is, the neutral wire in the deenergized circuit was mistakenly connected to the neutral wire of an energized circuit back at the junction box. Company officials believe that the victim had previously, inadvertently cross-wired the neutral conductors, which subsequently allowed electricity to flow from the live circuit on the completed side of the building through a light fixture, to the circuit on which the victim was working. When he handled the energized neutral conductor to make the connection, his body provided a path to the metal gridwork he was touching, which was connected to the building structural steel. The circuit breaker had to be closed, and the light switch in the "on" position on the live circuit to allow current to flow through the light fixture to the neutral line wire of the parallel circuit.

Adapted from: FACE 89-50. Accessed 9/11/08 at http://www.cdc.gov/niosh/face/In-house/full8950.html.

1. What could the employer have done to prevent this tragedy?
2. What could the worker have done to help prevent this tragedy?

CHAPTER 16

Chapter Outline

- **Introduction**
- **Article 110 Review**
- **Lockout/Tagout and an Electrically Safe Work Condition Review**
- **Article 130 Review**
- **Practice Examples Review**
- **Summary**
- **Lessons Learned**
- **Current Knowledge**

Learning Objectives

1. Understand *NFPA 70E*® Article 110 requirements.
2. Understand *NFPA 70E* Article 130 requirements that detail justification for energized work, minimum content of an Energized Electrical Work Permit, other precautions for personnel activities, and alerting techniques.
3. Understand *NFPA 70E* Article 130 requirements that address shock hazard analysis, including the approach boundaries for shock, arc flash hazard analysis, including the AFPB, incident energy analysis, and hazard/risk categories (HRCs) to determine arc flash protection, equipment labeling, and personal and other protective equipment.

Reference

1. *NFPA 70E*, 2009 Edition

Article 110 Review

Host and Contractor Responsibilities

Host and contractor responsibilities are contained in *NFPA 70E* Section 110.5. The host employer responsibilities include, but are not limited to, making sure that they make contractors aware of hazards related to the work they are performing, that they are provided with the information necessary to perform an electrical hazard assessment, and notifying the contractor if any of the contractor's employees are witnessed violating the provisions of *NFPA 70E*. Contractor responsibilities include, but are not limited to, making sure that their employees are trained in and follow the applicable provisions of both *NFPA 70E* and those of the host employer. The contractor is also responsible for ensuring that their employees are aware of and trained to recognize and avoid the hazards that are reported by the host employer. These concepts are examined in detail in Chapter 6.

Training Requirements

Training requirements in *NFPA 70E* are defined in Section 110.6. Those requirements are broken down into five major categories:

1. Safety training
2. Type of training
3. Emergency procedures
4. Qualified and unqualified person training
5. Documentation of training

Workers must be trained in the safety-related work practices and procedures that can help them avoid injury from the electrical hazards associated with the tasks they perform. *NFPA 70E* training requirements are primarily detailed in Chapter 8.

Electrical Safety Program

NFPA 70E electrical safety program requirements are detailed in Section 110.7 and are broken down into eight major segments. Addressed first is the general requirement for an employer to develop, document, and put into practice an electrical safety program that provides work practices and procedures that will protect workers from the voltage, energy level, and circuit conditions to which they will be exposed. The other segments that *NFPA 70E* requires as part of an electrical safety program include:

- Worker awareness and self-discipline
- Electrical safety program principles
- Controls
- Procedures

Figure 16.1 A job briefing is an essential part of an electrical safety program.

- A hazard/risk evaluation procedure
- Job briefing (**Figure 16.1**)
- Electrical safety auditing

Additional electrical safety program considerations are illustrated in *NFPA 70E* Annex E. These concepts are examined in detail in Chapter 7.

Working While Exposed to Electrical Hazards

NFPA 70E Section 110.8 describes requirements for working while exposed to electrical hazards. If it is infeasible or a greater hazard to deenergize and the lockout/tagout provisions are not implemented, other safety-related work practices must be put into practice (**Figure 16.2**). Appropriate safety-related work practices must be determined before a worker is exposed to electrical hazards by an electrical hazard analysis. An electrical hazard analysis consists of both a shock and an arc flash hazard analysis. These concepts are examined in detail in Chapter 6.

Figure 16.2 Other safety-related work practices must be used where it is infeasible or a greater hazard to deenergize.

Use of Equipment

NFPA 70E Section 110.9 details requirements for test instruments and equipment, portable electric equipment, ground-fault circuit interrupter (GFCI) protection devices, and overcurrent protection modification. Among other things, the following must be determined:

- Test instruments are rated for circuits to which they will be connected
- Proper operation of the test equipment is verified before and after a test
- Portable cord-and-plug-connected equipment is visually inspected for external defects and damage before each use and removed if warranted
- GFCI protection devices are tested per manufacturer's instructions

The requirements for use of equipment are examined in Chapter 6.

Figure 16.3 An adequately rated voltage detector must be used to test each phase conductor or circuit part to verify they are deenergized.

Lockout/Tagout and an Electrically Safe Work Condition Review

The Occupational Safety and Health Administration (OSHA) requires live parts to which an employee might be exposed to be deenergized before the employee works on or near them, unless the employer can demonstrate that deenergizing introduces additional or increased hazards or is infeasible due to equipment design or operational limitations. OSHA's lockout/tagout requirements are law and the first safety-related work practice to be implemented whenever possible.

NFPA 70E Article 120 and the process of achieving an electrically safe work condition is one example of how *NFPA 70E* can supplement OSHA's lockout/tagout requirements. This includes, but is not limited to, the six-step process outlined in Section 120.1 to verify that an electrically safe work condition exists after the provisions contained in *NFPA 70E* Section 120.2 have been taken into account. An electrically safe work condition is achieved when performed in accordance with the procedures of *NFPA 70E* Section 120.2 and verified by the following six steps outlined in Section 120.1:

1. Determine all possible sources of electrical supply to the specific equipment. Check applicable up-to-date drawings, diagrams, and identification tags.

2. After properly interrupting the load current, open the disconnecting device(s) for each source.

3. Wherever possible, visually verify that all blades of the disconnecting devices are fully open or that drawout-type circuit breakers are withdrawn to the fully disconnected position.

4. Apply lockout/tagout devices in accordance with a documented and established policy.

5. Use an adequately rated voltage detector to test each phase conductor or circuit part to verify they are deenergized (). Test each phase conductor or circuit part both phase-to-phase and phase-to-ground. Before and after each test, determine that the voltage detector is operating satisfactorily.

6. Where the possibility of induced voltages or stored electrical energy exists, ground the phase conductors or circuit parts before touching them. Where it could be reasonably anticipated that the conductors or circuit parts being deenergized could contact other exposed energized conductors or circuit parts, apply ground connecting devices rated for the available fault duty.

Article 130 Review

Justification

NFPA 70E Section 130.1(A) requires energized electrical equipment to be placed into an electrically safe work condition if a worker is to cross the **limited approach boundary (LAB)** associated with the energized electrical equipment. *NFPA 70E* Article 120 details the requirements for establishing an electrically safe work condition. Two fine print notes (FPNs) to Section 130.1(A) provide examples of infeasibility and increased or additional hazards that would permit work on energized equipment.

Increased or additional hazards according to *NFPA 70E* include, but are not limited to:

- Interruption of life support equipment
- Deactivation of emergency alarm systems

Figure 16.4 Safe work practices, including personal and other protective equipment, are required even if the employer has demonstrated that energized work is justified.

- Shutdown of hazardous location ventilation equipment

Work that might be infeasible to perform deenergized due to equipment design or operational limitations according to *NFPA 70E* include:

- Diagnostics and testing of circuits that can only be performed energized
- Work on circuits that form an integral part of a continuous process that would otherwise need to be completely shut down in order to permit work on one circuit or piece of equipment

Safe work practices, including personal and other protective equipment, are required even if the employer has demonstrated that deenergizing introduces additional or increased hazards or is infeasible due to equipment design or operational limitations (**Figure 16.4**).

These concepts are examined in detail in Chapter 10.

Energized Electrical Work Permit

NFPA 70E Section 130.1(B)(2) provides requirements as a means for employers to assess the task to be performed and determine appropriate PPE to protect against electrical hazards. Energized work is to be performed in accordance with the energized electrical work permit requirements of Section 130.1(B) if the equipment is not in an electrically safe work condition.

NFPA 70E Annex J.1 offers a sample permit. Annex J.1, like all *NFPA* annexes, is nonmandatory. *NFPA 70E* Section 130.1(B)(2), however, details the minimum information that must be documented in an Energized Electrical Work Permit. A sample permit is shown in **Figure 16.5**.

Energized Electrical Work Permit

1. **Description of the circuit and equipment to be worked on and their location:**

2. **Justification for why the work must be performed in an energized condition: (Check one and describe)**
   Infeasibility _____
   or
   Greater Hazard _____
   Describe _____

3. **Description of the safe work practices to be employed**

4. **Results of the shock hazard analysis**
   _____ Volts

5. **Determination of shock protection boundaries**
   Prohibited Approach Boundary: _____
   Restricted Approach Boundary: _____
   Limited Approach Boundary: _____

6. **Results of the arc flash hazard analysis:**
   Available short-circuit current: _____
   Fault clearing time: _____
   The Hazard/Risk Category is: _____
   or
   The Incident Energy is: _____ calories per square centimeter.

7. **The Arc Flash Protection Boundary:**

8. **Necessary personal protective equipment to safely perform the assigned task:**
   Note: PPE must meet the ANSI and ASTM standards that are listed in *NFPA 70E* Table 130.7(C)(8) and other protective equipment must conform to the standards in *NFPA 70E* Table 130.7(F).

9. **Means employed to restrict the access of unqualified persons from the work area:**

10. **Evidence of completion of a job briefing, including a discussion of any job-specific hazards:**

Energized work approval signature and date:

_____

Figure 16.5 The minimum that must be included in an Energized Electrical Work Permit is described in *NFPA 70E* Section 130.1(B)(2) and shown here.

NFPA 70E Section 130.1(B)(3) recognizes a limited number of examples where qualified persons inside the LAB can perform energized work without an Energized Electrical Work Permit as long as appropriate work practices and PPE are used.

Energized Electrical Work Permit requirements are examined in detail in Chapter 10.

Approach Boundaries

NFPA 70E Section 130.2(A) requires a shock hazard analysis. The shock hazard analysis must determine:

- The voltage to which workers will be exposed
- The shock protection boundaries
- The PPE required for shock protection

The results of the shock hazard analysis are among the information that is required to be documented on the Energized Electrical Work Permit to meet the requirements of *NFPA 70E* Section 130.1(B)(2).

There are three shock protection boundaries established by *NFPA 70E* (**Figure 16.6**) and Table 8.1 in Chapter 8:

- Limited approach boundary (LAB)
- **Restricted approach boundary**
- **Prohibited approach boundary**

These three shock protection boundaries apply when workers are exposed to energized electrical conductors or circuit parts. Table 130.2(C) establishes the distances associated within a range of nominal system voltages.

These concepts are applied to sample work permits later in this chapter and are examined in detail in Chapter 10.

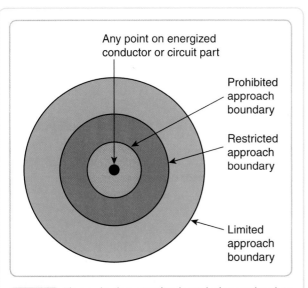

Figure 16.6 Three shock protection boundaries apply when workers are exposed to energized electrical conductors or circuit parts.

Arc Flash Hazard Analysis

An **arc flash hazard analysis** is required by *NFPA 70E* Section 130.3 to determine the AFPB and the PPE necessary for workers who cross that boundary. The analysis consists of three components:

- Determination of the AFPB
- Determination of protective clothing and other PPE
- Field labeling of the equipment

An exception recognizes that an arc flash hazard analysis is not required when all of the following apply:

- The circuit is rated 240 volts or less.
- The circuit is supplied by one transformer.
- The transformer supplying the circuit is rated less than 125 kilovolt-amperes.

Arc Flash Protection Boundary

The *NFPA 70E* Section 130.3(A) requirements for determining the AFPB are broken down into two voltage level categories: Between 50 and 600 volts and above 600 volts.

While it is often referred to as a "default" boundary, the 4 ft boundary is really a conditional boundary. Where detailed arc flash hazard analysis calculations are not performed, the AFPB is 4 ft for the product of clearing time of 2 cycles and an available bolted short-circuit (fault) current of 50 kiloamperes (or any combination not exceeding 100 kiloamperes). Regardless, the available bolted short-circuit current and clearing time must be known to determine the AFPB either by calculation or the use of the 4 ft conditional boundary.

Personal Protective Equipment

NFPA 70E Section 130.3(B) describes the second of three steps to fulfill what is required as part of an arc flash hazard analysis. Section 130.3(B) will determine the PPE required once the AFPB is crossed (**Figure 16.7**). Two options are available to determine the appropriate PPE for the work to be performed inside the AFPB; namely, incident energy calculation and using the tables provided in Section 130.7(C). Regardless, the available bolted short-circuit current and clearing time must be known to determine the PPE either by calculation or the use of the tables in Section 130.7(C).

Incident Energy Analysis

Incident energy analysis is one option permitted to determine the required PPE to cross the AFPB per *NFPA 70E* Section 130.3(B)(1). This analysis determines the incident energy exposure of the worker in calories per square centimeter (cal/cm^2). PPE must be selected and used based on the incident energy analysis.

Figure 16.7 All parts of the body that will be inside the AFPB must be protected before the AFPB is crossed.

Use of Hazard Risk Category Table in Article 130

NFPA 70E Section 130.3(B)(2) details another option permitted for the determination, selection, and use of the personal and other protective equipment required for a worker to be protected once he or she crosses the AFPB. The HRC associated with the task to be performed is determined from *NFPA 70E* Table 130.7(C)(9). Tables 130.7(C)(10) and 130.7(C)(11) are used once the level of PPE (the HRC) is determined from Table 130.7(C)(9).

Labeling of Equipment

NFPA 70E Section 130.3(C) details the third component of an arc flash hazard analysis. Electrical equipment must have a label affixed that documents either the results of the incident energy analysis (in cal/cm^2) or the HRC for the task to be performed.

Performing Testing Work

NFPA 70E Section 130.4 requires workers to be qualified to perform electrical testing on electrical equipment at voltage levels of 50 volts or more once they cross the LAB for that energized equipment.

Uninsulated Overhead Lines

NFPA 70E Section 130.5 addresses tasks performed within the LAB of uninsulated overhead lines (**Figure 16.8**). There are five categories of requirements that address uninsulated overhead lines located in *NFPA 70E* Section 130.5:

- Uninsulated and energized
- Deenergizing or guarding
- Employer and employee responsibility
- Approach distances for unqualified persons
- Vehicular and mechanical equipment

These requirements are, in many cases, similar to OSHA's requirements in 29 Code of Federal Regulations (CFR) 1910.333(c)(3). Additional requirements that address tasks performed within the LAB of uninsulated overhead lines are covered in Chapter 13 and in *NFPA 70E* Section 130.5.

Figure 16.8 *NFPA 70E* Section 130.5 addresses tasks performed within the LAB of uninsulated overhead lines within the scope of the standard.

Other Precautions for Personnel Activities

NFPA 70E Section 130.6 details numerous precautions required to be considered for work involving electrical hazards. The *NFPA 70E* precautions for personnel activities are broken down into 11 subsections:

- Alertness
- Blind reaching
- Illumination
- Conductive articles being worn
- Conductive materials, tools, and equipment being handled
- Confined or enclosed work spaces
- Housekeeping duties
- Occasional use of flammable materials
- Anticipating failure
- Routine opening and closing of circuits
- Reclosing circuits after protective device operation

Many of these requirements in *NFPA 70E* are similar to what OSHA requires in 29 CFR part 1910. Chapter 13 of this book and *NFPA 70E* Section 130.6 cover the 11 categories of "other precautions for personnel activities" in detail.

General Requirements for Personal and Other Protective Equipment

Like OSHA, *NFPA 70E* Section 130.7(A) assigns responsibility to employers and employees for protective equipment use. Workers must be provided with, and they must use, protective equipment that is designed for the use and adequate for the hazard when performing tasks where electrical hazards are present.

Equipment Care
NFPA 70E Section 130.7(B) requires protective equipment to be maintained, inspected, and stored properly (**Figure 16.9**). It must be maintained in a safe, reliable condition, visually inspected before each use, and stored in a manner to prevent damage.

Personal Protective Equipment

NFPA 70E Section 130.7(C)(1) requires all parts of the body inside the AFPB to be protected in accordance with *NFPA 70E* Section 130.3. There are generally two methods to determine the PPE that workers need to wear once they cross the shock and AFPB; namely, incident energy calculation or determination of the HRC and PPE by using the *NFPA 70E* Section 130.7(C) tables. Appropriate PPE is selected by the requirements of *NFPA 70E* Section 130.7(C) once the incident energy is determined per Section 130.3(B)(1). The HRC is permitted to be used to select PPE per Section 130.3(B)(2). PPE must conform to the standards provided in Table 130.7(C)(8).

Arc-rated flame-resistant (FR) clothing and other PPE is selected based on the incident energy exposure in accordance with the requirements of *NFPA 70E* Sections 130.7(C)(1) though (8) and 130.7(C)(12) though (16), unless the HRC (table method) is used, as permitted by Section 130.3(B)(2).

The HRC method is used for the selection of PPE in conjunction with the requirements of *NFPA 70E* Sections and Tables 130.7(C)(9), 130.7(C)(10), and 130.7(C)(11).

These concepts are examined in detail in Chapter 14.

Other Protective Equipment

NFPA 70E Section 130.7(D)(1) details considerations for nine types of other protective equipment.

Figure 16.9 Protective equipment must be visually inspected before each use.

Figure 16.10 Insulated tools are required when working inside the LAB of exposed energized electrical conductors or circuit parts where tools might make accidental contact.

- Insulated tools (Figure 16.10)
- Fuse or fuse holding equipment
- Ropes and handlines
- Fiberglass-reinforced plastic rods
- Portable ladders
- Protective shields
- Rubber insulating equipment
- Voltage rated plastic guard equipment
- Physical or mechanical barriers

These requirements are discussed in detail in Chapter 13.

Alerting Techniques

NFPA 70E Section 130.7(E)(1) requires that safety signs, safety symbols, or accident prevention tags be used to warn employees about electrical hazards. *NFPA 70E* Section 130.7(E)(2) requires barricades to be used in conjunction with safety signs where it is necessary to prevent or limit employee access. *NFPA 70E* Section 130.7(E)(3) requires an attendant for the following reasons:

- If signs and barricades do not provide sufficient warning and protection from electrical hazards
- To warn and protect employees
- To keep unqualified employees outside a work area where the unqualified employee might be exposed to electrical hazards
- To remain in the area as long as there is a potential for employees to be exposed

These concepts are examined in detail in Chapter 13.

Standards for Other Protective Equipment

Other protective equipment (Figure 16.11) required in Section 130.7(D) must conform to the standards provided in Table 130.7(F). Other protective equipment includes the following:

- Ladders
- Safety signs and tags
- Blankets
- Covers
- Line hoses
- Fiberglass tools
- Fiberglass ladders
- Plastic guards
- Temporary grounding
- Insulated hand tools

A.

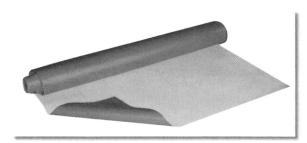

B.

C.

Figure 16.11   Other protective equipment such as line hose and covers (**A.**) and blankets (**B.** and **C.**) must conform to the standards provided in Table 130.7(F).

Practice Examples Review

Determination of justification for energized work, the AFPB, protective equipment, and completion of a sample Energized Electrical Work Permit to perform three tasks is illustrated in the examples that follow. Example 1 is a review of Test 1, example 2 is a review of Test 4, and example 3 is a review of Test 3, which were introduced in Chapter 2 and discussed in detail in Chapter 12. All three tests were conducted on the same electrical circuit set-up with an available bolted 3-phase, short-circuit current of 22,600 symmetrical root mean square (rms) amperes at 480 volts. In each case, an arcing fault was initiated in a size 1 combination motor controller enclosure with the door open, as if an electrician were working on energized equipment.

Example 1, Test 1

Determine the justification for energized work, to perform a task on the Test 1 setup as described in Chapter 12. The circuit has an available bolted 3-phase, short-circuit current of 22,600 symmetrical rms amperes at 480 volts. The task to be performed is to terminate a 10 American Wire Gauge (AWG) copper conductor on the load side of the motor controller for the exhaust fan (EF-4) for electrical closet 4 (**Figure 16.12**).

Justification

Determine the justification for energized work per *NFPA 70E* Section 130.1(A). The task of terminating a 10 AWG copper conductor on the load side of a motor controller for an exhaust fan for an electrical closet does not qualify as infeasible or a greater hazard. Therefore, the task must be performed under lockout/tagout.

Voltage testing must be performed to verify absence of voltage. Appropriate PPE must also be selected for the task. Testing for voltage to verify absence of voltage is a task that is infeasible to perform deenergized. The AFPB and appropriate PPE must be selected for the task and are covered in the next two examples.

Example 2, Test 4

Determine the justification for energized work, the AFPB and PPE, and complete each portion of a sample Energized Electrical Work Permit to perform Test 4 as described in Chapter 12. A 640-ampere circuit breaker with a STD setting of 6 cycles is protecting the circuit. The circuit has an available bolted 3-phase, short-circuit current of 22,600 symmetrical rms amperes at 480 volts (**Figure 16.13**). An arcing fault is initiated in a size 1 combination motor controller enclosure with the door open. The task being performed is voltage testing, and the equipment is energized. The 640-ampere circuit breaker clears the arcing fault in 6 cycles.

Justification

Determine the justification for energized work per *NFPA 70E* Section 130.1(A). The task is voltage testing. Testing for voltage to verify absence of voltage is a task that is infeasible to perform deenergized. The AFPB and appropriate protective equipment must be selected for the task.

Determine the AFPB for Test 4

NFPA 70E Section 130.3(A)(1) addresses determination of the AFPB for voltage levels between 50 and 600 volts. The conditional AFPB of 4 ft can be used where the clearing time does not exceed 2 cycles and an avail-

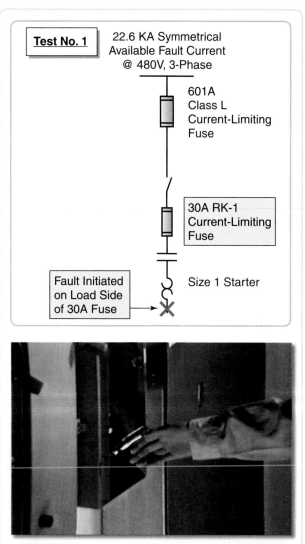

Figure 16.12   The circuit has an available bolted 3-phase, short-circuit current of 22,600 symmetrical rms amperes at 480 volts.

© 1997 Institute of Electrical and Electronics Engineers. Still photo extracted from the Cooper Bussmann *Safety BASICs Handbook* with permission from Cooper Bussmann.

able bolted short-circuit current does not exceed 50 kiloamperes (or any combination of clearing time and available bolted short-circuit current not exceeding 100 kiloamperes). The clearing time in Test 4 is 6 cycles, and the available bolted short-circuit current is 22,600 amperes.

Calculation:

Combination is 22.6 kiloamperes × 6 cycles = 135.6-kiloampere cycles

The product of the clearing time and available bolted short-circuit current for Test 4 exceeds the conditions where the 4 ft boundary can be used. Therefore, the AFPB must be calculated.

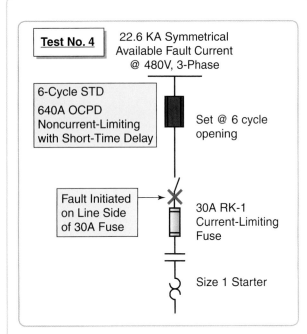

Test No. 4   22.6 KA Symmetrical
Available Fault Current
@ 480V, 3-Phase

6-Cycle STD
640A OCPD
Noncurrent-Limiting
with Short-Time Delay

Set @ 6 cycle
opening

Fault Initiated
on Line Side
of 30A Fuse

30A RK-1
Current-Limiting
Fuse

Size 1 Starter

Figure 16.13 A 640-ampere circuit breaker with a STD set to open at 6 cycles is protecting the circuit with an available bolted 3-phase, short-circuit current of 22,600 symmetrical rms amperes at 480 volts.
© 1997 Institute of Electrical and Electronics Engineers. Still photo extracted from the Cooper Bussmann *Safety BASICs Handbook* with permission from Cooper Bussmann.

Voltage testing does not require an Energized Electrical Work Permit, per *NFPA 70E* Section 130.1(B)(3). However, safe work practices and PPE are still required in accordance with *NFPA 70E* Chapter 1. Therefore, the Energized Electrical Work Permit will be used to document the necessary information to perform the task safely.

The AFPB for Test 4 was calculated in Chapter 12. The calculated AFPB per the basic formula in Institute of Electrical and Electronics Engineers (IEEE) 1584 is 47 in. This distance is information that must be entered on the Energized Electrical Work Permit (**Figure 16.14**). All parts of the body must be protected once the worker crosses the AFPB of 47 in.

Energized Electrical Work Permit

7. The Arc Flash Protection Boundary:

47 inches

Figure 16.14 The AFPB is information that must be entered on the Energized Electrical Work Permit.

Determine the PPE for Test 4

NFPA 70E Section 130.3(B) addresses determination of the PPE required once the AFPB is crossed. Two options are available to determine the appropriate PPE for the work to be performed inside the AFPB; namely, incident energy calculation and determination of the HRC.

HRC

The HRC associated with the equipment and the task to be performed is determined from *NFPA 70E* Table 130.7(C)(9). The equipment and the task must be listed in Table 130.7(C)(9) and be within the parameters of the specific notes to the table:

- The task is being performed in a size 1, 600-volt, combination motor controller enclosure.
- The category of equipment is listed in Table 130.7(C)(9) [**Table 16.1** includes working on energized parts in a size 1, 600-volt, combination motor controller, which falls under the category "Other 600V Class (277 V through 600 V, nominal) Equipment"]. According to the table, the HRC for this category is 2*, rubber insulating gloves are required, and insulated and insulating hand tools are required.
- The available bolted short-circuit current and clearing time must not exceed the parameters of Note 2 under the specific notes to the table: Maximum of 65 kiloamperes available; maximum of 0.03 sec (2 cycle) fault clearing time.

Test 4

Bolted short-circuit current = 22.6 kiloamperes
Clearing time = 6 cycles
Note: The feeder overcurrent protective device (OCPD) was a 640-ampere circuit breaker with STD setting of 6 cycles; the short-circuit current was cleared in 6 cycles.

Result:

No. Table 130.7(C)(9) cannot be used to select PPE, since the clearing time of 6 cycles exceeds the 2-cycle limit imposed by Note 2 of the table. Therefore, the incident energy must be calculated.

Table 16.1	*NFPA 70E* Table 130.7(C)(9), in part			
Tasks Performed on Energized Equipment		**Hazard/Risk Category**	**Rubber Insulating Gloves**	**Insulated and Insulating Hand Tools**
Other 600 V Class (277 V through 600 V, nominal) Equipment—Note 2 (except as indicated)				
Work on energized electrical conductors and circuit parts, including voltage testing		2*	Y	Y

Y = yes (required), N = no (not required)
Note 2 Maximum of 65 kA available; maximum of 0.03 sec (2 cycle) fault clearing time.

Reproduced with permission from *NFPA 70E*®-2009, *Electrical Safety in the Workplace*, Copyright © 2008, National Fire Protection Association. This reprinted material constitutes only a portion of the referenced table, is presented for educational purposes only, and is not the complete and official permission of the NFPA on the referenced subject, which is represented only by the standard in its entirety.

Incident Energy Analysis

Incident energy analysis is another option permitted to determine the required PPE to cross the AFPB per *NFPA 70E* Section 130.3(B)(1). This analysis determines the incident energy exposure of the worker in cal/cm². PPE must be selected and used based on the incident energy analysis. The incident energy for Test 4 was calculated in Chapter 12.

According to the basic formula in IEEE 1584, the calculated incident energy is 5.84 cal/cm² at 18 in. Note: IEEE 1584 requires the use of incident energy levels of 1.2, 8, 25, 40, and 100 cal/cm² as discussed in Chapter 12. Therefore, an incident energy of 8 cal/cm² is selected. The incident energy must be documented on the Energized Electrical Work Permit (**Figure 16.15**) and on a label on the motor controller enclosure.

Equipment Labeling

NFPA 70E Section 130.3(C) details the third component of an arc flash hazard analysis. Electrical equipment must have a label affixed that documents either the results of the incident energy analysis (in cal/cm²) or the HRC for the task to be performed.

The incident energy of 8 cal/cm² must be marked on a label on the motor controller enclosure for Test 4.

PPE Selection

Table 130.7(C)(9) cannot be used to select the HRC. When the incident energy analysis option is used, PPE must be selected according to *NFPA 70E* Sections 130.7(C)(1) through (C)(8) and (C)(12) through (C)(16). Summarized, in part:

Section 130.7(C)(1): All parts of the body need to be protected inside the AFPB.
Section 130.7(C)(2): FR clothing must cover all ignitable clothing.

Energized Electrical Work Permit

6. **Results of the arc flash hazard analysis:**
   Available short-circuit current: *22,600 amperes*
   Fault clearing time: *6 cycles*
   The Hazard/Risk Category is: *N/A*
   or
   The Incident Energy is: *8 calories per square centimeter.*

Figure 16.15   The incident energy is required to be documented on the Energized Electrical Work Permit.

Section 130.7(C)(3):
- Head
- Face
- Neck
- Chin

Section 130.7(C)(4): Safety glasses or goggles
Section 130.7(C)(5): FR clothing with an adequate arc rating is required for all parts of the body within the AFPB.
Section 130.7(C)(6):
- Rubber gloves rated for the voltage and leather protectors for shock protection if parts are exposed and energized
- Gloves must be inspected before use (including air test) and electrically tested at least every 6 months
- FR for arms (additionally required by Section 130.7(C)(5) for body protection)

Section 130.7(C)(7): Dielectric overshoes required if protection is required for step and touch potential.

Section 130.7(C)(8): PPE must meet the American National Standards Institute (ANSI) and American Society for Testing and Materials (ASTM) standards that are listed in Table 130.7(C)(8).

Section 130.7(C)(12):

- Clothing can be single layer or be layered.
- Where required, FR protection shall cover all parts of the body inside the AFPB.
- Where required, FR protection shall cover all flammable apparel (such as cotton, wool, and silk).
- If nonmelting flammables are worn under FR clothing, breakopen of FR protection must not cause flammables to ignite.
- Outer layer must be FR (such as rainwear, jackets, sweatshirt, etc.).
- Meltables (such as nylon, polyester, or spandex) worn as underlayers (such as underwear) are not permitted except for incidental amounts of elastic on flammables (e.g., cotton socks and underwear).
- Shirtsleeves must be fastened at the wrist.
- Shirts and jackets must be closed at the neck.
- Garments selected must provide enough protection but not interfere with the task.

Section 130.7(C)(13):

- Arc flash suit hoods must be rated for the exposure.
- Face shields must have an arc rating for the exposure.
- Leather gloves are required when rubber insulating gloves are used for shock protection.
- Heavy-duty leather work shoes must be worn for exposures equivalent to HRC 2 and higher.

Section 130.7(C)(14):

- Clothing must meet the requirements of ASTM F1506.

Section 130.7(C)(15):

- Hardhat liners, hairnets, and other clothing and apparel cannot be made of flammable synthetic material (except nonmelting flammables used as underlayers within limitations).

Section 130.7(C)(16):

- Inspect FR protective equipment before each use.
- Do not use FR protective equipment that is contaminated or damaged.
- Follow manufacturer instructions for care and maintenance.

Sample Energized Electrical Work Permit for Example 2

Figure 16.16 shows a sample Energized Electrical Work Permit for Example 2.

Other protective equipment is selected according to Section 130.7(D) and standards of Table 130.7(D).

Example 3, Test 3

Determine the justification for energized work, the AFPB and PPE, and complete the related portion of a sample Energized Electrical Work Permit to perform Test 3 as described in Chapter 2. The arcing fault is initiated on the line side of the motor branch circuit device (the fault is on the feeder circuit but within the controller enclosure) as illustrated in **Figure 16.17**. The circuit has an available bolted 3-phase, short-circuit current of 22,600 symmetrical rms amperes at 480 volts. Three 601-ampere (KRP-C-601SP) current-limiting fuses (Class L) protect the circuit. The fuses open the circuit in less than ½ cycle and limit the current. An arcing fault is initiated in a size 1 combination motor controller enclosure with the door open (Figure 16.17). The task being performed is voltage testing, and the equipment is energized.

Justification

Determine the justification for energized work per *NFPA 70E* Section 130.1(A). The task is voltage testing. Testing for voltage to verify absence of voltage is a task that is infeasible to perform deenergized. The AFPB and appropriate protective equipment must be selected for the task.

Determine the AFPB for Test 3

The *NFPA 70E* Section 130.3(A)(1) addresses determination the AFPB for voltage levels between 50 and 600 volts. The AFPB is 4 ft for the product of clearing time of 2 cycles and an available bolted short-circuit current of 50 kiloamperes (or any combination not exceeding 100 kiloamperes). The clearing time is 0.5 cycles, and the available bolted short-circuit current is 22,600.

- Bolted short-circuit current = 22.6 kiloamperes
- Clearing time = 0.5 cycles
- The feeder OCPD is a 601-ampere current-limiting Class L fuse.
- The 601-ampere current-limiting fuse clears the fault in ½ cycle (0.008 seconds)

Calculation:
Combination is 22.6 kiloamperes × 0.5 cycles = 11.3-kiloampere cycles

Result:
Yes, the 4-ft conditional AFPB can be used, since an 11.3-kiloampere cycle does not exceed 100-kilo-

Energized Electrical Work Permit

1. **Description of the circuit and equipment to be worked on and their location:**

   Equipment to be worked on is the starter controlling Exhaust Fan #4.

   The starter controlling Exhaust Fan #4 is located on the East wall in Electrical Closet 4.

   Exhaust Fan #4 (EF-4) is located on the roof above Electrical Closet 4.

   The circuit feeding Exhaust Fan #4 (EF-4) is fed by MDP-1 circuits 1, 3, and 5.

   MDP-1 is located on the North wall in Electrical Closet 1.

2. **Justification for why the work must be performed in an energized condition: (Check one and describe)**
   Infeasibility *X*

   or

   Greater Hazard_____

   Describe *Voltage Testing*

3. **Description of the safe work practices to be employed**

   Personal and other protective equipment will be used based on a Shock Hazard Analysis and an Arc Flash Hazard Analysis.

4. **Results of the shock hazard analysis**
   480 Volts

5. **Determination of shock protection boundaries**
   Prohibited Approach Boundary: *1 inches*
   Restricted Approach Boundary: *12 inches*
   Limited Approach Boundary: *42 inches*

6. **Results of the arc flash hazard analysis:**
   Available short-circuit current: *22,600 amperes*
   Fault clearing time: *6 cycles (MDP-1 circuits 1, 3, 5)*
   The Hazard/Risk Category is: _____

   or

   The Incident Energy is: *8* calories per square centimeter.

7. **The Arc Flash Protection Boundary:**
   47 inches

8. **Necessary personal protective equipment to safely perform the assigned task:**
   Note: PPE must meet the ANSI and ASTM standards that are listed in *NFPA 70E* Table 130.7(C)(8) and other protective equipment must conform to the standards in *NFPA 70E* Table 130.7(F).

   Hard hat
   Arc-rated face shield, minimum arc rating of 8
   Balaclava, minimum arc rating of 8
   FR shirt, minimum arc rating of 8
   FR pants, minimum arc rating of 8
   Safety glasses
   Heavy duty leather work boots
   Insulated tools
   Class 00 rubber insulating gloves
   Leather protectors
   Hearing protection*

   * The need for hearing protection is required for all HRCs in Table 130.7(C)(10) but not specifically by any other requirements of Section 130.7(C). Determine the need for hearing protection in accordance with OSHA regulations and standards and company safety program.

9. **Means employed to restrict the access of unqualified persons from the work area:**

   Alerting techniques including signs, barricades, and attendants per NFPA 70E Section 130.7(E).

10. **Evidence of completion of a job briefing, including a discussion of any job-specific hazards:**

    A job briefing per Section 110.7(G) was conducted. See attached completed Job Briefing Checklist Form.

Energized work approval signature and date:

John Q. Safety Officer   *Feb. 1, 2009*

Figure 16.16 Sample Energized Electrical Work Permit for Example 2.

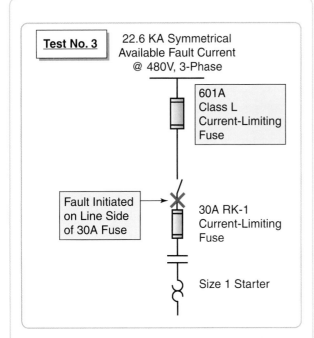

Test No. 3

22.6 KA Symmetrical
Available Fault Current
@ 480V, 3-Phase

601A
Class L
Current-Limiting
Fuse

Fault Initiated
on Line Side
of 30A Fuse

30A RK-1
Current-Limiting
Fuse

Size 1 Starter

Figure 16.17 The circuit is protected by 601-ampere current limiting fuses. An arcing fault is initiated in a size 1 combination motor controller enclosure with the door open.
© 1997 Institute of Electrical and Electronics Engineers. Still photo extracted from the Cooper Bussmann *Safety BASICs Handbook* with permission from Cooper Bussmann.

Energized Electrical Work Permit

7. The Arc Flash Protection Boundary:

48 inches

Figure 16.18 The AFPB must be entered on the Energized Electrical Work Permit.

Energized Electrical Work Permit

7. The Arc Flash Protection Boundary:

10 inches

Figure 16.19 The AFPB can also be calculated. All parts of the body within the AFPB must be protected.

---

ampere cycles. This distance must be entered on the Energized Electrical Work Permit (**Figure 16.18**). All parts of the body must be protected once the worker crosses the AFPB of 48 in.

However, the AFPB can also be calculated per Section 130.3(A)(1). The calculated AFPB is 10 in as covered in Chapter 12. All parts of the body within the AFPB must be protected. In this case, all parts of the body within 10 in must be protected, rather than 48 in, if the conditional boundary is used (**Figure 16.19**).

Voltage testing does not require the use of an Energized Electrical Work Permit per *NFPA 70E* Section 130.1(B)(3). However, safe work practices and PPE are still required in accordance with *NFPA 70E* Chapter 1.

Therefore, the Energized Electrical Work Permit is used to document the necessary information to perform the task safely.

Determine the PPE for Test 3

Section 130.3(B) will determine the PPE required once the AFPB is crossed. Two options are available to determine the appropriate PPE for the work to be performed inside the AFPB; namely, incident energy analysis and determination of the HRC.

The HRC associated with the task to be performed is determined from *NFPA 70E* Table 130.7(C)(9). The task must be listed in Table 130.7(C)(9) and be within the parameters of the specific notes to the table.

- The task is being performed in a size 1, 600-volt, combination motor controller enclosure.
- The category of equipment is listed in Table 130.7(C)(9) as "Other 600 V Class (277 V through 600 V, nominal) Equipment—Note 2" (**Table 16.2**). The table indicates that for work on energized electrical conductors and circuit parts, including voltage testing, the HRC is 2*, rubber insulating gloves are required, and insulated and insulating hand tools are required.
- The available bolted short-circuit current and clearing time must be within Note 2 under the specific notes to the table: Maximum of 65 kilo-amperes available; maximum of 0.03 sec (2 cycle) fault clearing time.

Table 16.2	*NFPA 70E* Table 130.7(C)(9), in part			
Tasks Performed on Energized Equipment		**Hazard/Risk Category**	**Rubber Insulating Gloves**	**Insulated and Insulating Hand Tools**
Other 600 V Class (277 V through 600 V, nominal) Equipment—Note 2 (except as indicated)				
Work on energized electrical conductors and circuit parts, including voltage testing		2*	Y	Y

Y = yes (required), N = no (not required)

Note 2 Maximum of 65 kA available; maximum of 0.03 sec (2 cycle) fault clearing time.

Reproduced with permission from *NFPA 70E*®-2009, *Electrical Safety in the Workplace*, Copyright © 2008, National Fire Protection Association. This reprinted material constitutes only a portion of the referenced table, is presented for educational purposes only, and is not the complete and official permission of the NFPA on the referenced subject, which is represented only by the standard in its entirety.

Test 3

Bolted short-circuit current = 22.6 kiloamperes

Note: The feeder OCPD was a 601-ampere current-limiting fuse with a clearing time of 0.5 cycles.

Result:

Yes, Table 130.7(C)(9) may be used to select PPE, since neither the maximum bolted short-circuit current of 65 kiloamperes nor the clearing time of 2-cycle exceed the limitations imposed by Note 2 of the table. Therefore, PPE may be selected by Table 130.7(C)(9), Table 130.7(C)(10), and Table 130.7(C)(11).

The HRC is 2*. Rubber insulating gloves and insulated and insulating hand tools are required, as indicated in Table 16.2. However, Note (f) to Table 130.7(C)(9) permits the HRC to be reduced by one number.

(f) For equipment protected by upstream current limiting fuses with arcing fault current in their current limiting range (½ cycle fault clearing time or less) the hazard/risk category required may be reduced by one number.

Table 130.7(C)(10) is then used to select the additional protective equipment based on the HRC of 2* reduced by one number to a HRC 2. According to that table, an HRC of 2 requires the following protective clothing and PPE:

- FR clothing, minimum arc rating of 8 (Note 1)
  ○ Arc-rated long-sleeve shirt (Note 5)
  ○ Arc-rated pants (Note 5)
  ○ Arc-rated coverall (Note 6)
  ○ Arc-rated face shield or arc flash suit hood (Note 7)
  ○ Arc-rated jacket, parka, or rainwear (AN)
- FR protective equipment
  ○ Hard hat
  ○ Safety glasses or safety goggles (SR)

  ○ Hearing protection (ear canal inserts)
  ○ Leather gloves (Note 2)
  ○ Leather work shoes

Note 1 indicates that an arc rating for a garment or system of garments is expressed in cal/cm^2 and that Table 130.7(C)(11) must be considered as well.

Note 2 indicates that additional leather or arc-rated gloves are not required if rubber insulating gloves with leather protectors are required by Table 130.7(C)(9) and that the combination of rubber insulating gloves with leather protectors satisfies the arc flash protection requirement.

Note 5 indicates that FR shirt and FR pants used for HRC 2 shall have a minimum arc rating of 8.

Note 6 indicates that an alternate is to use FR coveralls (minimum arc rating of 8) instead of FR shirt and FR pants.

Note 7 indicates that a face shield with a minimum arc rating of 4 for hazard/risk category 1 or a minimum arc rating of 8 for hazard/risk category 2, with wrap-around guarding to protect not only the face, but also the forehead, ears, and neck (or, alternatively, an arc-rated arc flash suit hood), is required.

Finally, SR indicates "selection required."

Note that the *NFPA 70E* Technical Committee unanimously accepted the phrase "may be reduced by one number" in its action on Proposal 70E-322 in the NFPA 2008 Annual Revision Cycle Report on Proposals. Some consider reducing HRC 2* by one number to be reducing the HRC from HRC 2* to HRC 1.

The HRC, protective clothing, and PPE must be documented on the Energized Electrical Work Permit (Figure 16.20) and on a label on the equipment to be worked on.

Protective equipment may also be selected by using the calculated incident energy. The incident energy was calculated for Test 3 in Chapter 12. The incident

Energized Electrical Work Permit

6. Results of the arc flash hazard analysis:
Available short-circuit current: *22,600 amperes*

Fault clearing time: *0.5 cycles*

The Hazard/Risk Category is: *2*

or

The Incident Energy is *N/A* calories per square centimeter.

8. Necessary personal protective equipment to safely perform the assigned task:
Note: PPE must meet the ANSI and ASTM standards that are listed in *NFPA 70E* Table 130.7(C)(8) and other protective equipment must conform to the standards in *NFPA 70E* Table 130.7(F).

Arc-rated long-sleeve shirt, minimum arc rating of 8

Arc-rated pants, minimum arc rating of 8

Arc-rated face shield, minimum arc rating of 8

Hard hat

Safety glasses

Hearing protection (ear canal inserts)

Leather work shoes

Insulated tools

Class 00 rubber insulating gloves

Leather protectors

Figure 16.20 The HRC and protective equipment must be documented on the Energized Electrical Work Permit.

energy value is used in conjunction with the requirements of Sections 130.7(C)(1) through (C)(8) and (C)(12) through (C)(16) to select the PPE. Other protective equipment is selected according to Section 130.7(D) and standards of Table 130.7(D).

Sample Energized Electrical Work Permit for Example 3

Figure 16.21 shows a sample Energized Electrical Work Permit for Example 3.

Summary

This chapter briefly revisited many of the *NFPA 70E* requirements that must be considered for work to be performed where workers are exposed to electrical hazards. Justification for energized work, shock hazard analysis, flash hazard analysis, and the Energized Electrical Work Permit are among the requirements that help protect workers if they are implemented. Engineering out the hazards and limiting energized work to only those tasks that are infeasible or would create a greater hazard if they were deenergized reduces workers' exposure to electrical hazards.

While many topics can be covered in a classroom setting, on-the-job training also helps to build workers' knowledge. Understanding site-specific hazards, demonstrating safe work practices such as lockout/tagout, conducting job briefings, and demonstrating skills are examples of on-the-job training results. Workers should demonstrate those skills on deenergized equipment. They should practice those skills while using the tools, wearing the PPE, and experiencing the environmental conditions to be encountered when the task will be performed energized (such as temperature, lighting, and work space).

Energized Electrical Work Permit

1. **Description of the circuit and equipment to be worked on and their location:**

   Equipment to be worked on is the starter controlling Exhaust Fan #4.

   The starter controlling Exhaust Fan #4 is located on the East wall in Electrical Closet 4.

   Exhaust Fan #4 (EF-4) is located on the roof above Electrical Closet 4.

   The circuit feeding Exhaust Fan #4 (EF-4) is fed by MDP-1 circuits 1, 3, and 5.

   MDP-1 is located on the North wall in Electrical Closet 1.

2. **Justification for why the work must be performed in an energized condition: (Check one and describe)**

   Infeasibility *X*

   or

   Greater Hazard_____

   Describe *Voltage Testing*

3. **Description of the safe work practices to be employed**

   Personal and other protective equipment will be used based on a Shock Hazard Analysis and an Arc Flash Hazard Analysis.

4. **Results of the shock hazard analysis**

   480 Volts

5. **Determination of shock protection boundaries**

   Prohibited Approach Boundary: *1 inches*

   Restricted Approach Boundary: *12 inches*

   Limited Approach Boundary: *42 inches*

6. **Results of the arc flash hazard analysis:**

   Available short-circuit current: *22,600 amperes*

   Fault clearing time: *0.5 cycles (MDP-1 circuits 1, 3, 5)*

   The Hazard/Risk Category is: _____*2*_____

   or

   The Incident Energy is: ____ calories per square centimeter.

7. **The Arc Flash Protection Boundary:**

   48 inches

8. **Necessary personal protective equipment to safely perform the assigned task:**

   Note: PPE must meet the ANSI and ASTM standards that are listed in *NFPA 70E* Table 130.7(C)(8) and other protective equipment must conform to the standards in *NFPA 70E* Table 130.7(F).

   Arc-rated long-sleeve shirt, minimum arc rating of 8

   Arc-rated pants, minimum arc rating of 8

   Arc-rated face shield, minimum arc rating of 8

   Hard hat

   Safety glasses

   Hearing protection (ear canal inserts)

   Leather work shoes

   Insulated tools

   Class 00 rubber insulating gloves

   Leather protectors

9. **Means employed to restrict the access of unqualified persons from the work area:**

   Alerting techniques including signs, barricades, and attendants per NFPA 70E Section 130.7(E).

10. **Evidence of completion of a job briefing, including a discussion of any job-specific hazards:**

    A job briefing per Section 110.7(G) was conducted. See attached completed Job Briefing Checklist Form.

    Energized work approval signature and date:

    John Q. Safety Officer   *Feb. 1, 2009*

Figure 16.21 Sample Energized Electrical Work Permit for Example 3.

Lessons *learned*

18-Year-Old Electrician's Apprentice Electrocuted

An 18-year-old electrician's apprentice was electrocuted when he attempted to work on an energized 277-volt lighting system at a newly constructed industrial park complex. The employer is an electrical contractor and employs 43 persons. The company was established in 1958, but it has been operating for the last 5 years under the present management. The company has no safety program, and its president is the only state-certified electrician employed by the company. All other employees received their electrical training in high school voc-tech classes or through on-the-job training. It is a common company practice to perform electrical work on energized electrical systems.

Two employees of the company (the victim and his supervisor) were in the process of relocating overhead junction boxes for a lighting system in a newly constructed industrial park complex. The initial electrical work at the complex was not done by this company. The two men were working in the same room while another contractor was installing dry wall in an adjacent room.

At approximately 9:30 a.m., the victim was standing on a fiberglass ladder and had just completed the connection on a 277-volt energized system. He proceeded to secure the connection with wire nuts, when he came in contact with the uninsulated wires. He descended the ladder, took three steps, and collapsed. The supervisor called for help, and the dry wall workers from the adjacent room immediately responded and began performing cardiopulmonary resuscitation (CPR) on the victim. The victim could not be revived. A burn wound on the victim's index finger suggested point of entry.

Adapted from: Fatality and Assessment and Control Evaluation (FACE) 86-08. Accessed 9/11/08 at http://www.cdc.gov/niosh/face/In-house/full8608.html.

Recommendations/Discussion

Recommendation 1: Electrical work should not be performed on energized systems.

Discussion: Work was being performed on the electrical system while it was energized. Disconnecting the source of the power supply before working on the system would have prevented this fatality. Working on energized systems is not recommended even if working conditions appear to be ideal. This was not the case. The only source of working light in the room was the sunlight. Glare from the sun could have hampered the victim. The dry wall work also created a considerable amount of dust that limited visibility.

Recommendation 2: The employer should develop a written safety policy and a safety program.

Discussion: The company does not provide training in safe work procedures nor are there any rules or written policies governing safe electrical installations. None of the benefits that a safety program would provide (i.e., training, hazard identification, personal protective equipment, and safe operating procedures) were utilized. This is evident by the poor safety record for 1985 (290 lost work days excluding this fatality).

Recommendation 3: Employers should determine the capabilities of employees prior to job assignment and employees should be assigned tasks that they are qualified to perform.

Discussion: The victim had been employed by the company for approximately three months. He had received no previous electrical training and on-the-job training was very limited. In spite of his limited training, he was expected to perform the same work tasks as more experienced employees within this company. Employers should determine employee capabilities prior to job assignment and only tasks within those capabilities should be assigned.

Questions

1. Who should be blamed for this fatality?
   - **A.** The electrician's apprentice himself
   - **B.** The supervisor
   - **C.** The owner of the company
   - **D.** The two drywall installers

2. Was the victim qualified to do the work?
   - **A.** Yes
   - **B.** No

3. What was the most important factor that could have prevented this electrocution?
   - **A.** Portable lighting should have been brought in.
   - **B.** The electrician's helper should have been wearing voltage-rated gloves.
   - **C.** The electrician's helper should have received more training.
   - **D.** The circuit where the work was to be performed should have been locked and tagged out.

4. Why might the apprentice have been assigned a task that required working on an energized circuit?
   - **A.** He was being supervised while undergoing on-the-job training.
   - **B.** He was the junior employee and had to serve his time with energized circuits.
   - **C.** He had participated in the original installation of the lighting circuit.
   - **D.** He had no family to support, so a lost-time injury would affect only the number of workers available to the contractor.

Ready for Review

- This chapter reviews many requirements related to electrical safety, and also provides examples of determining whether energized work is justified, determining the AFPB and protective equipment, and how the results of a shock hazard analysis and arc flash hazard analysis are integrated into a sample energized electrical work permit.
- Host and contractor responsibilities are contained in *NFPA 70E* Section 110.5. Host employers must ensure that contractors are aware of hazards. Contractors must ensure that employees are trained to recognize and avoid the reported hazards.
- Host employers must inform the contract employer about information about the installation that the contract employer needs to make the assessments required by *NFPA 70E* Chapter 1.
- Training requirements in *NFPA 70E* are addressed in Section 110.6. Those requirements are broken down into five major categories:
  - Safety training
  - Type of training
  - Emergency procedures
  - Qualified and unqualified person training
  - Documentation of training
- *NFPA 70E* electrical safety program requirements are detailed in Section 110.7. Employers must implement an electrical safety program to protect workers from the voltage, energy level, and circuit conditions. *NFPA 70E* Annex E covers additional electrical safety program considerations.
- *NFPA 70E* Section 110.8 describes requirements for working while exposed to electrical hazards. Appropriate safety-related work practices must be used, and are determined by an electrical hazard analysis, which includes both a shock and arc flash hazard analysis.
- *NFPA 70E* Section 110.9 details requirements for test instruments and equipment, portable electric equipment, GFCI protection devices, and overcurrent protection modification.
- OSHA requires live parts to which an employee might be exposed to be deenergized before the employee works on or near them, unless the employer can demonstrate that deenergizing introduces additional or increased hazards or is infeasible due to equipment design or operational limitations.
- OSHA's lockout/tagout requirements are law and the minimum safety-related work practice that must be implemented.
- *NFPA 70E* Section 120.1 outlines six steps to verify that an electrically safe work condition exists after the provisions contained in *NFPA 70E* Section 120.2 have been taken into account.
- *NFPA 70E* Section 130.1(A) requires energized electrical equipment to be placed into an electrically safe work condition if a worker is to cross the LAB. Two FPNs provide examples of infeasibility and increased or additional hazards that would permit energized work.
- Safe work practices, including personal and other protective equipment, are required even if the employer has demonstrated that energized work is justified.
- Energized work is generally required to be performed in accordance with the Energized Electrical Work Permit requirements of Section 130.1(B) if the equipment is not in an electrically safe work condition.
- *NFPA 70E* Annex J.1 offers a sample Energized Electrical Work Permit.
- *NFPA 70E* Section 130.1(B)(2) details the minimum information that must be documented in an energized electrical work permit.
- *NFPA 70E* Section 130.1(B)(3) recognizes a limited number of examples where energized work can be performed by qualified persons inside the LAB without an energized electrical work permit as long as appropriate work practices and PPE are used.
- *NFPA 70E* Section 130.2(A) requires a shock hazard analysis. The shock hazard analysis must determine the voltage to which workers will be exposed, shock protection boundaries, and PPE required for shock protection.
- The three shock protection boundaries established by *NFPA 70E* include the LAB, the re-

stricted approach boundary, and the prohibited approach boundary. These three shock protection boundaries apply when workers are exposed to energized electrical conductors or circuit parts.

- An arc flash hazard analysis is required by *NFPA 70E* Section 130.3 to determine the AFPB and the PPE necessary for workers who cross that boundary.

- The *NFPA 70E* Section 130.3(A) requirements for determining the AFPB are broken down into two voltage level categories: Between 50 and 600 volts and above 600 volts.

- Per *NFPA 70E* Section 130.3(A)(1), a conditional AFPB of 4 ft may be used where any combination of clearing time and available bolted short-circuit current not exceeding 100 kiloamperes.

- When the product of the clearing time and available bolted short-circuit current exceeds the conditions where the 4 ft boundary can be used, the AFPB must be calculated.

- The available bolted short-circuit current and clearing time must be known to determine the AFPB either by calculation or the use of the 4 ft conditional boundary.

- *NFPA 70E* Section 130.3(B) determines the PPE required once the AFPB is crossed. Two options are available to determine the appropriate PPE for the work to be performed inside the AFPB; namely incident energy calculation and the *NFPA 70E* HRC method.

- Incident energy analysis determines the incident energy exposure of the worker in calories per square centimeter. PPE must be selected and used based on the incident energy analysis.

- The incident energy must be documented on the Energized Electrical Work Permit and on a label on the enclosure.

- *NFPA 70E* Section 130.4 requires workers to be qualified to perform electrical testing on electrical equipment at voltage levels of 50 volts or more once they cross the LAB for that energized equipment.

- *NFPA 70E* Section 130.5 addresses tasks performed within the LAB of uninsulated overhead lines.

- *NFPA 70E* Section 130.6 details numerous precautions that must be considered for personnel activities.

- *NFPA 70E* Section 130.7(A) assigns responsibility to employers and employees for protective equipment use. Workers must be provided with, and they must use, protective equipment that is designed for the use and adequate for the hazard when performing tasks where electrical hazards are present.

- *NFPA 70E* Section 130.7(B) requires protective equipment to be maintained, inspected, and stored properly. It must be maintained in a safe, reliable condition; be visually inspected before each use; and be stored in a manner that prevents damage.

- *NFPA 70E* Section 130.7(C)(1) requires all parts of the body inside the AFPB to be protected in accordance with *NFPA 70E* Section 130.3.

- Appropriate PPE is selected by the requirements of *NFPA 70E* Section 130.7(C), once the incident energy or HRC is determined.

- PPE must conform to the standards provided in Table 130.7(C)(8).

- The HRC method is used for the selection of personal and other protective equipment in conjunction with the requirements of *NFPA 70E* Sections and Tables 130.7(C)(9), 130.7(C)(10), and 130.7(C)(11).

- The HRC associated with the equipment and the task to be performed is determined from *NFPA 70E* Table 130.7(C)(9). The equipment and the task must be listed in Table 130.7(C)(9) and be within the parameters of the specific notes to the table. If the equipment and task are not within these parameters, the incident energy must be calculated.

- If the HRC method is being used, Table 130.7(C)(9) determines the HRC and whether rubber insulating gloves and insulated and insulating hand tools are required.

- PPE is selected according to *NFPA 70E* Sections 130.7(C)(1) through (C)(8) and (C)(12) through (C)(16) in conjunction with an incident energy analysis.
- *NFPA 70E* Section 130.7(D)(1) details considerations for other protective equipment.
- *NFPA 70E* Section 130.7(E)(1) requires that safety signs, safety symbols, or accident prevention tags be used to warn employees about electrical hazards.
- *NFPA 70E* Section 130.7(E)(2) requires barricades to be used in conjunction with safety signs where it is necessary to prevent or limit employee access. *NFPA 70E* Section 130.7(E)(3) describes where the use of an attendant is warranted.
- *NFPA 70E* Section 130.7(E)(4) addresses considerations for look-alike equipment.
- Other protective equipment required in Section 130.7(D) must conform to the standards provided in Table 130.7(F).

Vocabulary

arc flash hazard analysis* A study investigating a worker's potential exposure to arc-flash energy, conducted for the purpose of injury prevention and the determination of safe work practices, arc flash protection boundary, and the appropriate levels of PPE.

arc flash protection boundary (AFPB)* When an arc flash hazard exists, an approach limit at a distance from a prospective arc source within which a person could receive a second degree burn if an electrical arc flash were to occur.

limited approach boundary* An approach limit at a distance from an exposed energized electrical conductor or circuit part within which a shock hazard exists.

prohibited approach boundary* An approach limit at a distance from an exposed energized electrical conductor or circuit part within which work is considered the same as making contact with the electrical conductor or circuit part.

restricted approach boundary* An approach limit at a distance from an exposed energized electrical conductor or circuit part within which there is an increased risk of shock, due to electrical arc over combined with inadvertent movement, for personnel working in close proximity to the energized electrical conductor or circuit part.

*This is the *NFPA 70E* definition.

OSHA Citation Examples

Example 1: Violation of 29 CFR 1926.21(b)(2)

U.S. Department of Labor

Occupational Safety and Health Administration

Inspection Number: 306541145

Inspection Date: 09/11/2003–09/11/2003

Issuance Date: 09/22/2003

Citation and Notification of Penalty

Company Name: XXXXXXXXXXXX

Inspection Site: XXXXXXXXXXXX

Citation 1 Item 2b   Type of Violation: **Serious**

29 CFR 1926.21(b)(2): The employer did not instruct each employee in the recognition and avoidance of unsafe condition(s) and the regulation(s) applicable to his work environment to control or eliminate any hazard(s) or other exposure to injury:

a.   On the site, employees were terminating welding conductors on a 120/208 energized panel thereby exposed to electrical hazards without personal protective equipment.

OR IN THE ALTERNATIVE:

Section 5(a)(1) of the Occupational Safety and Health Act of 1970: The employer did not furnish employment and a place of employment which were free from recognized hazards that were causing or likely to cause death or serious physical harm to employees in that employees were exposed to:

In Columbus, OH on 09/11/03: Employees were terminating welding conductors on a 120/208 energized electrical panel, thereby exposed to electrical hazards without personal protective equipment.

Among other methods, one feasible and acceptable means to correct the hazardous condition may include:

a.   Comply with NFPA 70E, Part II, paragraph 2-1.1.1 which requires de-energizing of live electrical parts.

b.   Comply with NFPA 70E, part II, paragraph 2-1.3.2 which requires work practices to be developed, based on a Flash Hazard Analysis to protect employees from live electrical parts.

Date By Which Violation Must Be Abated: 11/14/03

See pages 1 through ? of this Citation and Notification of Penalty for information on employer and employee rights and responsibilities.

U.S. Department of Labor

Occupational Safety and Health Administration

Worksheet

Fri Sep 19, 2003 4:07pm

				Inspection Number	306541145
				Opt. Insp. Number	Ex 7(e)

Establishment Name					
Type of Violation	S Serious	Citation Number	01	Item/Group	002 (b)
Number Exposed	2	No. Instances	1	REC	
Std. Alleged Vio.	1926.0021(b)(2)				

Abatement Period	MultiStep Abatements			Final Abatement	Action Type/Dates
	PPE Period	Plan	Report		
7					

Abatement Documentation Required		Date Verified	

Substance Codes	

AVD/Variable Information:	

29 CFR 1926.21(b)(2): The employer did not instruct each employee in the recognition and avoidance of unsafe condition(s) and the regulation(s) applicable to his work environment to control or eliminate any hazard(s) or other exposure to illness or injury:

     a. On the site, employees were terminating welding conductors on a 120/208 energized panel thereby exposed to electrical hazards without personal protective equipment.

OR IN THE ALTERNATIVE:

Section 5(a)(1) of the Occupational Safety and Health Act of 1970: The employer did not furnish employment and a place of employment which were free from recognized hazards that were causing or likely to cause death or serious physical harm to employees in that employees were exposed to:

In Columbus, OH on 09/11/03: Employees were terminating welding conductors on a 120/208 energized electrical panel, thereby exposed to electrical hazards without personal protective equipment.

Among other methods, one feasible and acceptable means to correct the hazardous condition may include:

a. Comply with NFPA 70E, Part II, paragraph 2-1.1.1 which requires de-energizing of live electrical parts.

b. Comply with NFPA 70E, part II, paragraph 2-1.3.2 which requires work practices to be developed, based on a Flash Hazard Analysis to protect employees from live electrical parts.

Penalty Calculations				Adjustment Factors			Proposed Adjusted Penalty
Severity	Probability	Gravity	GBP	Size	Good Faith	History	
H High	L Lesser	03	2500.00	40	0	0	0.00
Repeat Factor		0					

Employee Exposure:	
Occupation	
Nr of Employees	
Employee Name	
Address	Ex 7(c)
Occupation	

OSHA-1B/1BIHprint(Rev. 9/93)

Fri Sep 19, 2003 4:07pm

Inspection Nr. 306541145 Citation Nr. 01 Item/Group 002 (a)

---

a) Hazards-Operation/Condition-Accident: On the site, the employer did not ensure that an electrical panel was de-energized prior to employees working on it. The employees were installing the connection for the welder. The employees were not wearing personal protective equipment while working on the energized panel. A temporary light was connected to the panel and it was on.

b) Equipment: Electrical Panel

c) Location: On the site, Basement, Area B, Mechanical Room

d) Injury/Illness: Shock/Electrocution

e) Measurements: 120/208 - photo

21. Photo Number	22. Location on Video

23. Employer Knowledge: Yes, the employer knew the requirement to ensure the panel was de-energized prior to work being done. He was on the site and the violation was in plain view.

24. Comments (Employer, Employee, Closing Conference): Closing conference was held with Ex 7(c) The violation was discussed.

25. Other Employer Information:

26. Classification:				
Serious	Knowledge	S or O	Repeat?	Willful?
Y	Y	S	N	N

First Repeat	Second Repeat	Repeat Penalty

Event Date	Event Code	Action Code	Citation Type	Penalty	Abate Date	Final Order
	Z Add transaction	A Add	S Serious	1500.00		

OSHA-1B/1BIHprint(Rev. 9/93)

Example 2: Violation of 29 CFR 1926.416(a)(1)

U.S. Department of Labor

Occupational Safety and Health Administration

Inspection Number: 305430878

Inspection Date: 10/30/2002–10/30/2002

Issuance Date: 11/20/2002

Citation and Notification of Penalty

Company Name: XXXXXXXXXXXX

Inspection Site: XXXXXXXXXXXX

Citation 1 Item 3   Type of Violation: **Serious**

29 CFR 1926.416(a)(1): Employees were permitted to work in proximity to electric power circuits and were not protected against electric shock by deenergizing and grounding the circuits or effectively guarding the circuits by insulation or other means:

a.   In the first floor switch room where wiring was performed on an energized electrical panel, protection was not used during the process.

OR IN THE ALTERNATIVE:

Section 5(a)(1) of the Occupational Safety and Health Act of 1970: The employer did not furnish a place of employment which was free from recognized hazards that were causing or likely to cause death or physical harm to employees in that employees were exposed to:

1.   In the first floor main switch room, wiring was installed in an energized 277.480 volt panel.

   Among other methods, one feasible and acceptable means to correct the hazardous condition may include:

   a.   Comply with NFPA 70E, Part II, paragraph 2-1.1.1 which requires deenergizing of live electrical parts.

   b.   Comply with NFPA 70E, part II, paragraph 2-1.3.2 which requires work practices to be developed which protects employees from live electrical parts.

   Abatement certification is required by 29 CFR Part 1903.

Date By Which Violation Must Be Abated:	11/28/2002
Proposed Penalty:	$ 2500.00

Deborah J. Zubaly, Area Director

Columbus Area Office

See pages 1 through ? of this Citation and Notification of Penalty for information on employer and employee rights and responsibilities.

Citation and Notification of Penalty   Page 10 of 11   OSHA-2 (Rev. 9/93)

Example 3: Violation of 29 CFR 1926.95(a)(1)

U.S. Department of Labor
Occupational Safety and Health Administration

Inspection Number: 305430878
Inspection Date: 10/30/2002–10/30/2002
Issuance Date: 11/20/2002

Citation and Notification of Penalty

Company Name:
Inspection Site: XXXXXXXXXXXX

Citation 1 Item 1 Type of Violation: **Serious**

29 CFR 1926.95 (a)(1): Protective equipment was not used where it was necessary by reason of hazards encountered in a manner capable of causing injury or impairment in the function of any part of the body:

a.   In the first floor main switch room where live panels were wired, personal protection was not worn.

Abatement certification is required by 29 CFR Part 1903.

Date By Which Violation Must Be Abated:	11/28/2002
Proposed Penalty:	$ 1250.00

Citation 1 Item 2 Type of Violation: **Serious**

29 CFR 1926.403(i)(2)(i): Live parts of electrical equipment operating at 50 volts or more were not guarded against accidental contact by cabinets or other forms of enclosures, or by any of the following means: (A) by location in a room, vault, or similar enclosure that is accessible only to qualified persons; (B) by partitions or screens so arranged that only qualified persons will have access to the space within reach of the live parts; (C) by location on a balcony, gallery, or platform so elevated and arranged as to exclude unqualified persons; (D) by elevation of 8 feet or more above the floor or other working surface and so installed as to exclude unqualified persons:

a.   In the first floor main switch room where electrical work was performed, covers were not placed over the 277/480 volt energized panel.

Date By Which Violation Must Be Abated:	Corrected During Inspection
Proposed Penalty:	$ 1000.00

See pages 1 through ? of this Citation and Notification of Penalty for information on employer and employee rights and responsibilities.

Citation and Notification of Penalty   Page 9 of 11   OSHA-2 (Rev. 9/93)

Occupation	
Nr of Employees	
Employee Name	Ex 7(c)
Address	

Instance Description:	A. Hazard	B. Equipment	C. Location	D. Injury/Illness	E. Measurements

4. Date/Time
10-30-2002

20.   Instance Description - Describe the following:
   a) In the first floor switch room where wiring was performed on a 277/408 energized panel, the proper procedures were not implemented or used thereby exposing employees to a contact hazard. The employees were putting in new electrical service for the Drury Hotel. The foreman was connecting circuits to an energized panel without proper procedures or protection from energized circuits.
   b) Safety related work practices/Lockout
   c) First floor main switch room
   d) Fracture/death
   e) 277/480 volt main/tic tracer

21. Photo Number	22. Location on Video

23. Employer Knowledge: Yes, the foreman stated he knew the requirements and did not know why they were working on the equipment energized.

24. Comments (Employer, Employee, Closing Conference): CSHO explained hazard/requirement and suggested means for abatement

25. Other Employer Information:

26. Classification:				
Serious	Knowledge	S or O	Repeat?	Willful?
Y	Y	S	N	N

First Repeat	Second Repeat	Repeat Penalty

INSPECTION 305430878

Example 4: Violation of 29 CFR 1910.333(a)(2)

U.S. Department of Labor
Occupational Safety and Health Administration

Worksheet

Wed May 7, 2003 2:09pm

	Inspection Number	304363542
	Opt. Insp. Number	

Establishment Name						
Type of Violation	**S Serious**	Citation Number	**01**	Item/Group	006 (a)	
Number Exposed	12	No. Instances	1	REC		
Std. Alleged Vio.	**1910.0333(a)(2)**					

Abatement Period	MultiStep Abatements			Final Abatement	Action Type/Dates	
	PPE Period	Plan	Report			
45						

Abatement Documentation Required		N	Date Verified		

Substance Codes	**8870 - ELECTRICAL SHOCK**

AVD/Variable Information:

29 CFR 1910.333(A)(2): Where exposed live parts were not de-energized, safety related work practices were not used to protect employees who could be exposed to the electrical hazards involved:

a.   The employer had not developed electrical safety related work practices as required by this part to protect employees performing required duties involving work exposure to energized electrical circuits. Employees, such as the Maintenance manager, Maintenance mechanics and Maintenance technicians perform tasks with energized circuits involving 110V and 480V without wearing any electrical protective equipment, or without using any barriers, shields or insulating materials. Those tasks include preventive maintenance on compressors (using an "amp probe" to take voltage readings while the compressor motor is under load, i.e. energized circuits), checking relays in electrical panel boxes while energized to determine amperage readings and general troubleshooting to evaluate electrical equipment failure (such as the replacement of the ringblower wiring for the Civic high speed molding machine due to improper voltage). In addition, an arc flash analysis had not been performed to determine the potential hazards of the work and determine the necessity of personal protective equipment, specialized equipment, barriers, shields or insulating materials. Due to these practices, employees are exposed to related electrical hazards of shock, contact, electrocution, arc blast and arc flash.

OR IN THE ALTERNATIVE:

Section 5(a)(1) of the Occupational Safety and Health Act of 1970: The employer did not furnish a place of employment which was free from recognized hazards that were causing, or likely to cause, death or physical harm to employees in that employees were exposed to:

4.   Tasks performed on energized circuits of up to 480 Volts, including but not limited to, preventive maintenance on compressors (using an "amp probe" to take voltage readings while the compressor motor is under load, i.e. energized circuits), checking relays in electrical panel boxes while energized to determine amperage readings and general troubleshooting to evaluate electrical equipment failure (such as the replacement of the ringblower wiring for the Civic high speed molding machine due to improper voltage). The Maintenance manager, Maintenance mechanics and maintenance technicians perform these tasks without protection from the related electrical hazards of shock, electrocution, arc blast and arc flash.

Among other methods, one feasible and acceptable means to correct the hazardous condition may include:

a.   Comply with NFPA 70E, Part II, paragraph 2-1.1.1 which requires de-energizing of live electrical parts.

b.   Comply with NFPA 70E, part II, paragraph 2-1.3.2 which requires work practices to be developed which protects employees from live electrical parts.

OSHA-1B/1BIHprint(Rev. 9/93)

In general the employees knew to keep a "safe" distance and one always wore clean leather (not electrical insulating, i.e. rubber) gloves.

b) Equipment: Mongo Press, subpanel boxes, Atlas compressors
c) Location: throughout facilities
d) Injury/Illness electrocution, arc blast/flash
e) Measurement: 24 V, 110 V, 240V, 480 V

21. Photo Number	22. Location on Video

23. Employer Knowledge: employees had been sent to electrical work safety classes in the past per Ex 7(c) stated employees had been sent to outside classes but couldn't recall what the classes were, direct awareness of the National Electrical Code

24. Comments (Employer, Employee, Closing Conference):

25. Other Employer Information:

26. Classification:				
Serious	Knowledge	S or O	Repeat?	Willful?
Y	Y	S	N	N

Event Date	Event Code	Action Code	Citation Type	Penalty	Abate Date	Final Order
	Z Add transaction	**A Add**	**S Serious**	1875.00		

Example 5: Violation of 29 CFR 1910.335(a)(1)(i)

U.S. Department of Labor
Occupational Safety and Health Administration

Inspection Number: 306539271
Inspection Date: 06/25/2003–06/25/2003
Issuance Date: 07/15/2003

Citation and Notification of Penalty

Company Name: XXXXXXXXXXXX
Inspection Site: XXXXXXXXXXXX

Citation 1 Item 1   Type of Violation: **Serious**

29 CFR 1910.335(a)(1)(i): Employees working in areas where there were potential electrical hazards were not provided with, and/or did not use, electrical protective equipment that was appropriate for the specific parts of the body to be protected and for the work to be performed:

a.   In the Production Area, where an employee was installing switches, on an energized buss, proper protection was not used during the process.

OR IN THE ALTERNATIVE:

Section 5(a)(1) of the Occupational Safety and Health Act of 1970: The employer did not furnish a place of employment free from recognized hazards that were causing or likely to cause death or serious physical harm to employees in that employees were exposed to:

1.   In the Production Area, Switches were installed in an energized buss without the proper protective equipment such as face shields, gloves and long sleeves.

Among other methods, one feasible and acceptable method to correct the hazardous condition may include:

a.   Comply with NFPA 70E, Part II, paragraph 2-1.3.2 which requires deenergizing of live electrical parts.

b.   Comply with NFPA 70E, part II, paragraph 2-1.3.2 which requires work practices to be developed which protects employees from live electrical parts.

c.   Comply with NFPA 70E, part II, paragraph 2-1.3.3 which requires a flash hazard analysis before employees approach any electrical circuits.

Abatement certification is required by 29 CFR Part 1903.

Date By Which Violation Must Be Abated:	07/23/2003
Proposed Penalty:	$ 3500.00

Deborah J. Zubaty, Area Director
Columbus Area Office

See pages 1 through ? of this Citation and Notification of Penalty for information on employer and employee rights and responsibilities.

Instance Description:	A. Hazard	B. Equipment	C. Location	D. Injury/Illness	E. Measurements

4. Date/Time
6-25-03

20.   Instance Description - Describe the following:
   a) Hazards-Operation/Condition-Accident: An employee installing a large switch on an energized buss suffered burns when the switch arced and flashed creating a contact exposure to the employee. He had no face shield, gloves and was wearing short sleeves. He contracted burns to his face, arms, chest and hands.
   b) Equipment: face shield, gloves and long sleeves.
   c) Location: Production Area
   d) Injury/Illness   burns
   e) Measurement: na

21. Photo Number	22. Location on Video
see photos	

5(a)(1) Worksheet

Ex 7(e) and 5

23. Employer Knowledge: The employer was familiar with the operation and could easily observe the condition.

24. Comments (Employer, Employee, Closing Conference): Ex 7(d)

25. Other Employer Information: na

26.   Classification:

AGREEMENT ESTABLISHING A PARTNERSHIP

BETWEEN

THE OCCUPATIONAL SAFETY AND HEALTH ADMINISTRATION
U.S. DEPARTMENT OF LABOR—OSHA REGION V
COLUMBUS AREA OFFICE

AND

THE NATIONAL ELECTRICAL CONTRACTORS ASSOCIATION (NECA)
CENTRAL OHIO CHAPTER

AND

THE INTERNATIONAL BROTHERHOOD OF ELECTRICAL WORKERS (IBEW)
LOCAL UNIONS 683 & 1105

Excerpts from the Occupational Safety and Health Act of 1970: *29 USC 651.*

(a) The Congress finds that personal injuries and illnesses arising out of work situations impose a substantial burden upon, and are a hindrance to, interstate commerce in terms of lost production, wage loss, medical expenses, and disability compensation payments.

(b) The Congress declares it to be its purpose and policy, through the exercise of its powers to regulate commerce among the several States and with foreign nations and to provide for the general welfare, to assure so far as possible every working man and woman in the Nation safe and healthful working conditions and to preserve our human resources --

> *(1) by encouraging employers and employees in their efforts to reduce the number of occupational safety and health hazards at their places of employment, and to stimulate employers and employees to institute new and to perfect existing programs for providing safe and healthful working conditions;*

I. Introduction and Identification of Partners

In an effort to more fully realize the objectives of the Occupational Safety and Health Act of 1970 to provide a safe and healthful work environment for all workers engaged in the electrical construction and maintenance industry; The Central Ohio Chapter, National Electrical Contractors Association, Inc. (NECA) and International Brotherhood of Electrical Workers Local Unions 683 and 1105 (IBEW), an industry partnership, by this Charter enter into an agreement with the United States Department of Labor, Occupational Safety and Health Administration, Region V, Columbus Area Office (OSHA).

For over 100 years NECA/IBEW labor–management partnerships have provided its respective members and the entire construction industry with model programs designed to meet industry specific issues. These models are as diverse as award winning Columbus

State Community College credit joint training programs for the professional development of apprentices and journeymen. The Council on Industrial Relations which provides a means for labor and management to settle disputes without strikes or lockouts, to multiple trust agreements to manage benefit and other industry funds. It is because of these relationships that our business proudly bears the moniker of a "strike–proof industry," our issues are settled in house between peers.

The NECA/IBEW partnership is pleased to extend the hand of cooperation and consideration to the Columbus Area OSHA Office, not only in the construction industry but in general industry as well; once more building groundbreaking relationships and expanding the umbrella of partnership to the key organization within the safety community.

Our organizations have enjoyed a high level of cooperation though the year, the development of a national electrical worker safety curriculum is just one of many examples. This Charter builds on this important foundation, and denotes a new era of labor–management–agency cooperation and insight. And, of course, in the end meaningful increases in workplace health and safety benefits the central Ohio electrical worker, and their employers.

Employers who perform electrical construction and maintenance contracting services within the jurisdictional boundaries of, and are signatory to, International Brotherhood of Electrical Workers, Local Union 683 and Local Union 1105 and are a member in good standing of The Central Ohio Chapter, National Electrical Contractors Association, Inc. shall be eligible to apply to this charter.

II.   Purpose and Scope

Working as partners and associates the above parties are committed to achieve measurable, meaningful improvements in electrical worker safety using the following blueprint:

   a. It is the objective to design, and build an open, and continuous communication channel between OSHA, the NECA/IBEW partnership and participating firms. This channel will appreciate the unique role electrical workers and their employers play in today's construction industry and indeed society in general.

   b. Actively research, share and implement the top safety and health programs for electrical workers. This includes technology, innovations and best practices that provide measurable improvement in electrical worker safety. These will be shared with the NECA/IBEW Charter members at the quarterly meetings.

   c. Continuously develop, build and share improved, effective safety programs specifically for electrical workers.

   d. With help from the National Joint Apprenticeship and Training Committee (NJATC) develop and build improved, effective, meaningful safety training programs specific enough for our trade, yet broad enough to be effective in every facet of our diverse industry.

   e. Constantly recognize and promote electrical worker safety excellence.

   f. Through consequential and honest communication ensure that enforcement policies and practices are consistent, fair and effective.

   g. Understand that any and all parties to this Charter may withdraw from this agreement 30 days following submission of written notification of intent to withdraw.

III. Goals, Strategies and Measures

GOALS	STRATEGIES	MEASURES
1. Decrease in the participating employee injury illness rate over the life of the Charter.	a. Increase implementation of effective safety and health program through training, self-inspections, use of NFPA 70e *Electrical Safe Work Practices* and employee involvement.	i. Comparison of DART injury/illness rate to the most recently published BLS data.
2. Increased accessibility to quality safety training and education, thereby raising safety awareness for both employee and employer.	a. Provide quarterly meeting for Charter participants and stakeholders b. Collect local accident and near-miss data. c. The use of on the job tool box talks appropriate to the project will be encouraged	i. Document quarterly meetings ii. Use the collected data to focus training efforts in a meaningful manner to empirically identified topics.
3. All parties will work to promote and encourage new members	a. Charter will be promoted at industry events and functions.	i. Track the number of new employers who apply to this Charter. ii. Track the number of new employers who become members of this Charter

These outcomes will be obtained using the following methods:

- Adopt an industry standard checklist designed to exceed OSHA requirements that will be required to be used when working energized circuits. This policy will be based on the most recent National Fire Protection Association® (NFPA®) 70e Standard for Electrical Safety Requirements for Employee Workplaces which can be found in Appendix A of this partnership.
  - Part II, Appendix C of this standard provides a typical description of Electrical Safety Programs that are built around NFPA 70e. To quote:

C-1 Typical Electrical Safety Program Principles. *Electrical safety program principles can include, but are not limited to, the following:*

- *(a) Inspect/evaluate the electrical equipment*
- *(b) Maintain the electrical equipment's insulation and enclosure integrity*
- *(c) Plan every job and document first-time procedures*
- *(d) De-energize, if possible (see 2-1.1.3)*
- *(e) Anticipate unexpected events*
- *(f) Identify and minimize the hazard*
- *(g) Protect the employee from shock, burn, and blast, and other hazards that are due to the working environment*
- *(h) Use the right tools for the job*
- *(i) Assess people's abilities*
- *(j) Audit these principles*

C-2 Typical Electrical Safety Program Controls. *Electrical safety program controls can include, but are not limited to, the following:*

- *(a) Every electrical conductor or circuit part is considered energized until proven otherwise.*
- *(b) No bare-hand contact is to be made with exposed energized electrical conductors or circuit parts above 50 volts to ground, unless the "bare-hand method" is properly used.*

(c) *De-energizing an electrical conductor or circuit part and making it safe to work on is in itself a potentially hazardous task.*

(d) *Employer develops programs, including training, and employees apply them.*

(e) *Use procedures as "tools" to identify the hazards and develop plans to eliminate/control the hazards.*

(f) *Train employees to qualify them for working in an environment influenced by the presence of electrical energy.*

(g) *Identify/categorize tasks to be preformed on or near exposed energized electrical conductors and circuit parts.*

(h) *Use a logical approach to determine potential hazard of task.*

(i) *Identify and use precautions appropriate to the working environment.*

C-3 Typical Electrical safety program Procedures. *Electrical safety program procedures can include, but are not limited to, the following:*

(a) *Purpose of task*

(b) *Qualifications and number of employees to be involved*

(c) *Hazardous nature and extent of task*

(d) *Limits of approach*

(e) *Safe work practices to be utilized*

(f) *Personal protective equipment involved*

(g) *Insulating materials and tools involved*

(h) *Special precautionary techniques*

(i) *Electrical diagrams*

(j) *Equipment details*

(k) *Sketches/pictures of unique features*

(l) *Reference data*

- Mandatory OSHA 10 hour courses for all field employees and mandatory OSHA 30 hour courses for all field supervisors. All employees will receive refresher courses every three years.
- Site specific safety training for new hires.
- Regular safety program third party audits.
- Quarterly meetings discussing best industry practices.

All parties will be consulted on a continuing basis for feedback to assess the progression and potential amelioration of these goals.

IV. Performance Measures

1. The measurement system will use OSHA recordable fatalities and accidents identified.

2. Activity measures shall include:
- The applicable number of employers, supervisors, and employees trained.
- Number of job safety analyses conducted.
- Number of work-site audits.
- Increased employee involvement.
- Enhanced communication between management and employees.

3. Outcome measures will be gathered on an annual basis and will be comprised of data analyzing the number of hours worked, number of injuries, illnesses, and fatalities during the baseline year.

4. All Participants will monitor and submit an Incident Worksheet on all accident and near-misses. Forms will be submitted to the NECA office on quarterly basis utilizing the worksheet found in Appendix B. *All information will remain confidential as to company name and employee. The only purpose of this data is to assess performance and to target specific training requirements.*

V.   Annual Evaluation

The annual evaluation shall use the Strategic Partnership Annual Evaluation Format measurement system as specified in Appendix C of CSP 03-02-002, OSHA Strategic Partnership Program for Worker Safety and Health Directive.

All data for the annual evaluation shall be submitted to the NECA/IBEW partnership no later than August 15th of each year.

New applications and re-certifications shall be submitted by June15th to be considered for the upcoming year. Applications turned in past the June 15th deadline will not be considered for the partnership.

It will be the responsibility of the NECA/IBEW partnership to gather participant data to evaluate and track the overall results and success of the Charter. Data collected will be given to OSHA as well.

The OSHA liaison will use the collected data to write the annual evaluation and submit to the Area Director. The final draft will be made available to the NECA/IBEW partnership.

VI.   NECA/IBEW Partnership Management and Operation

Representative(s) from the NECA/IBEW partnership will administer this program as outlined herein and will serve as the primary safety resource, supporting the participating employers and employees. To fulfill this Charter the NECA/IBEW partnership will also provide the following services:

1. Act as a liaison for NECA/IBEW members with OSHA. Members will be able to call the NECA/IBEW Partnership with questions and the NECA/IBEW Partnership will contact OSHA for responses, if required. The OSHA primary contact will be appointed by the OSHA Area Director and the NECA/IBEW Partnership will be notified in writing.

2. In concert with The Electrical Trades Center, offer ongoing, quality training on topics of importance for members - specifically the focused areas of fall protection, electrical hazards, trenching, etc.

3. Provide up-to-date informational materials and brochures to NECA/IBEW Partnership members (from OSHA, OSHA Ohio On-Site Consultation Service, the Bureau of Workers' Compensation (BWC) and other appropriate organizations).

4. Organize and provide to participating employers OSHA's interpretations of major standards, as well as local inspection perspectives.

5. Develop and build written safety and health policies and programs for participating employers, including emphasis on employer/employee responsibilities.

6. Promote construction safety excellence through an annual NECA/IBEW Partnership Safety Recognition Program.

7. Administer the overall Charter, including but not limited to the initial evaluation of potential participating employer applications to determine whether the firm meets the criteria specified within the Charter. Information considered by the NECA/IBEW Partnership will include pertinent company information as referenced in Section VII (demonstrated safety and health program, training commitments, OSHA citation history, fatalities, injury/illness experience and similar factors). Any and all information garnered by the NECA/IBEW partnership will be held with the greatest confidentiality.

8. Notify OSHA on a regular and recurring basis with the name(s) of contractors which have met the partnership criteria.

9. Conduct periodic audits to determine the impact and effectiveness of this partnership with OSHA.

10. A representative(s) from the NECA/IBEW Partnership and a representative(s) from OSHA together will conduct random on-site verification of 10 percent or more of participating employers, to ensure that participating employers are fulfilling their commitment to the partnership. During these on site evaluations, if an OSHA representative observes serious hazards these hazards will be addressed in a comprehensive and systematic manner, which may include citations.

11. If necessary, terminate employers from the partnership, if findings indicate unacceptable performance or submission of falsified documentation. (Note: At the discretion of the NECA/IBEW Partnership and OSHA a participating employer may be permitted to correct deficiencies within 30 days of notification and apply to the NECA/IBEW Partnership for continued recognition as a partner in good standing before termination would take effect.)

12. Although providing mandatory fall protection at the 6′ level is not required by the current OSHA construction standards and is not mandated by OSHA as a requirement for participation in any OSHA partnership agreement, the contractors on this partnership are committed to providing a greater level of protection to the employees working at this site and will require protection at the 6′ level and above.

OSHA and the NECA/IBEW Partnership have the discretion to jointly veto employers from participating for just cause. The OSHA Area Director has the discretion to unilaterally veto employers from participating in this Charter, for just cause upon providing notification and explanation to the NECA/IBEW Partnership.

VII. OSHA's Commitment/Role

Upon acceptance into this program, OSHA will provide the following initiatives and benefits to participating employers:

1. OSHA will conduct quarterly meetings with the NECA/IBEW partnership and participating employers, to provide information on "what's hot" and to answer general and specific questions. Additionally these meetings will serve as evaluation sessions as provided in Section IX.

2. Participating employers will not be cited for other-than-serious items that are fixed "on the spot."

3. Unplanned inspections will occur only for imminent danger, national emphasis programs, local emphasis programs, fatalities/catastrophes or signed complaints (all other complaints will go through the phone/fax process).

4. Formal complaints will be handled by the phone/fax process, if the complainant agrees.

5. For inspections resulting from formal complaints, the inspection will only address the complaint item and those in plain view.

6. Participating employers will receive the maximum allowable good-faith discount (35%).

7. The OSHA Area Director will meet with participating employers to address their role in the partnership.

8. During inspections, if minor (ex: missing midrail) problems are found; OSHA may review the participating employer's records and provide limited on-site training as needed.

VIII. Participating Employer Verification

To become and take full advantage of participating employer status, a firm must:

1. Must submit application and be willing to submit documents to the NECA/IBEW

Partnership for review by OSHA (such as the OSHA Log 300 and the company's safety program).

2. In Ohio, must not have had willful or repeat violations in the last three years. Nationwide, must not have had serious violations that resulted in a fatally(s) or catastrophe(s).

3. Must have a written safety and health program that complies with ANSI A-10.38 recommendations or OSHA's 1989 Voluntary Guidelines. The program must include active employee involvement.

4. All new field employees must receive site-specific training before beginning work. And receive at least a 2-hour safety orientation within the first week of hire. Topics for the orientation must include fall protection, electrical, struck-by, trenching/caught-between and personal protective equipment.

5. Within a year, all field employees must attend an OSHA 10-Hour Training Course. Field employees must receive refresher training in the 10-Hour Course every three years.

6. Within a year, all field supervisors must attend an OSHA 30-Hour Training Course. Field supervisors must receive refresher training in the 30-Hour Course every three years.
   a. Note: A supervisor is defined as someone who directs/controls work.

7. Must have a Days Away Restricted and Transfer (DART) rate for work in Ohio at least 25 percent below the BLS national average for the company's North American Industry Classification Code (NAICS).
   a. The company will submit individual DART rates for the past three years, and will be evaluated on a three-year average.
   b. If a company fulfills all other requirements, but does not have a qualifying DART, it may appeal for inclusion in the partnership. OSHA and the NECA/IBEW partnership will review these appeals on a case by case basis, and may allow the company to participate if the improvement can be shown.

8. Must show evidence that both employees and supervisors are held accountable for safety.

9. Allow OSHA access to sites for inspection, if the employer has the authority to allow an inspection of the site. OSHA will follow the guidelines for inspections as outlined in the Field Inspection Reference Manual (FIRM).

10. The participating employers recognize that OSHA implements Local Emphasis Programs (LEP) and National Emphasis Programs (NEP) to better manage specified hazards. These specific programs will involve inspections. The NECA/IBEW Partnership will be informed by OSHA of all LEP's and NEP's and will pass this information to all participating employers.

11. Must participate in a site audit by an outside, independent source approved by the NECA/IBEW partnership. The audit must include an action plan to prevent future hazards, as well as methods to abate current hazards. These audits will be made available to OSHA upon request.

12. All work on energized circuits will be performed under an industry standard permit policy to be developed and built in accordance with the current National Fire Protection Association publication 70e. This policy will cover hazard/risk evaluation, and procedures including protective barriers and shields, communication, insulated tools and equipment, and personal protective equipment.

13. Must be a member in good standing of the National Electrical Contractors Association.

IX.   Employer/Employee Rights

This partnership does not preclude employees and/or employers from exercising any right provided under the OSH Act (or federal employees, 29 CFR 1960), nor does it abrogate any responsibility to comply with rules and regulations adopted pursuant to the Act.

Routine employee involvement in the day to day implementation of worksite safety and health programs is expected to be assured, including employee participation in employer self-audits, site inspections, job hazard analysis, safety and health program reviews and near miss investigations.

X.   Termination of this Charter

Any party may withdraw from this Charter by providing written notification to the other parties. Termination will be effective 30 days after receipt of said notification. Furthermore an individual participating employer may withdraw from this agreement by providing written notification to the NECA/IBEW partnership and OSHA, termination shall be effective 30 days after receipt of notification.

An individual participating employer's violation of this Charter shall not be grounds for OSHA to terminate this Charter.

The NECA/IBEW partnership and OSHA may terminate an individual employer from the partnership if the employer fails to meet the qualifications or otherwise violates the terms and conditions of this Charter.

Any party may propose modification or amendment to this Charter subject to concurrence by the every other party to the Charter.

Unless modified or superceded this Charter will remain in effect until 31st day of October, 2010

Agreed this 21st day of October, 2008

_____	_____
Deborah Zubaty	Mario Ciardelli
Area Director	Business Manager, Financial Secretary
U.S. Department of Labor, OSHA	Local Union 683, IBEW
Columbus Area Office	Columbus, Ohio
_____	_____
Brian Damant	William Hamilton
Chapter Manager	Business Manager, Financial Secretary
Central Ohio Chapter, NECA, Inc.	Local Union 1105, IBEW
Columbus, Ohio	Newark, Ohio

This material relates to Chapter 15 which covered the importance of maintenance for electrical safety. This appendix is reprinted with the permission of the InterNational Electrical Testing Association (NETA) and is extracted from Appendix B of ANSI/NETA MTS-07, *Standard for Maintenance Testing Specifications for Electrical Power Distribution Equipment and Systems.* Appendix D in this book provides an example of the maintenance and tests recommended for molded case circuit breakers and insulated case circuit breakers. For the complete standard, visit www.netaworld.org.

APPENDIX B
Frequency of Maintenance Tests

NETA recognizes that the ideal maintenance program is reliability-based, unique to each plant and to each piece of equipment. In the absence of this information and in response to requests for a maintenance timetable, NETA's Standards Review Council presents the following time-based maintenance schedule and matrix.

One should contact a NETA Accredited Testing Company for a reliability-based evaluation.

The following matrix is to be used in conjunction with Appendix B, Inspections and Tests. Application of the matrix is recognized as a guide only.

Specific condition, criticality, and reliability must be determined to correctly apply the matrix. Application of the matrix, along with the culmination of historical testing data and trending, should provide a quality electrical preventive maintenance program.

MAINTENANCE FREQUENCY MATRIX			
	EQUIPMENT CONDITION		
	POOR	AVERAGE	GOOD
EQUIPMENT RELIABILITY REQUIREMENT — LOW	1.0	2.0	2.5
MEDIUM	0.50	1.0	1.5
HIGH	0.25	0.50	0.75

Inspections and Tests Frequency in Months (Multiply These Values by the Factor in the Maintenance Frequency Matrix)				
Section	Description	Visual	Visual & Mechanical	Visual & Mechanical & Electrical
7.1	Switchgear & Switchboard Assemblies	12	12	24
7.2	Transformers			
7.2.1.1	Small Dry-Type Transformers	2	12	36

(continued)

Copyright © 2007 InterNational Electrical Testing Association. Reprinted with permission. Courtesy of InterNational Electrical Testing Association (NETA).

	Inspections and Tests Frequency in Months (Multiply These Values by the Factor in the Maintenance Frequency Matrix)			
Section	**Description**	**Visual**	**Visual & Mechanical**	**Visual & Mechanical & Electrical**
7.2.1.2	Large Dry-Type Transformers	1	12	24
7.2.2	Liquid-Filled Transformers	1	12	24
	Sampling	–	–	12
7.3	Cables			
7.3.2	Low-Voltage Cables	2	12	36
7.3.3	Medium- and High-Voltage Cables	2	12	36
7.4	Metal-Enclosed Busways	2	12	24
	Infrared Only	–	–	12
7.5	Switches			
7.5.1.1	Low-Voltage Air Switches	2	12	36
7.5.1.2	Medium-Voltage Metal-Enclosed Switches	–	12	24
7.5.1.3	Medium- and High-Voltage Open Switches	1	12	24
7.5.2	Medium-Voltage Oil Switches	1	12	24
7.5.3	Medium-Voltage Vacuum Switches	1	12	24
7.5.4	Medium-Voltage SF_6 Switches	1	12	24
7.5.5	Cutouts	12	24	24
7.6	Circuit Breakers			
7.6.1.1	Low-Voltage Insulated-Case/Molded-Case CB	1	12	36
7.6.1.2	Low-Voltage Power CB	1	12	36
7.6.1.3	Medium-Voltage Air CB	1	12	36
7.6.2	Medium-Voltage Oil CB	1	12	36
	Sampling	–	–	12
7.6.2	High-Voltage Oil CB	1	12	12
	Sampling	–	–	12
7.6.3	Medium-Voltage Vacuum CB	1	12	24
7.6.4	Extra-High-Voltage SF_6	1	12	12
7.7	Circuit Switchers	1	12	12
7.8	Network Protectors	12	12	24
7.9	Protective Relays			
7.9.1	Electromechanical and Solid State	1	12	12
7.9.2	Microprocessor-Based	1	12	12
7.10	Instrument Transformers	12	12	36
7.11	Metering Devices	12	12	36
7.12	Regulating Apparatus			
7.12.1.1	Step-Voltage Regulators	1	12	24
	Sample Liquid	–	–	12
7.12.1.2	Induction Regulators	12	12	24
7.12.2	Current Regulators	1	12	24
7.12.3	Load-Tap-changers	1	12	24
	Sample Liquid	–	–	12

Copyright © 2007 InterNational Electrical Testing Association. Reprinted with permission. Courtesy of InterNational Electrical Testing Association (NETA).

	Inspections and Tests Frequency in Months (Multiply These Values by the Factor in the Maintenance Frequency Matrix)			
Section	**Description**	**Visual**	**Visual & Mechanical**	**Visual & Mechanical & Electrical**
7.13	Grounding Systems	2	12	24
7.14	Ground-Fault Protection Systems	2	12	12
7.15	Rotating Machinery			
7.15.1	AC Induction Motors and Generators	1	12	24
7.15.2	Synchronous Motors and Generators	1	12	24
7.15.3	DC Motors and Generators	1	12	24
7.16	Motor Control			
7.16.1.1	Low-Voltage Motor Starters	2	12	24
7.16.1.2	Medium-Voltage Motor Starters	2	12	24
7.16.2.1	Low-Voltage Motor Control Centers	2	12	24
7.16.2.2	Medium-Voltage Motor Control Centers	2	12	24
7.17	Adjustable Speed Drive Systems	1	12	24
7.18	Direct-Current Systems			
7.18.1	Batteries	1	12	12
7.18.2	Battery Chargers	1	12	12
7.18.3	Rectifiers	1	12	24
7.19	Surge Arresters			
7.19.1	Low-Voltage Devices	2	12	24
7.19.2	Medium- and High-Voltage Devices	2	12	24
7.20	Capacitors and Reactors			
7.20.1	Capacitors	1	12	12
7.20.2	Capacitor Control Devices	1	12	12
7.20.3.1	Reactors, Dry-Type	2	12	24
7.20.3.2	Reactors, Liquid-Filled	1	12	24
	Sampling	–	–	12
7.21	Outdoor Bus Structures	1	12	36
7.22	Emergency Systems			
7.22.1	Engine Generator	1	2	12
	Functional Testing	–	–	2
7.22.2	Uninterruptible Power Systems	1	12	12
	Functional Testing	–	–	2
7.22.3	Automatic Transfer Switches	1	12	12
	Functional Testing	–	–	2
7.23	Telemetry/Pilot Wire SCADA	1	12	12
7.24	Automatic Circuit Reclosers and Line Sectionalizers			
7.24.1	Automatic Circuit Reclosers, Oil/Vacuum	1	12	24
	Sample	–	–	12
7.24.2	Automatic Line Sectionalizers, Oil	1	12	24
	Sample	–	–	12
7.27	EMF Testing	12	12	12

Copyright © 2007 InterNational Electrical Testing Association. Reprinted with permission. Courtesy of InterNational Electrical Testing Association (NETA).

APPENDIX D

Maintenance Tests for Molded-Case and Insulated-Case Circuit Breakers

This material relates to Chapter 15 which covered the importance of maintenance for electrical safety. This appendix is reprinted with the permission of the InterNational Electrical Testing Association (NETA) and is extracted from ANSI/NETA MTS-07, *Standard for Maintenance Testing Specifications for Electrical Power Distribution Equipment and Systems.* These specifications cover the suggested field tests and inspections that are available to assess the suitability for continued service and reliability of electrical power distribution equipment and systems. The purpose of these specifications is to assure that tested electrical equipment and systems are operational and within applicable standard and manufacturer's tolerances and that the equipment and systems are suitable for continued service. The following is extracted as an example of the maintenance and tests recommended for molded case circuit breakers and insulated case circuit breakers. Appendix C in this book provides guidelines on the Frequency of Maintenance. For the complete standard visit www.netaworld.org.

7. INSPECTION AND TEST PROCEDURES
7.6.1.1 Circuit Breakers, Air, Insulated-Case/Molded-Case

1. **Visual and Mechanical Inspection**
    1. Inspect physical and mechanical condition.
    2. Inspect anchorage and alignment.
    3. Prior to cleaning the unit, perform as-found tests, if required.
    4. Clean the unit.
    5. Operate the circuit breaker to insure smooth operation.
    6. Inspect bolted electrical connections for high resistance using one of the following methods:
        1. Use of a low-resistance ohmmeter in accordance with Section 7.6.1.1.2.
        2. Verify tightness of accessible bolted electrical connections by calibrated torque-wrench method in accordance with manufacturer's published data or Table 100.12.
        3. Perform a thermographic survey in accordance with Section 9.
    7. Inspect operating mechanism, contacts, and arc chutes in unsealed units.
    8. Perform adjustments for final setting in accordance with coordination study provided by end user.
    9. Perform as-left tests.
2. **Electrical Tests**
    1. Perform resistance measurements through bolted connections with a low-resistance ohmmeter, if applicable, in accordance with Section 7.6.1.1.1.
    2. Perform insulation-resistance tests for one minute on each pole, phase-to-phase and phase-to-ground with the circuit breaker closed, and across each open pole. Apply voltage in accordance with manufacturer's published data. In the absence of manufacturer's published data, use Table 100.1.
    3. Perform a contact/pole-resistance test.

Copyright © 2007 InterNational Electrical Testing Association. Reprinted with permission. Courtesy of InterNational Electrical Testing Association (NETA).

*4. Perform insulation-resistance tests on all control wiring with respect to ground. The applied potential shall be 500 volts dc for 300-volt rated cable and 1000 volts dc for 600-volt rated cable. Test duration shall be one minute. For units with solid-state components, follow manufacturer's recommendation.

5. Determine long-time pickup and delay by primary current injection.

6. Determine short-time pickup and delay by primary current injection.

7. Determine ground-fault pickup delay by primary current injection.

8. Determine instantaneous pickup current by primary injection.

*9. Test functions of the trip unit by means of secondary injection.

10. Perform minimum pickup voltage test on shunt trip and close coils in accordance with Table 100.20.

11. Verify correct operation of auxiliary features such as trip and pickup indicators, zone interlocking, electrical close and trip operation, trip-free, antipump function, and trip unit battery condition. Reset all trip logs and indicators.

12. Verify operation of charging mechanism.

3. Test Values

3.1 Test Values—Visual and Mechanical

1. Compare bolted connection resistance values to values of similar connections. Investigate values which deviate from those of similar bolted connections by more than 50 percent of the lowest value. (7.6.1.1.1.6.1)

2. Bolt–torque levels should be in accordance with manufacturer's published data. In the absence of manufacturer's published data, use Table 100.12. (7.6.1.1.1.6.2)

3. Results of the thermographic survey shall be in accordance with Section 9. (7.6.1.1.1.6.3)

4. Settings shall comply with coordination study recommendations. (7.6.1.1.1.8)

3.2 Test Values—Electrical

1. Compare bolted connection resistance values to values of similar connections. Investigate values which deviate from those of similar bolted connections by more than 50 percent of the lowest value.

2. Insulation-resistance values should be in accordance with manufacturer's published data. In the absence of manufacturer's published data, use Table 100.1. Values of insulation resistance less than this table or manufacturer's recommendations should be investigated.

3. Microhm or dc millivolt drop values should not exceed the high levels of the normal range as indicated in the manufacturer's published data. If manufacturer's data is not available, investigate values that deviate from adjacent poles or similar breakers by more than 50 percent of the lowest value.

4. Insulation-resistance values of control wiring should be comparable to previously obtained results but not less than two megohms.

5. Long-time pickup values should be as specified, and the trip characteristic should not exceed manufacturer's published time–current characteristic tolerance band, including adjustment factors. If manufacturer's curves are not available, trip times should not exceed the value shown in Table 100.7. (Circuit breakers exceeding specified trip time shall be tagged defective.)

6. Short-time pickup values should be as specified, and the trip characteristic should not exceed manufacturer's published time-current tolerance band. (Circuit breakers exceeding specified trip time shall be tagged defective.)

7. Ground fault pickup values should be as specified, and the trip characteristic should not exceed manufacturer's published time-current tolerance band. (Circuit breakers exceeding specified trip time shall be tagged defective.)

8. Instantaneous pickup values of molded-case circuit breakers should fall within manufacturer's published tolerances. In the absence of manufacturer's published tolerances, refer to Table 100.8. (Circuit breakers exceeding specified trip time shall be tagged defective.)

9. Pickup values and trip characteristics should be within manufacturer's published tolerances. (Circuit breakers exceeding specified trip time shall be tagged defective.)

10. Minimum pickup voltage on shunt trip and close coils should be in accordance with manufacturer's published data. In the absence of manufacturer's published data, refer to Table 100.20.

11. Breaker open, close, trip, trip-free, antipump, and auxiliary features should function as designed.

12. Charging mechanism shall function as designed.

* Optional

Copyright © 2007 InterNational Electrical Testing Association. Reprinted with permission. Courtesy of InterNational Electrical Testing Association (NETA).

Tables Referenced by 7.6.1.1

TABLE 100.1   Insulation-Resistance Test Values Electrical Apparatus and Systems

Nominal Rating of Equipment (Volts)	Minimum Test Voltage (DC)	Recommended Minimum Insulation Resistance in Megohms
250	500	25
600	1,000	100
1,000	1,000	100
2,500	1,000	500
5,000	2,500	1,000
8,000	2,500	2,000
15,000	2,500	5,000
25,000	5,000	20,000
34,500 and above	15,000	100,000

In the absence of consensus standards dealing with insulation-resistance tests, the NETA Standards Review Council suggests the above representative values.
See Table 100.14 for temperature correction factors.
Test results are dependent on the temperature of the insulating material and the humidity of the surrounding environment at the time of the test.
Insulation-resistance test data may be used to establish a trending pattern. Deviations from the baseline information permit evaluation of the insulation.

TABLE 100.7   Molded-Case Circuit Breakers Inverse Time Trip Test
(At 300% of Rated Continuous Current of Circuit Breaker)

Range of Rated Continuous Current (Amperes)	Maximum Trip Time in Seconds For Each Maximum Frame Rating [a]	
	≤ 250 V	251–600V
0–30	50	70
31–50	80	100
51–100	140	160
101–150	200	250
151–225	230	275
226–400	300	350
401–600	- - - - -	450
601–800	- - - - -	500
801–1000	- - - - -	600
1001–1200	- - - - -	700
1201–1600	- - - - -	775
1601–2000	- - - - -	800
2001–2500	- - - - -	850
2501–5000	- - - - -	900
6000	- - - - -	1000

Derived from Table 5-3, NEMA Standard AB 4-2000, *Guidelines for Inspection and Preventative Maintenance of Molded-Case Circuit Breaker Used in Commercial and Industrial Applications.*
a. Trip times may be substantially longer for integrally-fused circuit breakers if tested with the fuses replaced by solid links (shorting bars).

Copyright © 2007 InterNational Electrical Testing Association. Reprinted with permission. Courtesy of InterNational Electrical Testing Association (NETA).

TABLE 100.8   Instantaneous Trip Tolerances for Field Testing of Circuit Breakers

Breaker Type	Tolerance of Settings	Tolerances of Manufacturer's Published Trip Range	
		High Side	Low Side
Adjustable [a]	+40% −30%	- - - - -	- - - - -
Nonadjustable [b]	- - - - -	+25%	−25%

Reproduction of Table 5-4 from NEMA publication AB4-2000, *Guidelines for Inspection and Preventive Maintenance of Molded-Case Circuit Breakers Used in Commercial and Industrial Applications.*
NEMA AB4-2000 *Guidelines for Inspection and Preventative Maintenance of Molded-Case Circuit Breaker Used in Commercial and Industrial Applications, Table 5-4.*
a. Tolerances are based on variations from the nominal settings.
b. Tolerances are based on variations from the manufacturer's published trip band (i.e., −25% below the low side of the band; +25% above the high side of the band.)

TABLE 100.12.1   Bolt-Torque Values for Electrical Connections
US Standard Fasteners[a]
Heat-Treated Steel–Cadmium or Zinc Plated[b]

Grade	SAE 1&2	SAE 5	SAE 7	SAE 8
Head Marking	◯	◯	◯	◯
Minimum Tensile (Strength) (lbs/in^2)	64K	105K	133K	150K
Bolt Diameter (Inches)	**Torque (Pound-Feet)**			
1/4	4	6	8	8
5/16	7	11	15	18
3/8	12	20	27	30
7/16	19	32	44	48
1/2	30	48	68	74
9/16	42	70	96	105
5/8	59	96	135	145
3/4	96	160	225	235
7/8	150	240	350	380
1.0	225	370	530	570

a. Consult manufacturer for equipment supplied with metric fasteners.
b. Table is based on national coarse thread pitch.

Table 100.12.2   US Standard Fasteners [a]
Silicon Bronze Fasteners[bc]
Torque (Pound-Feet)

Bolt Diameter (Inches)	Nonlubricated	Lubricated
5/16	15	10
3/8	20	15
1/2	40	25
5/8	55	40
3/4	70	60

a. Consult manufacturer for equipment supplied with metric fasteners.
b. Table is based on national coarse thread pitch.
c. This table is based on bronze alloy bolts having a minimum tensile strength of 70,000 pounds per square inch.

Copyright © 2007 InterNational Electrical Testing Association. Reprinted with permission. Courtesy of InterNational Electrical Testing Association (NETA).

TABLE 100.12.3   US Standard Fasteners[a]
Aluminum Alloy Fasteners[bc]
Torque (Pound-Feet)

Bolt Diameter (Inches)	Lubricated
5/16	10
3/8	14
1/2	25
5/8	40
3/4	60

a. Consult manufacturer for equipment supplied with metric fasteners.
b. Table is based on national coarse thread pitch.
c. This table is based on aluminum alloy bolts having a minimum tensile strength of 55,000 pounds per square inch.

TABLE 100.12.4   US Standard Fasteners[a]
Stainless Steel Fasteners[bc]
Torque (Pound-Feet)

Bolt Diameter (Inches)	Uncoated
5/16	15
3/8	20
1/2	40
5/8	55
3/4	70

a. Consult manufacturer for equipment supplied with metric fasteners.
b. Table is based on national coarse thread pitch.
c. This table is to be used for the following hardware types:
Bolts, cap screws, nuts, flat washers, locknuts (18-8 alloy)
Belleville washers (302 alloy).

Tables in 100.12 are compiled from Penn-Union Catalogue and Square D Company, Anderson Products Division, *General Catalog: Class 3910 Distribution Technical Data, Class 3930 Reference Data Substation Connector Products.*

TABLE 100.20.2   Rated Control Voltages and their Ranges for Circuit Breakers
Solenoid-Operated Devices

Rated Voltage	Closing Voltage Ranges for Power Supply
125 dc	90–115 or 105–130
250 dc	180–230 or 210–260
230 ac	190–230 or 210–260

Some solenoid operating mechanisms are not capable of satisfactory performance over the range of voltage specified in the standard; moreover, two ranges of voltage may be required for such mechanisms to achieve an acceptable standard of performance.

The preferred method of obtaining the double range of closing voltage is by use of tapped coils. Otherwise it will be necessary to designate one of the two closing voltage ranges listed above as representing the condition existing at the device location due to battery or lead voltage drop or control power transformer regulation. Also, caution should be exercised to ensure that the maximum voltage of the range used is not exceeded.

Copyright © 2007 InterNational Electrical Testing Association. Reprinted with permission. Courtesy of InterNational Electrical Testing Association (NETA).

Glossary

29 CFR Part 1910 The part of the Code of Federal Regulations that covers occupational safety and health standards for general industry.

29 CFR Part 1926 The part of the Code of Federal Regulations that covers safety and health regulations for construction.

affected employee§ An employee whose job requires him/her to operate or use a machine or equipment on which servicing or maintenance is being performed under lockout or tagout, or whose job requires him/her to work in an area in which such servicing or maintenance is being performed.

American National Standard An accreditation provided by the American National Standard Institute (ANSI), an organization that approves standards and ensures that they can be used internationally.

ampacity* The current, in amperes, that a conductor can carry continuously under the conditions of use without exceeding its temperature rating.

arc blast‡ (also called arc flash) Dangerous condition caused by the release of energy in an electric arc, usually associated with electrical distribution equipment.

arc flash‡ (also called arc blast) Dangerous condition caused by the release of energy in an electric arc, usually associated with electrical distribution equipment.

arc flash hazard* A dangerous condition associated with the possible release of energy caused by an electric arc.

arc flash hazard analysis* A study investigating a worker's potential exposure to arc-flash energy, conducted for the purpose of injury prevention and the determination of safe work practices, arc flash protection boundary, and the appropriate levels of PPE.

arc flash protection boundary (AFPB)* When an arc flash hazard exists, an approach limit at a distance from a prospective arc source within which a person could receive a second degree burn if an electrical arc flash were to occur.

arc rating* The value attributed to materials that describes their performance to exposure to an electrical arc discharge. The arc rating is expressed in cal/cm^2 and is derived from the determined value of the arc thermal performance value (ATPV) or energy of breakopen threshold (E_{BT}) (should a material system exhibit a breakopen response below the ATPV value) derived from the determined value of ATPV or E_{BT}.

> FPN: *Breakopen* is a material response evidenced by the formation of one or more holes in the innermost layer of flame-resistant material that would allow flame to pass through the material.

arcing fault‡ Fault characterized by an electrical arc through the air.

article A component of *NFPA 70E* that appears second (after chapter) and bolded in this example: **110**.6.(D)(1)(a).

assured equipment grounding conductor program One of the methods recognized by OSHA to provide ground-fault protection to protect employees.

authorized employee§ A person who locks out or tags out machines or equipment in order to perform servicing or maintenance on that machine or equipment. An affected employee becomes an authorized employee when that employee's duties include performing servicing or maintenance covered under this section.

capable of being locked out§ An energy isolating device is capable of being locked out if it has a hasp or other means of attachment to which, or through which, a lock can be affixed, or it has a locking mechanism built into it. Other energy isolating devices are capable of being locked out, if lockout can be achieved without the need to dismantle, rebuild, or replace the energy isolating device or permanently alter its energy control capability.

confined space§ A space that is large enough and so configured that an employee can bodily enter and perform assigned work; has limited or restricted means for entry or exit (for example, tanks, vessels, silos, storage bins, hoppers, vaults, and pits are spaces that may have limited means of entry); and is not designed for continuous employee occupancy.

current-limiting overcurrent protective device† A device that, when interrupting currents in its current-limiting range, reduces the current flowing in the faulted circuit to a magnitude substantially less than that obtainable in the same circuit if the device were replaced with a solid conductor having comparable impedance.

electrical safety program A subset of an overall safety program, ideally tailored to address the specific issues in a particular workplace, and components of which can include policies and procedures, site assessment, task assessment, PPE requirements, hazardous boundaries and hazard/risk analysis, administration, lockout/tagout, training, auditing and recordkeeping, and budgeting.

electrically safe work condition＊ A state in which an electric conductor or circuit part has been disconnected from energized parts, locked/tagged in accordance with established standards, tested to ensure the absence of voltage, and grounded if determined necessary.

enclosed space§ A working space, such as a manhole, vault, tunnel, or shaft, that has a limited means of egress or entry, that is designed for periodic employee entry under normal operating conditions, and that under normal conditions does not contain a hazardous atmosphere, but that may contain a hazardous atmosphere under abnormal conditions.

energized＊ Electrically connected to, or is, a source of voltage.

energy source§ Any source of electrical, mechanical, hydraulic, pneumatic, chemical, thermal, or other energy.

energy-isolating device§ A mechanical device that physically prevents the transmission or release of energy, including but not limited to the following: A manually operated electrical circuit breaker; a disconnect switch; a manually operated switch by which the conductors of a circuit can be disconnected from all ungrounded supply conductors, and, in addition, no pole can be operated independently; a line valve; a block; and any similar device used to block or isolate energy. Push buttons, selector switches and other control circuit type devices are not energy isolating devices.

equipment＊ A general term, including material, fittings, devices, appliances, luminaires, apparatus, machinery, and the like used as a part of, or in connection with, an electrical installation.

exposed movable conductor A condition in which the distance between the conductor and a person is not under the control of the person. The term is normally applied to overhead line conductors supported by poles.

flame-resistant (FR)＊ The property of a material whereby combustion is prevented, terminated, or inhibited following the application of a flaming or non-flaming source of ignition, with or without subsequent removal of the ignition source.

> FPN: Flame resistance can be an inherent property of a material or it can be imparted by a specific treatment applied to the material.

ground-fault circuit interrupter (GFCI)* A device intended for the protection of personnel that functions to deenergize a circuit or portion thereof within an established period of time when a current to ground exceeds the values established for a Class A device.

> FPN: Class A ground-fault circuit-interrupters trip when the current to ground is 6 mA or higher and do not trip when the current to ground is less than 4 mA. For further information, see UL 943, *Standard for Ground-Fault Circuit Interrupters*.

hazard/risk evaluation* An analytical tool consisting of a number of discrete steps intended to ensure that hazards are properly identified and evaluated, and that appropriate measures are taken to reduce those hazards to a tolerable level (adapted from ANSI/ASSE Z244.1).

hazard/risk procedure* A comprehensive review of the task and associated foreseeable hazards that use event severity, frequency, probability, and avoidance to determine the level of safe practices employed.

impedance‡ Ratio of the voltage drop across a circuit element to the current flowing through the same circuit element, measured in ohms (Ω).

incident energy* The amount of energy impressed on a surface, a certain distance from the source, generated during an electrical arc event. One of the units used to measure incident energy is calories per centimeter squared (cal/cm^2).

interrupting rating (IR)* The highest current at rated voltage that a device is intended to interrupt under standard test conditions.

> FPN: Equipment intended to interrupt current at other than fault levels may have its interrupting rating implied in other ratings, such as horsepower or locked rotor current.

knife-blade fuses A type of fuse that contains blades that are inserted into fuse clips; it is important not to rely on test measurements by touching the end caps of the fuse with probes. Touch the probes to the fuse blades.

let-go threshold The electrical current level at which the brain's electrical signals to muscles can no longer overcome the signals introduced by an external electrical system. Because these external signals lock muscles in the contracted position, the body may not be able to let go when the brain is telling it to do so.

level 1 subdivision The fourth component of *NFPA 70E*, bolded in this example: 110.6.**(D)**(1)(a).

level 2 subdivision The fifth component of *NFPA 70E*, bolded in this example: 110.6.(D)(**1**)(a).

level 3 subdivision The sixth component of *NFPA 70E*, bolded in this example: 110.6.(D)(1)(**a**).

limited approach boundary* An approach limit at a distance from an exposed energized electrical conductor or circuit part within which a shock hazard exists.

listed† Equipment, materials, or services included in a list published by an organization that is acceptable to the authority having jurisdiction (AHJ) and concerned with evaluation of products or services, that maintains periodic inspection of production of listed equipment or materials or periodic evaluation of services, and whose listing states that the equipment, material, or services either meets appropriate designated standards or has been tested and found suitable for a specified purpose.

lockout§ The placement of a lockout device on an energy isolating device, in accordance with an established procedure, ensuring that the energy isolating device and the equipment being controlled cannot be operated until the lockout device is removed.

lockout device§ A device that utilizes a positive means such as a lock, either key or combination type, to hold an energy isolating device in the safe position and prevent the energizing of a machine or equipment. Included are blank flanges and bolted slip blinds.

National Electrical Code® A standard published by the National Fire Protection Association (NFPA) primarily addressing electrical installations.

nominal voltage* A nominal value assigned to a circuit or system for the purpose of conveniently designating its voltage class (e.g., 120/240 volts, 480Y/277 volts, 600 volts). The actual voltage at which a circuit operates can vary from the nominal within a range that permits satisfactory operation of equipment.

Occupational Safety and Health Administration (OSHA) The U.S. government agency that sets standards for worker safety and health protection, including outlining requirements for electrical safety for employers and employees.

Ohm's law‡ Mathematical relationship between voltage, current, and resistance in an electric circuit. Ohm's law states that the current flowing in a circuit is proportional to electromotive force (voltage) and inversely proportional to resistance: $I = E/R$.

overcurrent* Any current in excess of the rated current of equipment or the ampacity of a conductor. It may result from overload, short circuit, or ground fault.

> FPN: A current in excess of rating may be accommodated by certain equipment and conductors for a given set of conditions. Therefore, the rules for overcurrent protection are specific for particular situations.

overcurrent protective device (OCPD)‡ General term that includes both fuses and circuit breakers.

overload* Operation of equipment in excess of normal, full-load rating, or of a conductor in excess of rated ampacity that, when it persists for a sufficient length of time, would cause damage or dangerous overheating. A fault, such as a short circuit or ground fault, is not an overload.

performance language Wording that indicates that something must be done, but not how to do it. For example, "the authorized employee shall have knowledge of the type and magnitude of the energy, the hazards of the energy to be controlled, and the method or means to control the energy."

prohibited approach boundary* An approach limit at a distance from an exposed energized electrical conductor or circuit part within which work is considered the same as making contact with the electrical conductor or circuit part.

qualified person* One who has skills and knowledge related to the construction and operation of the electrical equipment and installations and has received safety training to recognize and avoid the hazards involved.

restricted approach boundary* An approach limit at a distance from an exposed energized electrical conductor or circuit part within which there is an increased risk of shock, due to electrical arc over combined with inadvertent movement, for personnel working in close proximity to the energized electrical conductor or circuit part.

safety program An organized effort to reduce injuries.

section The third component of *NFPA 70E*, bolded in this example: 110.**6**.(D)(1)(a).

selective coordination† Localization of an overcurrent condition to restrict outages to the circuit or equipment affected, accomplished by the choice of overcurrent protective devices and their ratings or settings.

shock hazard* A dangerous condition associated with the possible release of energy caused by contact or approach to energized electrical conductors or circuit parts.

short-circuit (fault) current‡ Overcurrent resulting from a short circuit due to a fault or an incorrect connection in an electric circuit.

short-circuit current rating† The prospective symmetrical fault current at a nominal voltage to which an apparatus or system is able to be connected without sustaining damage exceeding defined acceptance criteria.

short-circuit study An analysis of an electrical distribution system that determines the available short-circuit current at various points in the system; the purpose is to verify compliance with the *NEC* and *NFPA 70E* and increase electrical safety.

tagout§ The placement of a tagout device on an energy isolating device, in accordance with an established procedure, to indicate that the energy isolating device and the equipment being controlled may not be operated until the tagout device is removed.

tagout device§ A prominent warning device, such as a tag and a means of attachment, which can be securely fastened to an energy isolating device in accordance with an established procedure, to indicate that the energy isolating device and the equipment being controlled may not be operated until the tagout device is removed.

Type 1 protection# A form of motor controller protection that requires that, under short-circuit conditions, the contactor or starter shall cause no danger to persons or installation and might not be suitable for further service without repair and replacement of parts.

Type 2 protection# A form of motor controller protection that requires that, under short-circuit conditions, the contactor or starter shall cause no danger to persons or installation and shall be suitable for further use. The risk of contact welding is recognized, in which case the manufacturer shall indicate the measures to be taken regarding the maintenance of the equipment.

unqualified person* A person who is not a qualified person.

working on (energized electrical conductors or circuit parts)* Coming in contact with energized electrical conductors or circuit parts with the hands, feet, or other body parts, with tools, probes, or with test equipment, regardless of the personal protective equipment a person is wearing. There are two categories of "working on":

> *Diagnostic (testing)* is taking readings or measurements of electrical equipment with approved test equipment that does not require making any physical change to the equipment, *repair* is any physical alteration of electrical equipment (such as making or tightening connections, removing or replacing components, etc.).

zone-selective interlocking A system of communications between line-side and load-side circuit breakers allowing the line-side circuit breaker to open without delay for short-circuits located between the circuit breakers and, for purposes of obtaining selective coordination, allowing the line-side circuit breaker to delay operation for short-circuits on the load side of the load-side circuit breaker.

*This is the *NFPA 70E* definition.
†This is the *National Electrical Code* (*NEC*®) definition.
‡This definition is from the National Fire Protection Association's (NFPA's) *Pocket Dictionary of Electrical Terms*.
§This is the OSHA definition.
#This definition is from International Electrotechnical Commission (IEC) 60947-4-1.

Index

Adjustable instantaneous trip setting, 73
Affected employees, 152, 159
AFPB. *See* Arc flash protection boundary
Alerting techniques, 248, 312
Alertness precautions, 243, 311
American National Standard, 89, 90
American Society for Testing and Materials (ASTM), 246, 247–248, 268
Ampacity, 192
Application of control devices, 159–160
Approach boundaries. *See also* Limited approach boundary (LAB)
    hazard analysis and, 178–180, 309
    prohibited, 106, 178, 309
    restricted, 106, 178, 309
    training on, 140
Arc blast, 4, 24
Arc flash
    defined, 23–24
    maintenance switching to reduce, 74–76
    NFPA 70E requirements, 106–107
    safety training for, 136
    short-circuit currents and, 190–191
    staged tests of, 25–27
Arc flash hazard analysis, 210–237, 309
    AFPB determination, 218–222
    arcing current and, 213
    current-limiting fuses and, 216–217
    documentation of, 180–182
    electronic trip molded case circuit breakers and, 215–216
    incident energy determination for, 222–231
    information updating for, 283–284
    low-voltage power circuit breakers and, 216
    maintenance and, 290–291
    marking equipment for, 281–283
    methods available, 217–218
    NFPA 70E requirements, 106
    OCPD clearing time and, 65, 213–214
    overcurrent protective devices and, 215–216, 231
    PPE selection and, 181, 258, 264–268
    required information for, 213
    thermal magnetic molded case circuit breakers and, 214–215
Arc flash protection boundary (AFPB), 309
    arc flash hazard analysis to determine, 211, 305
    conditional value and, 218–219
    determination of, 218–222
    IEEE 1584 method of determining, 220–222
    LAB vs., 179, 181
    maintenance switching and, 74
    moving people outside of, 281
    NFPA 70E method of determining, 220–222
    PPE and, 265

Ralph Lee method of determining, 220
    remote racking and, 77
    requirements for, 106
    short-circuit currents and, 189
Arc flash suits, 264
Arcing current, 213
Arcing faults, 23–29
    arc blast, 24
    arc flash, 23–24
    basics of, 23
    defined, 4
    injuries from, 7, 24–25
    OCPDs and, 27–29
    staged arc flash tests, 25–27
Arc rating
    of arc flash suits, 23
    of FR clothing, 23, 271
    of PPE, 73, 223, 268
Arc-resistant medium voltage switchgear, 77
Arm protection, 260–262
Articles of *NFPA 70E*, 94
Assured equipment grounding conductor program, 39
ASTM. *See* American Society for Testing and Materials
Attendants, 248
Auditing of electrical safety program, 125–126
Authorized employees, 152
Awareness, 18–35
    of arcing faults, 23–29
    in culture of electrical safety, 3–8
    electrical safety program for, 120
    of electrical shock, 20–22
    of workplace hazards, 19–20

Baclawski, Vince A., 280
Barricades, 242, 248
Blind reaching, 243, 311
Body protection, 259–260
Boundaries. *See* Approach boundaries
Burns from arcing faults, 24–25
Bus differential relays, 287

Capable of being locked out, 152–153
Capelli-Schellpfeffer, Mary, 5–7, 136–137
Cardiopulmonary resuscitation (CPR), 3, 4, 22, 138
Casini, Virgil, 134, 141–142
Central Ohio IBEW/NECA/OSHA Charter, 126, 337–344
Chin protection, 259
Circuit breakers. *See also* Overcurrent protective devices (OCPDs)
    electronic trip molded case, 215–216
    incident energy analysis for, 60–64
    instantaneous trip setting adjustments, 286

Circuit breakers (Cont.)
    interrupting ratings of, 284
    low-voltage power, 216
    maintenance of, 280, 292
    resetting, 279–280
    safety enhancement with, 66
    thermal magnetic molded case, 214–215
Circuits
    closing of, 78, 245
    deenergized, 41–42
    downsizing of, 73
    isolation of, 70
    opening of, 78, 245
    switches for circuits 800 amperes and above, 67
Closing of circuits, 78, 245
Communication
    alerting techniques, 248, 312
    labeling of hazards, 182, 279, 281–283, 310
    of lockout/tagout, 157–159
    safety signs and tags, 248
Computer software for short-circuit calculations, 202–203
Conditional value, 218–219
Conductive articles, materials, tools, and equipment, 311
Conductive heat burns, 24
Conductive materials, tools, and equipment, 243
Conductors, 39, 193–194
Confined spaces, 47, 243–244, 311
Contractor responsibilities, 105, 161, 306
Control of hazardous energy, 148–173
    electrically safe work condition and, 163–165
    lockout/tagout for, 151–163
Convective heat burns, 24
Cooper Bussmann
    Arc Flash Incident Energy Calculator, 225
    knife-blade fuses, 281
    Short-Circuit Calculator, 202
Cooper Bussmann Safety BASICs Handbook for Electrical Safety, 225
CPR. See Cardiopulmonary resuscitation
Culture of electrical safety, 1–17
    appropriate priorities in, 9
    choices in, 9–10
    costs of energized work vs. shutting down, 9
    employee qualification and, 10
    hazards and, 3–8
    historical perspective on, 2–3
    implementation choices, 10–11
    requirements for, 5–7
    time pressures and, 9–10
    training and personal responsibility, 9
    workplace pressure and, 2
Current-limiting overcurrent protective devices, 58–64, 66, 216–217

Deenergized equipment, 41–42, 45–46. See also Electrically safe work condition
Design considerations, 56–87
    overcurrent protective devices, 58–64
    safety measures and devices, 65–78
Dielectric overshoes, 262

Doughty, R. L., 222
Doughty/Neal/Floyd method, 222

Electrical burns, 24
Electrically safe work condition
    defined, 5
    with lockout/tagout, 150, 163–165
    NFPA 70E requirements, 106
Electrical safety program, 116–131
    auditing of, 125–126
    for awareness, 120
    cost considerations, 119
    establishment of, 118–119
    hazard/risk evaluation in, 120–124
    job briefing in, 124–125
    NFPA 70E and, 119–126, 306
    objectives, 118–119
    principles, controls, and procedures in, 120
    requirements for, 120
The Electrical Safety Program Book (Jones, Jones, & Mastrullo), 118, 133, 135
Electrical shock
    hazard awareness, 20–22
    injuries from, 5–7, 21–22
    prevention of, 22, 45–49
    skin resistance and, 20–21
Electronic trip molded case circuit breakers, 215–216
Emergency procedures
    CPR, 3, 4, 22, 138
    training for, 7, 137–138
Employees. See also Qualified person; Unqualified person
    lockout/tagout protection of, 154
    outside personnel and lockout/tagout, 161
    protection from hazards, 41–42
    qualification of, 10
Enclosed spaces, 47, 243–244, 311
Energized electrical work permit, 125, 176–178, 308–309
Energized equipment
    costs of shutting down vs. working with, 9
    justification of, 175–176, 307–308
    lockout/tagout and, 153
    NFPA 70E requirements, 107
    OSHA requirements on, 47
    qualified persons and, 139–140
    requirements for, 306
    safety training for, 136–137
    testing of, 239–240
Energy control program, 153
Energy-isolating device, 153
Energy source, 153
Equipment. See also Energized equipment; Insulated equipment; Personal protective equipment (PPE)
    grounding of, 242
    labeling of, 182
    for lockout/tagout, 155–157
    maintenance of, 278–303
    marking for arc flash hazards, 281–283
    overview, 311–312
    requirements for, 106, 307
Exposed movable conductors, 179

Eye protection, 259

FACE program. *See* Fatality Assessment and Control Evaluation
Face protection, 259, 264–265
Failure anticipation, 244–245, 311
Fatality Assessment and Control Evaluation (FACE), 2, 12, 134–135
Fault currents. *See* Short-circuit currents
Fiberglass-reinforced plastic rods, 247–248
Finger-safe products, 69–70
First aid training, 137–138
Flame-resistant (FR) clothing
    arc flash protection of, 267–268
    care and maintenance of, 266–268
    characteristics of, 266
    defined, 267
    electrical shock injuries and, 8
    evaluating and comparing, 267
    foot protection, 266
    need for, 267
    PPE and, 258
    prohibited apparel for, 266
Flammable materials, 244, 311
Floyd, H. Landis, 10, 222
Foot protection, 262, 266
FR clothing. *See* Flame-resistant clothing
Fuses
    arc flash hazard reduction and, 286
    current-limiting, 59–60, 216–217
    insulated holding equipment, 246–247
    maintenance requirements for, 292
    misapplied fuse application, 291
    rejection style, 66–67
    replacement of, 279–280, 281
    testing and maintenance of, 280–281

Gloves, 260–262, 265, 268–271
Ground-fault circuit interrupters (GFCI), 5, 39–41, 107–108
Group lockout/tagout, 161–162

Handlines, insulated, 246–247
Hand protection, 260–262, 265
Hazard analysis, 178–182. *See also* Arc flash hazard analysis
    approach boundaries, 178–180
    arc flash, 180–182
    in electrical safety program, 120–124
    emergency procedure training and, 137–138
    maintenance and, 293–295
    NFPA 70E on, 92
Hazard awareness, 18–35
    arcing faults, 23–29
    in culture of electrical safety, 3–8
    electrical shock, 20–22
    workplace hazards, 19–20
Hazard risk categories (HRCs)
    arc flash hazard analysis and, 182, 229–230
    PPE selection and, 260, 271, 310
    protective clothing characteristics and, 271–272
    rubber insulating gloves and tools and, 268–271
Head protection, 259

Hearing protection, 271
Heinrich, H. W., 118–119
Host employer responsibilities, 105, 306
Housekeeping precautions, 244, 311
HRCs. *See* Hazard risk categories

IEEE 1584 method
    of AFPB determination, 23, 213, 220–222
    of incident energy analysis, 223–228
    resources and tools for, 225
Illumination precautions, 243, 311
Impedance, 23, 191
Impedance-grounded systems, 71
Incident energy analysis
    for arc flash hazards, 181–182, 222–231
    arcing faults and, 23
    circuit breakers and, 60–64
    Doughty/Neal/Floyd method of, 222
    HRC method, 229–230
    IEEE 1584 method of, 223–228
    NFPA 70E method of, 223–228
    PPE and, 105, 260, 309–310
    Ralph Lee method of, 222–223
    short-circuit currents and, 190
Injuries
    from arcing faults, 7, 24–25
    from electrical shock, 21–22
    trauma following electrical events, 5–7
"In-sight" fusible disconnects, 70
Instantaneous trip setting, 286
Insulated equipment
    case circuit breakers, 348–352
    fuse/fuse holding equipment, 246–247
    gloves, 268–271
    handlines, 246–247
    overview, 246–247, 248
    protective equipment, 246–247
    ropes, 246–247
    tools, 268–271
International Brotherhood of Electrical Workers (IBEW), 126, 337–344
Interrupting rating (IR), 189, 284
Isolation of circuits, 70, 159

Justification of hazards, 175–176, 307–308

Knife-blade fuses, 280–281
Kowalski-Trakofler, Kathleen, 288

LAB. *See* Limited approach boundary
Labeling of equipment
    for arc flash hazards, 182, 281–283, 310
    NEC requirements, 279
    safety signs and tags, 248
Ladders, portable, 247–248
Layering of PPE, 263–264
Lee, Ralph, 220, 222–223
Legal repercussions for maintenance, 295–296
Let-go threshold, 21
Level 1 subdivisions of *NFPA 70E*, 94

Level 2 subdivisions of *NFPA 70E*, 94
Level 3 subdivisions of *NFPA 70E*, 94
Liggett, Danny, 125
Limited approach boundary (LAB)
    AFPB vs., 179, 227
    defined, 178
    energized work and, 307–308
    NFPA 70E requirements, 106
    qualified persons and, 139–140, 178
    shock hazard analysis and, 309
    for uninsulated overhead lines, 241–242
    working inside of, 106–107
Lipster, Stephen M., 126
Listed equipment, 58. *See also* Underwriters Laboratories (UL)
Lockout/tagout, 151–163, 307
    application of, 159–160
    communication of, 157–159
    defined, 20, 153
    devices, 153
    electrically safe work condition with, 163–165
    employee protection with, 154
    group lockout/tagout, 161–162
    OSHA requirements on, 46–47
    outside personnel and, 161
    procedure for, 154–155, 163
    protective materials and hardware for, 155–157
    release of, 160–161
    requirements for, 307
    stored energy and, 160
    training for, 157–159
Look-alike equipment, 248
Low-voltage power circuit breakers, 216
Lubrication maintenance, 289–290

Main OCPDs, 71
Maintenance, 287–296
    of circuit breakers, 280, 292
    of equipment, 278–303
    of flame-resistant (FR) clothing, 266–268
    of fuses, 280–281
    hazard analysis and, 288–289, 293–295
    host employer responsibilities for, 105
    legal repercussions for, 295–296
    lubrication, 289–290
    misapplied fuse application and, 291
    of PPE, 43, 258, 262
    programs for, 292–293
    of protective equipment, 246
    requirements for, 291–292
    safety and, 290–291
    switching to reduce arc flash, 74–76
Maintenance Frequency Matrix, 293, 345–347
Marking equipment
    for arc flash hazards, 182, 281–283, 310
    safety signs and tags, 248
Material Safety Data Sheet (MSDS), 19
Mechanical barriers, 242, 248
Misapplied fuse application, 291
Molded-case circuit breakers, 292, 348–352
Monitoring, remote, 76

Motor controllers
    "in-sight" fusible disconnects for, 70
    "no damage" protection for, 68
    short-circuit calculations for, 191
MSDS (Material Safety Data Sheet), 19

National Electrical Code (NEC)
    arc flash labeling requirements, 279
    on GFCIs, 39
    on "in-sight" fusible disconnects, 70
    NFPA 70E and, 89–90, 91
National Electrical Contractors Association (NECA), 126, 337–344
National Electrical Safety Code (NESC), 134–135
National Fire Protection Association. *See NFPA 70E*
National Institute for Occupational Safety and Health (NIOSH)
    FACE program, 2, 12, 134–135
    on maintenance, 288
National Institute of Standards and Technology (NIST), 295
Neal, T. E., 222
NEC. *See* National Electrical Code
NECA (National Electrical Contractors Association), 126, 337–344
Neck protection, 259
Neitzel, Dennis, 66, 141, 180
NESC (National Electrical Safety Code), 134–135
Neuropsychological effects of electrical shock, 5–7
NFPA 70E, 88–115
    AFPB determination method of, 220–222
    on alerting techniques, 312
    on approach boundaries, 309
    on arc flash hazard analysis, 309
    on arc flash labeling, 279
    on contractor responsibilities, 105, 306
    definitions in, 103
    electrically safe work condition defined in, 4
    on electrical safety program, 105–106, 119–126, 306
    on energized electrical work permits, 308–309
    on equipment labeling, 310
    on equipment use, 107–108, 307
    evolution of, 90–91
    history of, 89–90
    on host employer responsibilities, 105, 306
    incident energy analysis method of, 223–228
    layout of, 91–92
    on limited approach boundary, 307–308
    on maintenance requirements, 292
    numbering style of, 92–95
    organization of, 92, 102–103
    on PPE requirements, 8, 309–310, 311
    purpose of, 91, 104–105
    on retraining, 141, 158
    scope of, 91–92, 103–104
    on testing, 310
    on training requirements, 105, 306
    on uninsulated overhead lines, 310
    on working while exposed to electrical hazards, 106–107, 306
NIOSH. *See* National Institute for Occupational Safety and Health
NIST (National Institute of Standards and Technology), 295

"No damage" protection for motor controllers, 68
Nominal voltage, 178
Non-current-limiting overcurrent protective devices, 58

Occupational Safety and Health Act of 1970, 37, 38, 101
Occupational Safety and Health Administration (OSHA), 36–55
    citation examples, 327–336
    on control of hazardous energy, 149
    electrical safety program and, 126, 337–344
    on employee protection from hazards, 41–42
    on ground-fault circuit protection, 39–41
    history and purpose of, 37–49
    on lockout/tagout procedures, 150, 163, 307
    NFPA and, 89–90
    on PPE, 42
    on prevention, 45–49
    requirements of, 38
    on training, 42–44
    29 CFR Part 1910 and, 42, 162–163
    29 CFR Part 1926 and, 38–39
OCPDs. *See* Overcurrent protective devices
Ohm's law, 21, 194
Opening of circuits, 78, 245
OSHA. *See* Occupational Safety and Health Administration
Outside personnel and lockout/tagout, 161
Overcurrent, 245
Overcurrent protective devices (OCPDs)
    arc flash hazard analysis and, 180–181, 231
    arcing faults and, 23
    clearing time, 213–214
    consistency over time of, 65
    current-limiting, 58–64
    design considerations, 57, 58–64, 231
    evaluation of, 284–285
    maintenance requirements for, 290–291, 292
    non-current-limiting, 58
    safety role of, 27–29
Overhead lines, uninsulated, 240–242, 310
Overload, 245

Performance language, 38, 95
Personal protective equipment (PPE), 256–277
    arc flash hazard analysis and, 23–24
    for arc flash hazards, 181, 258, 264–268
    arc rating for, 73
    arm protection, 260–262
    body protection, 259–260
    care of, 258
    chin protection, 259
    costs of, 10–11
    coverage of, 264
    eye protection, 259
    face protection, 259, 264–265
    fit of, 264
    flame-resistant clothing and, 258
    foot protection, 262
    hand protection, 260–262
    head protection, 259
    HRC classifications and, 268–272
    interference of, 264
    layering of, 263–264
    maintenance requirements, 262
    movement requirements for, 259
    neck protection, 259
    NFPA 70E on, 8
    OSHA on, 42
    overview, 309–310, 311
    requirements for, 258–259
    selection of, 258, 263
    visibility requirements for, 259
Physical barriers, 242, 248
Point-to-point calculation of short-circuit currents, 194–201
Portable electrical equipment, 107
Portable ladders, 247–248
PPE. *See* Personal protective equipment
Prevention. *See also* Electrical safety program
    of electrical shock, 22, 45–49
    OSHA on, 45–49
    safety program elements for, 134
Prohibited approach boundary
    defined, 178
    NFPA 70E requirements, 106
    shock hazard analysis and, 309
Protective clothing, 271–272
Protective equipment, 246–249. *See also* Personal protective
    equipment (PPE)
    alerting techniques, 248
    attendants, 248
    barricades, 248
    care for, 246
    fiberglass-reinforced plastic rods, 247–248
    insulated, 246–247
    look-alike equipment, 248
    mechanical barriers, 248
    physical barriers, 248
    portable ladders, 247–248
    requirements for, 246
    rubber insulating equipment, 248
    safety signs and tags, 248
    shields, 247–248
    standards for, 248–249, 312
    types of, 246
    voltage rated plastic guard equipment, 248
Psychological effects of electrical shock, 5–7

Qualified person
    approach to energized work by, 179–180
    job briefing and, 125
    for maintenance testing, 294–295
    NFPA 70E requirements, 106
    testing by, 239–240
    training of, 138–141

Racking, remote, 77
Radiant heat burns, 24
Ralph Lee method, 220, 222–223
Rejection style class J, T, R, and L fuses, 57, 66–67
Relaying schemes, 287
Remote monitoring, 76
Remote opening and closing, 78

Remote racking, 77
Restricted approach boundary
    defined, 178
    NFPA 70E requirements, 106
    shock hazard analysis and, 309
Retraining requirements, 141, 158
Risk evaluation. *See also* Arc flash hazard analysis
    in electrical safety program, 120–124
    emergency procedure training and, 137–138
    NFPA 70E on, 92
Ropes, insulated, 246–247
Routine opening and closing of circuits, 245
Rubber insulating equipment, 248, 260–262, 265, 268–271

Safety. *See also* Electrical safety program
    maintenance and, 290–291
    measures and devices, 65–78
    signs and tags for, 248
    training on, 136–137
Safety glasses or goggles, 271
Scannel, Gerard F., 90
SCCR (Short-circuit current rating), 189
Sections of *NFPA 70E*, 94
Selective coordination, 70
Self-discipline training, 120
Shields, protective, 247–248
Shock hazard analysis, 106, 178
Shock protection. *See also* Personal protective equipment (PPE)
    boundaries for, 178–179
    for hands and arms, 260–262
Short-circuit current rating (SCCR), 189
Short-circuit currents
    basic calculations for, 191–194
    calculation of, 188–209
    conductors effect on, 193–194
    effect on arc flash hazards, 190–191
    factors in, 192
    point-to-point calculation of, 194–201
    requirements for calculation of, 190
    software for calculation of, 202–203
    sources of, 191–192
    transformers effect on, 193
Short-circuit study, 189, 212
Short-time delays, elimination of, 73–74
Skin resistance and electrical shock, 20–21
Software for calculation of short-circuit currents, 202–203
Stored energy and lockout/tagout, 160
Subdivisions of *NFPA 70E*, 94
Switches
    for circuits 800 amperes and above, 67
    shunt-trip, 67

Tagout. *See* Lockout/tagout
Terminal covers, 69–70
Testing
    of circuit breakers, 292, 348–352

of energized equipment, 178
of fuses, 280–281
historical voltage testing methods, 3
lockout/tagout and, 161
NFPA 70E equipment requirements, 107
requirements for, 48, 239–240, 310
staged arc flash, 25–27
Thermal contact burns, 24
Thermal magnetic molded case circuit breakers, 214–215
Time pressures, 9–10
Tools
    conductive, 243
    insulated, 268–271
Training, 132–147
    in culture of electrical safety, 9
    documentation of, 141
    for electrical testing, 240
    on emergency procedure, 137–138
    for lockout/tagout, 157–159
    OSHA on, 42–44
    of qualified employees, 138–141
    requirements, 135–141, 306
    retraining requirements, 141, 158
    on safety, 134, 136–137
    types of, 137
    of unqualified employees, 138
Transformers, 193, 287
Trauma. *See* Injuries
29 CFR Part 1910, 42, 162–163
29 CFR Part 1926, 38–39
Type 1 protection, 68
Type 2 protection, 68

Underwriters Laboratories (UL), 59, 281
Uninsulated overhead lines, 240–242, 310
Unqualified person
    approach to energized work by, 180
    NFPA 70E requirements, 106
    training of, 138

Ventricular fibrillation, 21–22
Visibility requirements for PPE, 259
Voltage
    historical testing methods, 3
    measuring on energized equipment, 178
    nominal, 178
Voltage rated plastic guard equipment, 248

White, Jim, 287
Work permits, energized electrical, 176–178
Workplace hazard awareness, 19–20
Workplace pressures, 2

Zone-selective interlocking, 74

Credits

Lessons Learned Photo courtesy of Salisbury Electrical Safety, L.L.C.

Chapter 1

1-2 © 1997 Institute of Electrical and Electronics Engineers; 1-3 © Charles Stewart & Associates

Chapter 2

2-3 (A-G), 2-4 (A-G), 2-5 (A-F) © 1997 Institute of Electrical and Electronics Engineers

Chapter 3

3-2 © Iconos/age fotostock; 3-3A Courtesy of Siemens Energy & Automation, Inc.; 3-3B Courtesy of Pass & Seymour/Legrand; 3-3C, 3-4 Courtesy of Hubbell Wiring Device-Kellems; 3-5 Photo courtesy of National Electrical Contractors Association (NECA); 3-7 Courtesy of Salisbury Electrical Safety, L.L.C.; 3-9 Courtesy of Square D/Schneider Electric; 3-11, 3-12 (A-B) Courtesy of Salisbury Electrical Safety, L.L.C.

Chapter 4

4-16 Courtesy of Cooper Bussmann; 4-17 Courtesy of Eaton Corporation; 4-18 (A-D) Courtesy of Cooper Bussmann; 4-19 Courtesy of Boltswitch; 4-22 (A-F) Courtesy of Cooper Bussmann; 4-28 (A-B), 4-32, 4-33 (A-B), 4-34, 4-35, 4-36 Courtesy of Eaton Corporation

Chapter 6

6-1 Courtesy of Westex; 6-2 Courtesy of Tyndale; 6-3 Courtesy of Salisbury Electrical Safety, L.L.C.; 6-6 Courtesy of Fluke Corporation

Chapter 7

Opener Photo courtesy of National Electrical Contractors Association (NECA); 7-6 Courtesy of Salisbury Electrical Safety, L.L.C.

Chapter 8

8-1 Courtesy of Square D/Schneider Electric; 8-2 Courtesy of Salisbury Electrical Safety, L.L.C.; 8-4 Photo courtesy of National Electrical Contractors Association (NECA); 8-9 Courtesy of Salisbury Electrical Safety, L.L.C.

Chapter 9

Opener Photo courtesy of National Electrical Contractors Association (NECA); 9-1, 9-2, 9-3, 9-4, 9-5, 9-6, 9-7, 9-8, 9-9, 9-10, 9-11, 9-12 Courtesy of Ideal Industries; 9-13 Photo courtesy of National Electrical Contractors Association (NECA); 9-14 Courtesy of Ideal Industries; 9-15, 9-16, 9-17 Photo courtesy of National Electrical Contractors Association (NECA); 9-18, 9-19, 9-20, 9-21, 9-22 Courtesy of Ideal Industries; 9-23, 9-24, 9-25 Photo courtesy of National Electrical Contractors Association (NECA); 9-26 Courtesy of Ideal Industries

Chapter 10

Opener Photo courtesy of National Electrical Contractors Association (NECA); 10-7 Photo courtesy of Thomas & Betts Corporation

Chapter 11

11-3, 11-4, 11-5, 11-11, 11-12, 11-13, 11-14 Courtesy of Cooper Bussmann

Chapter 12

Opener Photo courtesy of National Electrical Contractors Association (NECA); 12-6, 12-7, 12-8 © 1997 Institute of Electrical and Electronics Engineers; 12-9, 12-10 Courtesy of Cooper Bussmann

Chapter 13

Opener Photo courtesy of National Electrical Contractors Association (NECA); 13-6 Courtesy of Salisbury Electrical Safety, L.L.C.

Chapter 14

Opener Photo courtesy of National Electrical Contractors Association (NECA); 14-1, 14-3, 14-4, 14-5, 14-6, 14-7, 14-8, 14-9, 14-10, 14-12, 14-13, 14-14 Courtesy of Salisbury Electrical Safety, L.L.C.

Chapter 15

15-2 Courtesy of DuPont; 15-4 Courtesy of Cooper Bussmann; 15-13, 15-14, 15-16 (A-B) Courtesy of James R. White, Shermco Industries; 15-17A Courtesy of H. Landis Floyd/ DuPont; 15-17B, 15-18 (A-D) Courtesy of James R. White, Shermco Industries; L15-1, L15-2, L15-3 Courtesy of James R. White, Shermco Industries

Chapter 16

Opener Photo Courtesy of Thomas & Betts Corporation; 16-01, 16-02, 16-03, 16-04, 16-07, 16-08, 16-09, 16-10, 16-11 (A-C) Courtesy of Salisbury Electrical Safety, L.L.C.; 16-12B, 16-13B, 16-17B © 1997 Institute of Electrical and Electronics Engineers

Unless otherwise indicated, all photographs are under copyright of Jones and Bartlett Publishers, LLC, the National Joint Apprenticeship and Training Committee, and the National Fire Protection Association.